Stress Analysis for Creep

Stress Analysis for Creep

J.T. BOYLE and J. SPENCE

Department of Mechanics of Materials,
University of Strathclyde,
Glasgow, Scotland

Butterworths
London - Boston - Durban - Singapore - Sydney - Toronto - Wellington

First published, 1983

British Library Cataloguing in Publication Data

Boyle, J.T.
Stress analysis for creep.
1. Materials–Creep 2. Strains and stresses
I. Title II. Spence, J.
620.1'1233 TA418.22

ISBN 0-408-01172-6

Typeset by Scribe Design, Gillingham, Kent
Printed in England by The Camelot Press Ltd., Southampton

Preface

The object in writing this book has been to provide a textbook which thoroughly introduces the reader to the basic methods of stress analysis for creep. It is not intended as a complete state-of-the-art review and should not be judged as such. The book is different from previous texts on engineering creep mechanics since it does not avoid any of the difficulties associated with stress analysis by concentrating on elementary problems or the special case of isothermal creep under constant loads. Rather, in all of the worked examples, it is presumed that, for a complete understanding of the analysis, the reader will have access to computing facilities. Also, the examples are given in sufficient detail to allow the reader to make use of common solution routines found in most computer libraries. In fact, *all* the examples presented here have been especially solved for this purpose and results are not simply reproduced from published analyses. To appreciate creep analysis, the reader must compute. If he does not, his understanding of the subject will be the poorer. However, one of the aims of this book is to develop and describe simplified methods of analysis, suitable for the designer, which do not require extensive computing. Nevertheless, it is the main premise of this text that, to appreciate fully the simplifications involved, a sound practical background in the methods of stress analysis for creep is of crucial importance.

An increasingly large amount of equipment in electric power generation, aeronautics and petrochemical plant contains metallic components which must be designed and constructed to operate at high temperatures for increased thermal efficiency. Indeed, we can expect this requirement for high-temperature design to extend well into the future as the demand for new energy sources – geothermal power and fusion reactors – grows. The safety of this equipment and its economic operation will depend on accurately assessing the possibility of structural failure. Consequently, much attention must be paid to the design of these components. Unfortunately, the engineer's familiar elastic analyses – and its attendant simple design procedures based on the use of tables and formulae – is inadequate in the high-temperature regime since the material will display a measurable amount of *creep*, the time-dependent slow deformation that occurs in many metals within certain ranges of temperature and stress. The engineer faced with creep problems will need to acquaint himself with more complex stress analysis procedures.

The arrangement of the subject matter is as follows. The text is mainly concerned with engineering mechanics, so that the physics of creep is discussed only briefly and a phenomenological approach is adopted. We begin with a description of the phenomenon of creep as it is commonly observed in the laboratory, and develop a number of uniaxial mathematical models of the material behaviour which have found application in stress analysis. The phenomenon of creep rupture is also described (Chapter 2). Following this, we demonstrate the basic behaviour of creeping structures under constant load, constant displacement, variable loading and non-isothermal conditions with reference to simple examples which require only a uniaxial model (Chapter 3). Some of the more useful simplifications to the creep behaviour are also introduced at this stage.

We then examine in detail methods of stress analysis for the special case of steady creep where the stress and rates of change of deformation remain constant in time. Analysis of this particular behaviour will be found useful in formulating approximate methods for the more general case of transient creep. A convenient multiaxial constitutive relationship is developed in detail and there follows a number of examples where a solution can be obtained in closed form. In two of these examples, the reader is introduced to the concept of generalised models for shells and plates. It is commented that the ability to obtain a closed-form solution is rare in creep mechanics; in general, some sort of numerical technique is required. We describe the most common of these techniques with reference to a simple example and suggest a method – known as quasinlinearisation – which is particularly useful in steady creep analysis. Two further examples of this method are also given. However, the most common methods of analysis for steady creep are based on the energy theorems and these, together with associated methods, are developed in full with a number of examples. We then describe how a simplification of the stress analysis of structures such as frameworks or piping and thin shells and plates can be obtained through the use of approximate generalised models (Chapters 4 and 5).

The following chapter also deals with steady creep, and with a particular difficulty associated with it – the inadequacy of the basic creep data, which can lead to unreliable predictions of component behaviour. It is shown how this can be overcome by relating component behaviour to uniaxial behaviour through the 'reference stress method' and how a simplified method – particularly useful in design – also results (Chapter 6).

Having dealt with the steady creep problem, we turn to the more general phenomenon of transient creep. Initially, the type of mathematical equations and the numerical methods required for their solution are developed. A more rigorous and general approach to that commonly found in other books on creep is adopted and it is shown how this can be specialised to particular problems. Specifically, we also show how this can be applied to the finite element method, which is the most common analysis tool for creep analysis (Chapter 7). We then describe how the solution of the transient creep problem can be simplified. We first consider the technique of characterising the solution in terms of known solutions through the use of comparison theorems. (This should simplify much recent work done on bounding theorems in England into a form that is more accessible to the reader.) The problems of constant load, non-isothermal creep, variable and cyclic loading and creep relaxation are considered. Secondly, we consider the simplification of the problem for particular structures which can be thought of as mechanically simpler, through the development of generalised models for transient creep. This has particular application to the analysis of shell structures (Chapter 8).

We then examine two possible structural failure modes caused by creep: creep rupture and creep buckling. In the case of creep rupture (Chapter 9), we describe the process of deterioration in a material and suitable mathematical models for its representation. Material deterioration leads to the study of a relatively new subject known as continuum damage mechanics. Two examples of problems in damage mechanics, and the difficulties which arise in their analysis, are described. Finally, a number of simple estimates for the lifetime of a deteriorating structure are developed. In the case of creep buckling (Chapter 10), we examine the initiation of buckling through loss of stability due to creep for two simple examples and extend their behaviour to structural creep in the presence of large deformations.

Finally, we turn to the principal concern of design for creep (Chapter 11). Here we take cognisance of the fact that this area is a wide and controversial one on which the body of opinion is changing rapidly. Unfortunately, in this area much ground must be covered merely to get to the point of saying very little about the true nature and problems of design for creep. The first concern is with material data requirements and constitutive modelling for design, and we consider the presentation of basic creep data, suitable methods of curve fitting and the problem of extrapolation. Secondly, we describe the need for 'benchmarking' a particular method of stress analysis, and reproduce a number of well known comparisons between experiment and theoretical predictions. The main task of this chapter is to present current design methodologies for dealing with time-dependent material behaviour. In relation to this task, two design procedures are contrasted and their respective design criteria described.

We have not followed the conventional practice of citing detailed references to the literature, with the exception of Chapters 1 and 11. This literature is now quite extensive and it would be physically impossible to record all of the original sources of the ideas presented here. Most of these ideas are, of course, attributable to others. The references or further reading lists supplied at the end of each chapter should allow the interested reader to trace the work back to its source and to uncover many related studies. The presentation of these ideas is largely our own and we take full responsibility for any omissions, errors in analysis or inaccuracies. To begin with, the presentation is slowly paced, but accelerates towards the end. It is therefore assumed, for example in Chapter 9 on creep rupture, that all of the previous chapters have been read and understood. Nevertheless, we have tried to include nothing inherently difficult; but, on the other hand, little has been omitted as being beyond the grasp of the reader (if he has followed the rest of the text). Above all, we have attempted to include nothing that later must be unlearnt as incorrect or speculative.

The book is principally aimed at research and design engineers employed in the design, construction and verification of high-temperature plant operating under creep conditions, who need both an *introduction* to the problems associated with creep and a *single* text which described the basic methods of analysis currently in use. Parts of the book can be used as a graduate text for mechanical engineers specialising in creep aspects of pressure-vessel design, and should be useful to specialists in mechanics or applied mathematics who wish a rigorous treatment of the subject.

In conclusion, it is a pleasure to acknowledge support given to this work through the various resources of the University of Strathclyde and through the United Kingdom Science and Engineering Research Council who provided financial assistance to one of the authors (J.T.B.) for continued research in creep mechanics. The more tedious aspects of the preparation of this book were undertaken during a pleasant visit to the Division of Solid Mechanics of Chalmers University of Technology, Gothenburg, Sweden, where much of it was presented by one of the authors (J.T.B.) as a course in creep mechanics. We are grateful to Professor Jan Hult for the arrangements which made this possible, and for his hospitality, enthusiasm and encouragement. Also, thanks are due to Mrs Mary Lauder, who typed the manuscript with speed and precision. Finally, the senior author (Professor John Spence) would like to record clearly that the major part of the text is the work of his colleague, to whom appropriate credit is due.

J.T. Boyle and J. Spence
1983

Contents

Chapter 1

Introduction

To begin with, we should define what we mean by 'creep'. Loosely, creep is a *time-dependent* deformation which occurs when a material is subjected to load for a prolonged period of time. The familiar load–deflection diagram for the linear elastic response of a tensile specimen (*Figure 1.1*) signifies that for a given load a fixed deflection will result. But, if the material creeps, the deflection will continue to increase, even though the load remains the same. The variation in the deflection with time is shown in *Figure 1.2* – this type of plot is usually

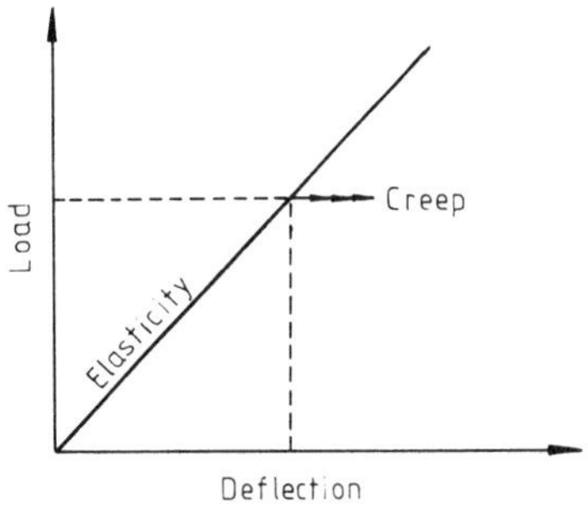

Figure 1.1 Load–deflection diagram

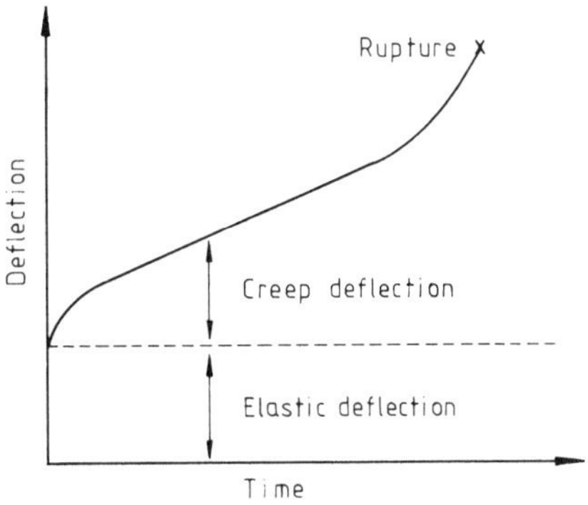

Figure 1.2 Standard creep curve

called the 'standard creep curve'. Of course, the tensile specimen cannot elongate indefinitely and ultimately creep will lead to fracture, or 'creep rupture'. A moment's thought and examination of the standard creep curve should suggest that a different approach to the design of engineering components will be necessary. If material creep is present, then we should design to a *finite lifetime* in order to avoid excessive distortion or rupture during the expected service life of a component. To do this, we must be able to analyse the stresses in a creeping structure.

1.1 The occurrence of creep in mechanical engineering components

The concept of material creep is not a new one; the phenomenon may be observed in the flow of ice, snow and rock and soil masses. Indeed, its significance was appreciated throughout antiquity in some manufactured wooden objects and in cloth[1]. Creep can occur in many materials used in modern engineering construction – bitumen, ceramics, concrete and, of course, metals. In general, the creep curves for these materials are similar in form, to the standard creep curves, *Figure 1.2* (although there may be some distortion); two useful summaries can be found in refs 2 and 3 given at the end of the chapter. In this book we will be concerned almost exclusively with the phenomenon of creep in metals. Although metal creep must have been encountered throughout the history of technology, particularly in furnace equipment and steam boilers, its true character and significance was not realised until the findings of the polymath, E.N. da Costa Andrade, were published in 1910[4]. The key was that metal creep is a *temperature-dependent* phenomenon.

From a practical viewpoint, what sort of engineering components are likely to experience creep? In the following, a representative selection of examples from a number of industries will be described briefly:

1.1.1 Power generation

Many creep problems occur in power generation equipment. In order to accommodate a higher thermal efficiency, the operating temperatures of simple steam engines gradually increased during the nineteenth century, but were brought into the material creep range with the growing use of superheated steam in the first quarter of the present century.

Nevertheless, engineering design problems resulting from creep did not become serious until the invention of rapidly rotating turbines. Steam turbines are still responsible for a significant proportion of the electric power generated[5]. A typical unit is configured as a boiler supplying a single turbo-generator with superheated steam in the range 538–565°C. The main components that need to operate in the creep range must usually be designed to avoid failure during the plant life. Such failure could occur through creep rupture of boiler tubes or turbine blades, although gross distortion of turbine casing due to creep deformation could also have disastrous consequences. In fossil-fired plant, such components have been developed for service in the creep regime over many years of operating experience and evaluation to the extent that they can be safely designed using simple empirical formulae and charts. Moreover, they allow a level of in-service inspection and subsequent maintenance. On the other hand, nuclear power generation, and in particular the liquid-metal fast breeder reactor (LMFBR), presents some novel design problems[6] in view of the public demand for additional structural integrity and the obvious difficulties of in-service inspection. A typical arrangement for the heat transport system in a fast reactor consists of the reactor core supplying hot sodium (with a mean temperature of about 540° C) to an intermediate heat exchanger by way of the 'primary' coolant piping. The heat exchanger transfers heat to steam generators through the 'secondary' piping circuit; the maximum secondary circuit steam temperature is about 510°C. Finally, the steam generators are linked to the turbines in the conventional manner. In order to avoid any alarming sodium–water reaction, the whole of the primary circuit (core, piping and heat exchanger) is contained in a single vessel. The principal components subject to creep in the primary circuit are the above-core structure, where sodium pressure is low but there is a severe thermal environment, and the intermediate heat exchanger. The design criterion is usually that rupture should not occur within the expected plant life.

In the secondary piping circuit, high thermal transients lead to the need for relatively thin-walled sections and the additional problem of buckling induced by time-dependent creep deformations needs to be taken into consideration. In order to ensure a high degree of operational reliability, a greater understanding of material creep behaviour and a subsequent stress analysis of the various components is demanded. This has been responsible for the considerable recent growth in interest in creep mechanics.

1.1.2 Aircraft gas-turbine engines

The need for cost effectiveness in the competitive airline construction industry is well known. The consequent demand for aircraft gas-turbine engines of minimum size and weight together with a good fuel economy and increased service life has led to the highest tolerable operating temperatures: gas temperatures can be as high as 1300° C! The principal failure mode due to creep is that of blade rupture[7], although there is an obvious need to maintain critical dimensions such as the clearance between rotor blades and casing. Although the stringent demands of operational safety cannot be understated, these are offset by the extensive testing and development of prototypes before a new design can be licensed to enter service. Thus, aircraft turbine design can achieve a considerable level of sophistication through development, in stark contrast to the design problems associated with nuclear technology which needs to be 'right first time'.

1.1.3 Process industries

High operating temperatures can be found in a number of processes in the petroleum and petrochemical industries. These occur mainly in such large-scale steam reformer processes used in the production of town gas or in the synthesis of ammonia. In a steam reformer, heat is supplied from hot furnace gases, at temperatures of about 950° C, flowing parallel to tubes containing the pressurised process gases. There is a resulting temperature gradient of about 100° C through the tube wall. The approach to design is *different* here from that of the previous examples cited. Rather than designing to avoid the possibility of a creep failure, the consequences of any such failure are accepted, and the design formulated accordingly[8]. This approach arises from two factors: first there is an incomplete knowledge of tube material properties and actual operational temperatures, thus ruling out any detailed stress analysis. Secondly, the economic penalties of plant shutdown can be very high, so that continuous operation must take precedence. Consequently, the design problem is the ability to isolate any failed tubes from the process until such time as replacement is practicable.

We have described three applications where creep is a design problem, although we have seen that their approaches to design are not the same. Other examples could be given, for example creep is a factor in the design of metallic heat shields for the NASA Space Shuttle thermal protection systems[9], and in coal liquefaction and gasification plants. It is likely that creep will remain a problem in high-temperature technology in the foreseeable future: it may be expected to occur in the heat transport systems associated with alternative energy sources such as geothermal or solar power plants or in fusion reactors. The need for methods of stress analysis for creep is clear.

1.2 Background to stress analysis for creep

In this book we will be concerned primarily with the techniques of stress analysis for creep. No detailed treatment of the underlying physical processes will be given, nor will any attempt be made to provide a comprehensive summary of the wide variety of alternative means of describing the phenomenon of creep. On these matters we can only give an introduction to these topics rather than deal with them in any depth. Unfortunately, neither can we provide any historical perspective on the development of creep analysis; happily much of this can be found in previously published textbooks and monographs. Since these texts have influenced the arguments and presentation of the present book, it is worthwhile briefly to summarise them for further reference.

Much of the early published work on creep was concerned with the description of time-dependent material behaviour. Among the first exhaustive studies of the aims and techniques of stress analysis and design was the work published by the Englishman, R.W. Bailey, in the 1930s[10]. The aims that he described, and the methods used, are surprisingly similar to those still used in practice today. A dramatic increase in research on creep occurred in the late 1950s and a comprehensive review of the work up to this time can be found in the first book dealing with stress analysis by I. Finnie and W. Heller[11]. Published in 1959 it is, despite its age, still a worthwhile text for engineers seeking their first acquaintanceship with creep. Following this, several other texts on creep appeared during the 1960s. Sadly, two of the best, by the Swedish authors F.K.G. Odqvist and J. Hult[12] and by the Soviet author L.M. Kachanov[13], were never published in English editions (although the latter is available in translation, but with many misprints). Both Odqvist[14], one of the earliest investigators of creep mechanics, and Hult[15] later produced separate books. The former gives a rigorous mathematical treatment of steady creep and creep rupture, while the latter remains one of the best short introductions to the subject. The considerable Soviet contribution to creep analysis has been summarised by Yu N. Rabotnov[16]. This extensive monograph is elegantly written and translated but is hard going for the uninitiated; it is however, worth the effort. The book by R.K. Penny and D.L. Marriott[17] is specialised, mostly descriptive and contains few examples. Nevertheless, it does deal with a number of simplifying concepts, which have largely stemmed from the UK and have enhanced our understanding of creep and its effect on structures as well as our ability to condense creep information into forms suitable for the designer. (We will discuss these later in this book.) The most recent text by H. Kraus[18] provides a very broad introduction to the design problems posed by material creep by way of elementary examples. In addition to these books, there are a few compilations of papers by different authors [19-21] which aim to introduce various aspects of the creep phenomenon and which deserve mention here. Creep is also the active subject of several international conferences, notably the IUTAM symposia on creep in structures[22-24] and the IMechE/ASTM collaborations[25-28] as well as numerous short conferences held under the auspices of these institutions and the ASME.

1.3 General-purpose computer programs for creep analysis

The type of examples that we will give in this book have been chosen to allow duplication by the reader. Any numerical solution which will be required (and there will be many) will be carried out using a 'special-purpose' computer program. (The reader will thus find things a bit easier if he has some background in elementary computer programming and numerical analysis. There are many texts to which the reader can refer, but that by Conte and de Boor[29] is especially recommended.) Nevertheless, it is more likely that an analyst will want to use one of the available general-purpose finite element programs which can handle material creep. There are several suitable programs in use today; below we cite those proprietary codes which are widely found in practice:

ABAQUS	Hibbit, Karlsson & Sorensen Inc., USA
ADINA	ADINA Engineering Inc., USA
ANSYS	Swanson Analysis Systems Inc., USA
ARGUS	Merlin Technologies Inc., USA
ASAS-NL	Atkins Research & Development, UK
ASKA	J.H. Argyris, Institut für Statik und Dynamic der Luft- und Raumfahit Konstruktionen, Stuttgart, Germany
BERSAFE	Central Electricity Generating Board, UK
MARC	MARC Analysis Research Corporation, USA
PAFEC	PAFEC Ltd, UK

There are of course many other in-house or non-proprietary programs in circulation, too many to mention here; a recent survey of American and European programs has been given by Noor[30]. The interested reader is also referred to the book by D.R.J. Owen and E. Hinton[31] in which a computer program for the finite element solution of elasto–viscoplastic (plasticity and creep) problems in two dimensions is developed and listed in its entirety.

Most of the general-purpose programs described above deal not only with creep but also with other structural nonlinearities (such as plasticity or large deformations). In general, the cost of nonlinear analysis substantially reduces the size and precision of a finite element solution.

Because of this, the analyst should be aware that it is not often necessary, or indeed desirable, to use such a powerful analytical tool, particularly when the problem in hand is otherwise simple. A greater understanding of the problem can usually be gained if time is available to develop a special-purpose program. It is the intent of this book not only to provide the analyst with the necessary background to do so but also to make more efficient use of, and to be more critical of, the general-purpose programs that he will meet in practice.

References

1. Gordon, J.E., *Structures: Or Why Things Don't Fall Down,* Pelican Books, 1978
2. Pomeroy, C.D. (Ed.), *Creep of Engineering Materials*, A Journal of Strain Analysis Monograph, Mechanical Engineering Publ., 1978.
3. Kelly, A. *et al.* (Eds), Creep of engineering materials and of the Earth, 1978. *Phil. Trans. R. Soc. A,* **288**, 1–136.
4. Andrade, E.N. da C., The viscous flow in metals and allied phenomena, *Proc. R. Soc. Lond., A*, **84**, 1, 1910.

5. Wyatt, L.M., *Materials of Construction for Steam Power Plant,* Applied Science Publ., 1976.
6. Holmes, J.A.G., High temperature problems associated with the design of the commercial fast reactor, *IUTAM Symp. on Creep in Structures,* Eds A.R.S. Ponter and D.R. Hayhurst, Springer-Verlag, 1981, p. 279.
7. Jenkins, R.E., Creep and aero gas turbine design, *Proc. 2nd Int. Conf. on Engineering Aspects of Creep,* I. Mech. E., 1980, Paper C33/80.
8. Jones, W.G., Design and operating experience with high temperature fired tubes in petrochemical plants, *Conf on Failure of Components Operating in the Creep Range,* I. Mech. E., 1976, Paper C108/76.
9. Harris, H.G., and Norman, K.N., *Creep of Metallic Thermal Protection Systems,* NASA TM X-2273, Vol. II, NASA, 1972.
10. Bailey, R.W., The utilisation of creep test data in engineering design, *Proc. I. Mech. E.,* **131**, 131, 1935.
11. Finnie, I. and Heller, W.R., *Creep of Engineering Materials*, McGraw-Hill, 1959.
12. Odqvist, F.K.G. and J. Hult., *Creep Strength of Metallic Structures* (in German), Springer-Verlag, 1962.
13. Kachanov, L.M., *The Theory of Creep*, Fizmatgiz, 1960 (transl. UK Natl Lending Library for Science and Technology, 1967).
14. Odqvist, F.K.G., *Mathematical Theory of Creep and Creep Rupture*, Clarendon Press, 1st Edn, 1966; 2nd Edn, 1974.
15. Hult, J., *Creep in Engineering Structures*, Blaisdell, 1966.
16. Rabotnov, Yu N., *Creep Problems in Structural Members,* North-Holland, 1969.
17. Penny, R.K. and Marriott, D.L., *Design for Creep,* McGraw-Hill, 1971.
18. Kraus, H., *Creep Analysis*, Wiley, 1980.
19. Smith, A.I. and Nicolson, A.M. (Eds), *Advances in Creep Design : the A.E. Johnson Memorial Volume*, Applied Science Publ., 1971.
20. Bernasconi, G. and Piatti, G., (Eds), *Creep of Engineering Materials and Structures,* Applied Science Publ., 1979.
21. Bressers, J. (Ed.), *Creep and Fatigue in High Temperature Alloys,* Applied Science Publ. 1981.
22. Hoff, N.J., (Ed.), *Creep in Structures, 1960; Proc. IUTAM Colloq., Stanford, 1960,* Springer-Verlag, 1962.
23. Hult, J., (Ed.), *Creep in Structures, 1970; Proc. IUTAM Symp., Gothenburg, 1970,* Springer-Verlag, 1972.
24. Ponter, A.R.S., and Hayhurst, D.R. (Eds), *Creep in Structures, 1980; Proc. IUTAM Symp., Leicester, 1980,* Springer-Verlag, 1981.
25. ASTM/I. Mech.E., *Joint Int. Conf. on Creep.*, I. Mech. E., 1963.
26. I Mech. E., *Thermal Loading and Creep in Structures and Components,* I.Mech.E., 1964.
27. ASTM/I.Mech. E., *Int. Conf. on Creep and Fatigue,* ASTM, 1973; I.Mech.E., 1974.
28. I.Mech. E., *Engineering Aspects of Creep,* I. Mech.E., 1980.
29. Conte, S.D. and de Boor, C. *Elementary Numerical Analysis: An Algorithmic Approach,* 3rd Edn McGraw-Hill/Kogakusta Int. Student Edn, 1980.
30. Noor, A.K., Survey of computer programs for solution of nonlinear structural and solid mechanics problems, *Comp. Struct.,* **13**, 425–465, 1981.
31. Owen, D.R.J. and Hinton, E. *Finite Elements in Plasticity : Theory and Practice*, Pineridge Press, 1980.

Chapter 2

A phenomenological description of creep

2.1 The phenomenon of creep

We begin by considering what happens if we load a uniaxial tensile specimen at a constant load for a period of time at a constant temperature sufficiently elevated to cause creep. The typical strain response with time is shown in *Figure 2.1*. Notice that if we repeat the experiment at different load levels (L_1 to L_3) the response may be quite different. Conventionally, the curve obtained is divided into several parts. There is an initial (linear) elastic time-independent

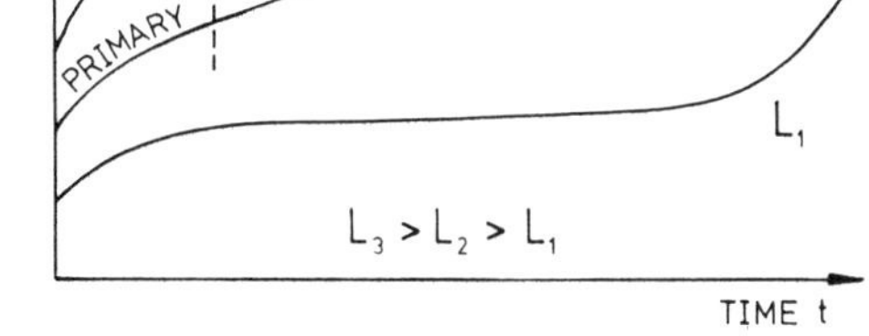

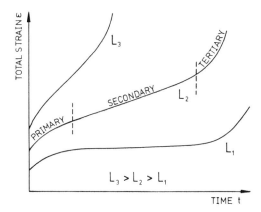

Figure 2.1 Basic creep curves for different loads, L_1, L_2, L_3

response (we will neglect for the moment situations where the load is high enough to cause initial plastic deformation). Thereafter, the creep curve is described in three parts: a part of decreasing creep strain rate called *primary creep*, a part of reasonably constant creep strain rate called *secondary* or *steady-state creep* and a final portion of increasing strain rate called *tertiary creep* preceding fracture. These definitions become more apparent if we plot creep strain rate against time (*Figure 2.2*). In primary and secondary creep, the test is also at constant stress. However, in tertiary creep, the increase in strain rate is largely the result of the change in the dimension of the cross-section of the specimen as deformation proceeds. Further, material deterioration, in the form

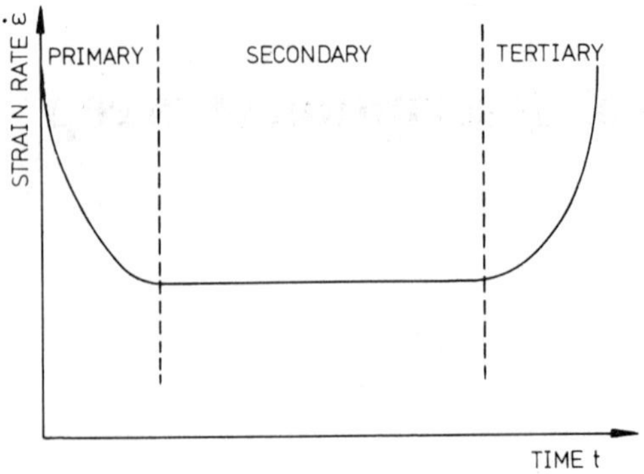

Figure 2.2 Typical plot of strain rate in time

of internal cavitation, may have altered the effective load-bearing cross-section. Hence, care must be taken in describing material behaviour in the tertiary phase.

The information from these basic creep curves can usually be presented in different forms which are more enlightening for different situations. The two most useful of these are the so-called *isochronous* (*Figure 2.3*) and *isostrain* (*Figure 2.4*) forms. In the former, contours of constant time are plotted on a log–log plot of strain rate versus stress. This sort of plot is useful in determining the stress–strain behaviour of the material. In the latter, contours of constant strain are plotted on a log–log plot of stress versus time. For a given

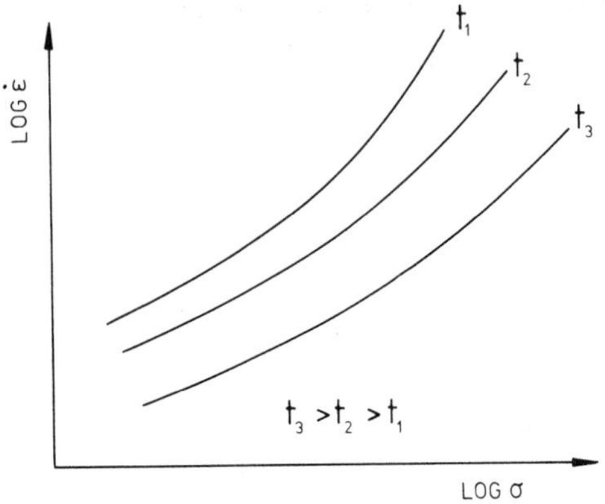

Figure 2.3 Isochronous creep curves

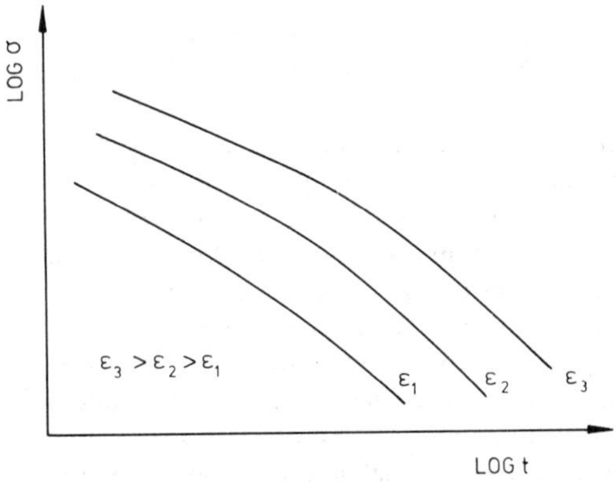

Figure 2.4 Isostrain creep curves

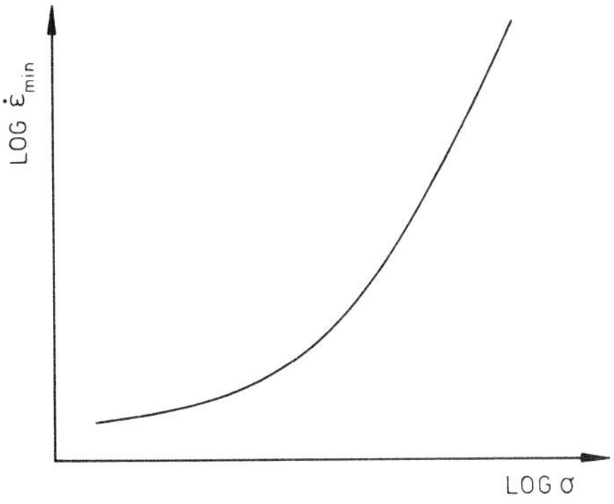

Figure 2.5 Typical plot of log(minimum strain rate) against log(stress)

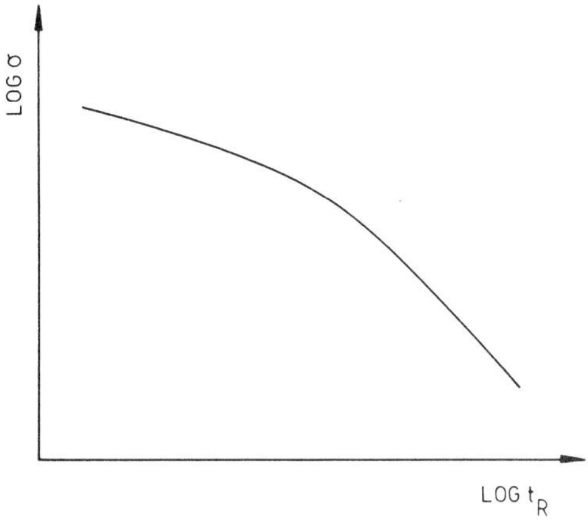

Figure 2.6 Typical plot of log(initial stress) against log(rupture time)

stress level, this sort of plot would tell us the time at which a given strain was reached. From these plots we can single out two which are particularly useful – that of minimum creep rate against stress (*Figure 2.5*) and that of rupture time against initial stress (*Figure 2.6*). The plot of minimum creep strain rate against stress is sometimes more or less straight and of positive slope, i.e. proportional to some power of stress, but for tests at lower load there is usually a transition to a lower slope. The plot of stress against rupture time suggests that the latter is inversely proportional to some power of stress, but again in longer-term tests there is a transition to a different slope. The former high-stress, short-time portion is often associated with ductile type fracture while the latter is of a brittle type taking place at low strains. Since by their very nature creep data require a long time to accumulate, it is natural that designers will wish to extrapolate short-term data to longer times, or to other load levels. Because of the shape of *Figures 2.5* and *2.6*, this is fraught with difficulty.

We can easily envisage the complexities that may arise when varying loads or temperatures are considered. The effect of different temperatures can readily be appreciated. If we plot minimum creep strain rate against temperature (*Figure 2.7*), it can be seen that the creep rate increases exponentially with temperature. Thus it is possible for a small increase in temperature to double the creep rate! (This exponential dependence is an example of what is known as Arrhenius's law, which has wide generality, applying not only to the creep of metals but also to other physical and biological processes.)

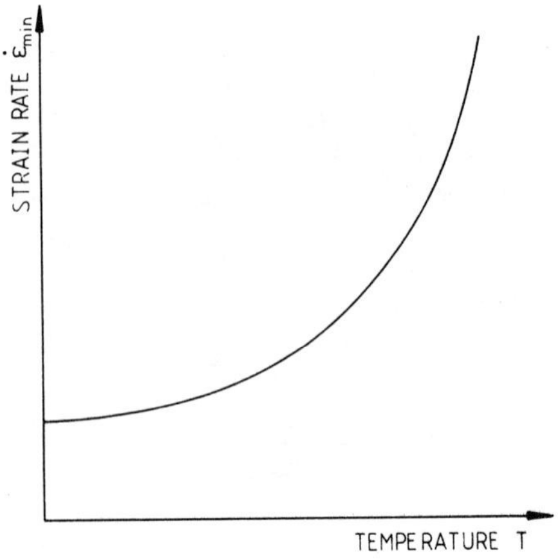

Figure 2.7 Variation of minimum creep rate with temperature

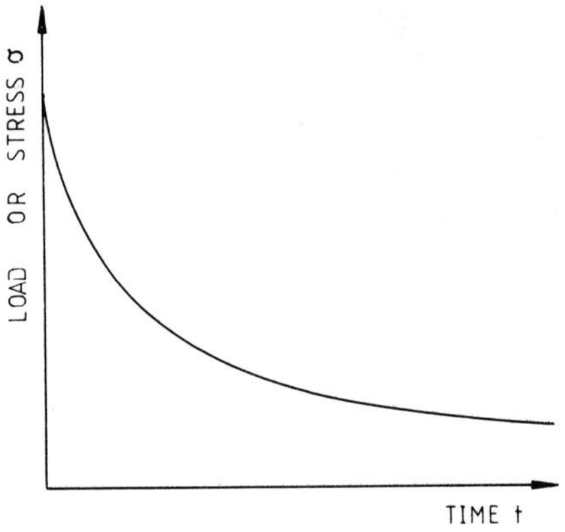

Figure 2.8 Typical relaxation curve

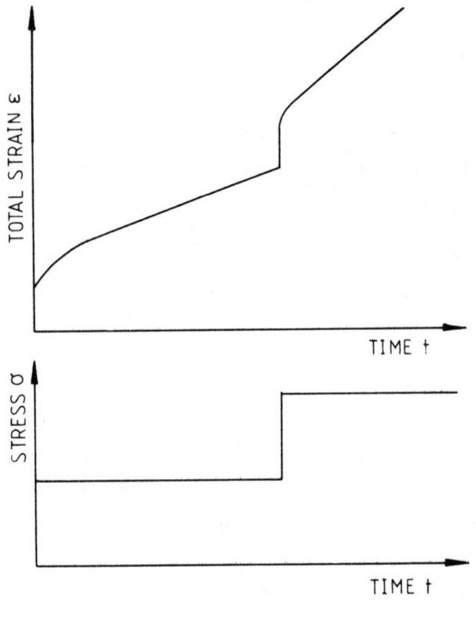

Figure 2.9 Typical creep response to step load

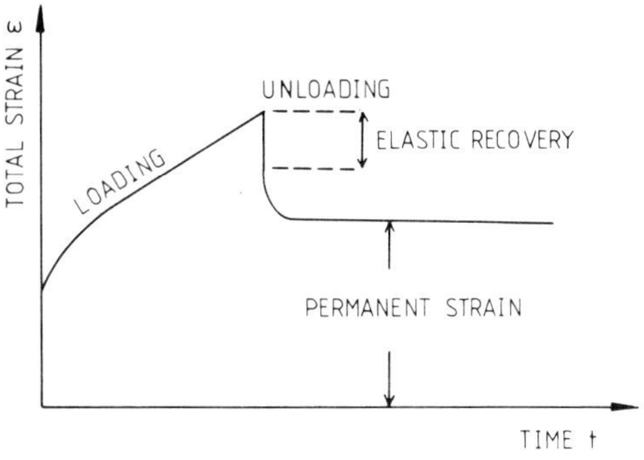

Figure 2.10 Effect of unloading on creep

The simplest example of non-constant loads is the so-called *relaxation* test wherein the elongation of a tensile specimen is held constant, resulting in a relaxation of the resultant load (*Figure 2.8*). Relaxation behaviour in metals can usually be predicted adequately using constant-load tests (although this is not true for some materials like polymers). One can give some idea of the complexity of material creep response to varying loads by examining the results of step loading tests (*Figure 2.9*) and unloading tests (*Figure 2.10*) which lead to the phenomenon of creep *recovery*. To combine these for cyclic loading obviously leads to complex behaviour and it should be clear that the simple constant-load tests will prove inadequate in attempts to establish a comprehensive model of material creep.

Finally, it should be emphasised that smooth curves, as depicted here, are not obtained as there is also the business of *scatter* in the experimental data, particularly in the cross-plots from different tests. Scatter arises not only from expected batch variations in specimens but also because of the sensitivity of creep strain to small changes in stress and temperature. Thus, in creep testing, the permitted tolerances of these control variables has to be small and usually this is difficult to achieve in practice.

2.2 The physical mechanisms of creep

The next step is to use the preceding experimental-type information to develop a mathematical model of the material behaviour for use in stress analysis. Here we will adopt a phenomenological approach to creep modelling. This means that our characterisation of the material behaviour is based on macroscopic observations rather than on the underlying physical processes. In stress analysis and materials engineering this is the usual approach. Nevertheless, there is a considerable understanding of the mechanisms of creep and a brief description follows. There are two dominant mechanisms called *dislocation creep* and *diffusional creep*. In the former, defects, known as dislocations, in the crystalline lattice structure of the metal can overcome the natural stiffness of the crystal, and other obstacles introduced in an alloy to prohibit creep, to move through the lattice. At low stresses, this dislocation motion stops or slows down, but creep can continue through the bulk movement of atoms – diffusional flow – from one atomic site under compression to another in tension. Dislocation creep has a high nonlinear dependence on stress, of the type discussed in the preceding section. Diffusional creep at low stress exhibits a more or less linear

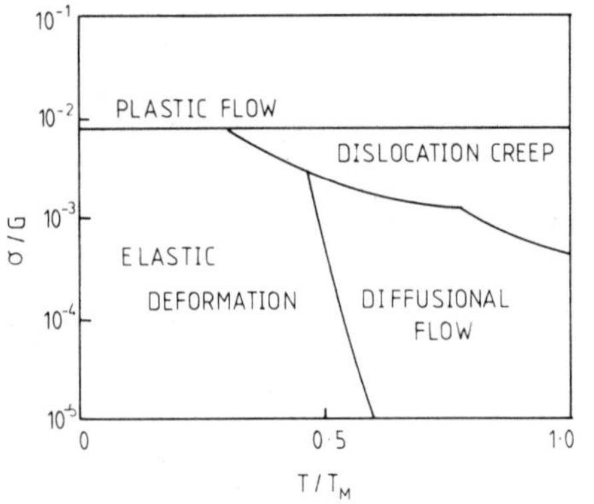

Figure 2.11 Deformation mechanism map: G = shear modulus, T_m = melting temperature

viscous dependence on stress. Indeed, the most important mechanism in most engineering structures is dislocation creep.

The relative importance of these mechanisms in particular stress and temperature ranges can be conveniently summarised on a *deformation mechanism map* (*Figure 2.11*). This is an idealised diagram, for a particular material, which also shows regions where instantaneous plastic flow occurs and where the deformation is simply elastic. It can be seen that creep is significant at temperatures above about $0.3T_m$, where T_m is the melting temperature of the material.

2.3 Convenient uniaxial constitutive relationships

Many 'simplified' uniaxial constitutive relations have been proposed to describe the standard creep curves. We will restrict our attention for the time being to primary and secondary creep, ignoring the tertiary phase, and describe a few relations which have found particular application in stress analysis.

The first step common to nearly all approaches is to separate the elastic and inelastic parts of strain

$$\epsilon = \epsilon_E + \epsilon_C$$

which can also be thought of as the definition of creep strain. In the broadest sense, the creep strain resulting from a constant-load test may be written as a function of stress σ, time t and temperature T as

$$\epsilon_C = f(\sigma, t, T)$$

which is usually assumed to be separable into

$$\epsilon_C = f_1(\sigma) f_2(t) f_3(T)$$

Some suggestions for the stress dependence are

$f_1(\sigma)$	$= B\sigma^n$	Norton
$f_1(\sigma)$	$= C \sinh(\alpha\sigma)$	Prandtl
$f_1(\sigma)$	$= D \exp(\beta\sigma)$	Dorn
$f_1(\sigma)$	$= A[\sinh(\gamma\sigma)]^n$	Garofalo
$f_1(\sigma)$	$= B(\sigma - \sigma)^n$	Friction stress

where symbols other than σ are material constants. The Garofalo relation contains the Norton, Prandtl and Dorn relations as special cases and predicts the 'bent' plot of minimum creep strain rate against stress (*Figure 2.5*). However, the Norton power law is predicted from physical arguments and is particularly useful in stress analysis.

Suggestions for the time dependence are

$f_2(t) = t$		Secondary creep
$f_2(t) = bt^m$		Bailey
$f_2(t) = (1 + bt^{1/3})e^{kt}$		Andrade
$f_2(t) = \sum_j a_j t^{m_j}$		Graham and Walles

According to Arrhenius's law, the temperature dependence is given as

$$f_3(T) = A \exp(-\Delta H/RT)$$

where ΔH is the activation energy, R is Boltzmann's constant and T is the absolute temperature.

The simplest of the above would take the form

$$\epsilon_C = C \exp(-\Delta H/RT) t^m \sigma^n$$

and for isothermal conditions

$$\epsilon_C = B\, t^m \sigma^n \tag{2.1}$$

which is very common in creep analysis.

The preceding development is, of course, for constant stress and is simply an attempt to model mathematically the basic creep curves for primary and secondary creep. For varying stress, we examine the *rate form*: for example, for the relation (2.1) we have

$$\dot{\epsilon}_C = \frac{d\epsilon_C}{dt} = mBt^{m-1}\sigma^n \tag{2.2}$$

This can be written in a form independent of time if we eliminate t between equations (2.1) and (2.2); then

$$\dot{\epsilon}_C = mB^{1/m}\sigma^{n/m}/\epsilon_C{}^{(1-m)/m} \tag{2.3}$$

The choice of the rate form is an attempt to model mathematically the primary creep phase of decreasing creep strain rate; this process is often called *hardening*. The relation (2.2) is usually called *time hardening* since the hardening phase is modelled using the time parameter. The relation (2.3) is similarly called a *strain hardening* model for the same reason. Of course, time and strain hardening formulations can be developed for functions other than the simple form (2.1).

These two results are obviously the same for constant stress. However, if we *assume* that they can be applied to varying stress conditions despite their derivation, we find that they give different results. They are compared for a step loading in *Figure 2.12*. Equation (2.2) says that the strain rate at any time depends on the elapsed *time* whereas (2.3) indicates that it depends on the accumulated creep *strain*. This can best be understood graphically from *Figure 2.12*. The curve 0AC represents the actual response to the step load test; the

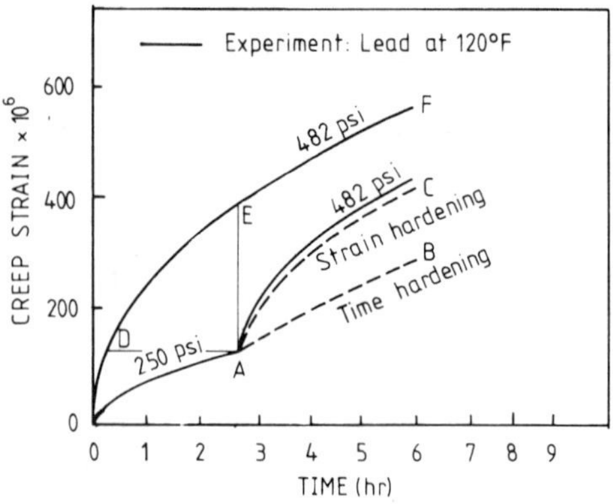

Figure 2.12 Experimental comparison of time and strain hardening theories (see Finnie, I. and Heller, W.R., *Creep of Engineering Materials*, McGraw-Hill, 1959, Chapter 6, Figure 6.2)

curve 0F represents a creep test held at a constant load equal to the final step load. The time hardening prediction for the step load, curve AB, is obtained by moving that portion EF of the curve 0F down to the point A. On the other hand, the strain hardening prediction, curve AC, is obtained by moving that portion DF of the curve 0F across to the point A. The time hardening prediction is clearly poorer. Fortunately, the differences are less marked for more gradual load changes and relation (2.2) is easier to use in stress analysis.

The representation of material creep behaviour described above by the time and strain hardening models is intended for use in the primary and secondary phases. But they do not account for all the creep strain observed, and do not represent well the important recovery phenomenon for unloading. A more complex form is required. Many possibilities have been suggested in the literature, but there is some agreement that an adaptation of the above hardening models which uses the concept of 'hidden' or 'internal' variables to describe the hardening should prove sufficient. Such formulations have two particular advantages: they are capable of representing a wider range of material behaviour including conventional plasticity and creep, and they are particularly useful in stress analysis. There are a number of these internal-variable theories, but they can mostly be expressed in a common form. In the simple time and strain hardening models, the history of the material behaviour at some instant was introduced into the constitutive relation between strain rate and stress by way of the time parameter and the current creep strain, respectively. They are used to describe the current state of the material and can also be called *state variables.* The underlying concept of a state-variable theory is to choose other 'hidden' state variables. We introduce here a model with two state variables, denoted by α and R, such that

$$\dot{\epsilon}_C = f(\sigma, \alpha, R) \tag{2.4}$$

(although we have used creep strain here, more general inelastic strains are also included). This is called the *equation of state*. The state variable α is usually called the 'rest' stress and the variable R the 'drag' stress. Initially,

$$t = 0: \qquad \epsilon_C = 0 \qquad \alpha = 0 \qquad R = R_0$$

and if the rest and drag stresses do not alter in time then the material exhibits only secondary creep for constant stress. In order to model more complex behaviour, we must also specify growth laws for the state variables. Two competing mechanisms are assumed to govern their growth: hardening and recovery (or 'softening') such that

$$\begin{aligned} \dot{\alpha} &= f_1(\sigma, \alpha, R)\dot{\epsilon}_C - f_2(\sigma, \alpha, R)\alpha \\ \dot{R} &= g_1(\sigma, \alpha, R)|\dot{\epsilon}_C| - g_2(\sigma, \alpha, R)(R - R_0) \end{aligned} \tag{2.5}$$

The first term in each expression is due to hardening, the second to softening; it is assumed that in a steady state the two mechanisms cancel and the state variables are subsequently constant.

This model has an analogy in conventional plasticity where α can be interpreted as the centre of the yield surface and R as the radius of the yield surface; the equation of state (2.4) is then piecewise defined. The above is a generalisation of these concepts to time-dependent material behaviour.

The functions f, f_1, f_2, g_1 and g_2 assume different forms for different theories. We briefly describe three of these below for illustration.

(i) (*Bailey–Orowan*). This theory is probably the seminal work for internal-variable models of creep. The rest stress is identically zero so that equations (2.4) and (2.5) assume the simpler forms

$$\dot{\epsilon}_C = \text{sgn}(\sigma) f(|\sigma| - R)$$

$$\dot{R} = h(R)|\dot{\epsilon}_C| - r(R)$$

where the function f is piecewise defined according to

$$f \geqslant 0 \qquad |\sigma| = R$$

$$f = 0 \qquad |\sigma| < R$$

The functions $h(R)$ and $r(R)$ have the suggested forms

$$r(R) = k_1 |R|^{n-m} \qquad 1/h(R) = k_2 |R|^m$$

In the steady state $\dot{R} = 0$, $|\sigma| = R$ and it follows that

$$f = r/h = k_1 k_2 \sigma^n$$

that is, the Norton power law.

In this model, there are four material constants, k_1, k_2, m and n, to be determined, some of which will depend on temperature.

(ii) (*Robinson*). In this theory, the drag stress is constant for all time with a fixed value R_0. The equations (2.4) and (2.5) take the special forms

$$\dot{\epsilon}_C = f(\sigma - \alpha)$$

$$\dot{\alpha} = f_1(\alpha)\dot{\epsilon}_C - f_2(\alpha)\alpha$$

where the function f is piecewise defined as

$$f = \begin{cases} \mu F^{(n-1)/2}(\sigma - \alpha) & F > 0 \text{ and } \sigma(\sigma - \alpha) > 0 \\ 0 & \end{cases}$$

with

$$F = \frac{(\sigma - \alpha)^2}{R_0^2} - 1$$

The functions f_1 and f_2 are also piecewise defined

$$f_1(\alpha) = \begin{cases} c_1 \left(\dfrac{\alpha}{R_0}\right)^l & \alpha > \alpha_0 \text{ and } \sigma\alpha \leqslant 0 \\ c_1 \left(\dfrac{\alpha_0}{R_0}\right)^l & \text{otherwise} \end{cases}$$

$$f_2(\alpha) = \begin{cases} c_2 \left(\dfrac{\alpha}{R_0}\right)^{m-l-1} & \alpha > \alpha_0 \text{ and } \sigma\alpha \leqslant 0 \\ c_2 \left(\dfrac{\alpha_0}{R_0}\right)^{m-l-1} & \text{otherwise} \end{cases}$$

Again for the case of steady creep, $\alpha \equiv 0$, we reduce to the Norton power law.

There are eight material constants μ, n, R_0, α_0, c_1, c_2, m and l to be determined; again some of these will depend on temperature.

(iii) (*Hart*). Finally, we describe a slightly more complex theory which has received some attention from stress analysts. In this, an alternative internal variable – the 'hardness' σ^* – is introduced while the drag stress does not appear. The equations (2.4) and (2.5) become

$$\dot{\epsilon}_C = c_1 (\sigma - \alpha)^m$$

$$\dot{\alpha} = c_2 [\dot{\epsilon}_C - f_2(\alpha, \sigma^*)\alpha]$$

where

$$f_2(\alpha, \sigma^*) = \frac{\dot{\epsilon}^*}{\alpha} \left[\ln\left(\frac{\sigma^*}{\alpha}\right)\right]^{-1/\lambda}$$

$$\dot{\epsilon}^* = c_3(\sigma^*)^n \exp(-\Delta H/RT)$$

The growth equation for the hardness is given as

$$\dot{\sigma}^* = f_2(\alpha, \sigma^*)\alpha\, \sigma^* \Gamma(\sigma^*, \alpha)$$

where the hardening function Γ assumes different forms for different materials.

As a special case for steady loading $\dot{\alpha} \simeq 0$, and the above takes the related form

$$\dot{\epsilon}_C = \dot{\epsilon}^* [\ln(\sigma^*/\sigma)]^{-1/\lambda}$$

$$\dot{\sigma}^* = \dot{\epsilon}_C\, \sigma^* \Gamma(\sigma^*, \sigma)$$

This model requires at least six material constants, c_1, c_2, m, n, λ, c_3

It should be obvious that a greater number of material constants are required by these models and subsequently an extensive testing programme must be followed. Indeed, in order to identify the constants, tests other than simple constant-load tests are necessary; for example, relaxation and constant-strain tests. With variable temperature, the testing programme could prove impracticable. For this reason, the designer often chooses to use the simpler time hardening model of the creep behaviour; who can blame him?

2.4 Creep rupture

In the previous section, we considered only primary and secondary creep. Here we examine the problem of determining the critical time in which a tensile specimen subject to constant force will fracture, and of how to describe material creep behaviour in the tertiary phase. It has already been mentioned that two possible mechanisms can occur. The specimen can rupture due to an infinite elongation, assuming that the flow of the material is uniform, or material deterioration will result in internal cavitation eventually also leading to fracture. The former is called *ductile rupture* and the latter *brittle rupture.*

In ductile rupture the tensile specimen will elongate at a certain rate, causing the cross-sectional area to decrease and the tensile strain to increase. The rate of elongation will subsequently increase at an ever-increasing rate, leading to fracture. Consider the tensile specimen shown in *Figure 2.13*; the initial length

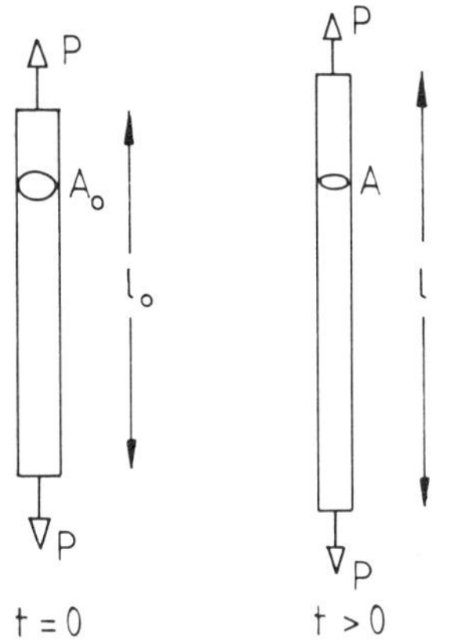

Figure 2.13 Cylindrical bar under constant load

and cross-sectional area are denoted by l_0 and A_0, respectively. The current magnitudes are denoted by l and A at time t. Since large strains are developed in ductile rupture, a logarithmic definition of strain is used so that

$$\epsilon = \ln(l/l_0)$$

and since creep deformation takes place at constant volume

$$A_0 l_0 = Al$$

The stress will be written as

$$\sigma = P/A \qquad \sigma_0 = P/A_0$$

If we ignore elastic strains and consider secondary creep alone described by a power law, then combining the above we have

$$\dot{\epsilon} = \frac{\dot{l}}{l} = -\frac{\dot{A}}{A} = B\sigma^n = B\left(\frac{P}{A}\right)^n = B\sigma_0^n\left(\frac{A_0}{A}\right)^n$$

and therefore

$$A^{n-1}\,\mathrm{d}A = -B\sigma_0^n - A_0^n\ \mathrm{d}t$$

After integrating,

$$nB\sigma_0^n t = 1 - (A/A_0)^n$$

and we see that A reduces to zero in a finite time – the rupture time

$$t_R^D = \frac{1}{nB\sigma_0^n} \tag{2.6}$$

On a log–log plot of lifetime against initial stress (*Figure 2.14*), this gives a straight line. This theory of ductile rupture does not in practice describe the mechanism too well since elastic effects and primary creep are neglected. Although the theory does appear to be adequate if there is a long secondary stage, in general equation (2.6) tends to overestimate the lifetime. This theory

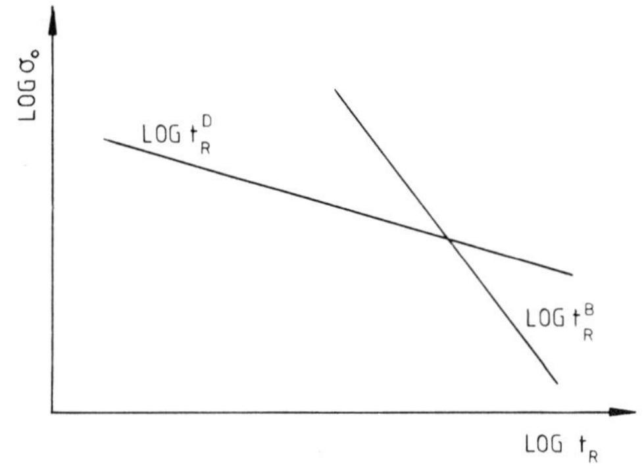

Figure 2.14 Ductile and brittle creep rupture curves

also predicts that for any finite load a finite lifetime will result; in fact, experimental observation is that there is a load at which the specimen will rupture instantaneously. Since large strains are assumed, instantaneous plasticity may also be present. Using a total deformation theory for illustration, we have

$$\epsilon = \epsilon_P + \epsilon_C \qquad\qquad \epsilon_P = B_0\sigma^{n_0}$$

again ignoring elastic strains. With the logarithmic definition of strain and using incompressibility, it follows that

$$-\frac{\dot{A}}{A}\left[1 - n_0 B_0\,\sigma_0^{n_0}\left(\frac{A_0}{A}\right)^{n_0}\right] = B\sigma_0^n\left(\frac{A_0}{A}\right)^n$$

and rupture occurs when $\dot{A} \to \infty$ at a finite value of area given by

$$A_R = (n_0 B_0)^{1/n_0} P$$

For instantaneous rupture at a load P_0, we should have $A_R = A_0$, i.e.

$$P_0 = \frac{A_0}{(n_0 B_0)^{1/n_0}}$$

In the theory of brittle rupture, at low stress these geometry changes are neglected and it is assumed that during creep small cavities appear and grow in time, reducing the effective load-bearing area of the specimen. To describe this phenomenon, a *damage parameter* ω is introduced so that the effective area is given by $(1-\omega)A_0$ and the *net stress* by

$$\sigma = \frac{P}{A} = \frac{P}{A_0(1-\omega)} = \frac{\sigma_0}{1-\omega}$$

where ω increases from zero in the virgin state to unity at rupture. This damage parameter is a further example of a hidden variable introduced in the previous section, for which we should specify a suitable growth law. It is assumed that the rate of damage creation depends on the current value of net stress

$$\dot{\omega} = D\sigma^k$$

This can be rearranged on eliminating σ to obtain the rupture time as

$$t_R^B = \int_0^1 \frac{d\omega}{D[\sigma_0/(1-\omega)]^k} = \frac{1}{(1+k)D\sigma_0^k} \tag{2.7}$$

which again gives a straight line on a log–log plot (*Figure 2.14*). Examination of *Figure 2.14* shows that $k < n$ and $D > B$. The rupture condition in this brittle model is $\omega = 1$, that is, the area reduces to zero. As in the case of ductile rupture. brittle rupture can occur at a finite area when the damage parameter is less than unity. One way of incorporating this into our model is to introduce the concept of instantaneous damage creation such that the total damage is the sum of an instantaneous part and an evolving part. Again for illustration purposes, we assume that the instantaneous part is a power law in the net stress such that

$$\dot{\omega} = \frac{d}{dt}(D_0\,\sigma^{k_0}) + D\sigma^k$$

Then

$$\dot{\omega}\left[1 - D_0\left(\frac{\sigma_0}{1-\omega}\right)^{k_0}\frac{k_0}{1-\omega}\right] = D\sigma^k$$

and damage increases without bound, $\dot{\omega} \to \infty$, when

$$\omega = 1 - [k_0(D_0\sigma_0^{k_0})]^{1/(k_0+1)} < 1$$

By definition, the damage on loading is $\omega_0 = D_0\sigma_0^{k_0}$ and the load which gives instantaneous brittle rupture can be found by solving the equation

$$\omega_0 = 1 - (k_0\omega_0)^{1/(k_0+1)}$$

The above analyses are not particularly useful for modelling more complex structural behaviour. They predict a sudden transition from ductile to brittle behaviour, which of course is not the case – rather, there is a smooth transition. Moreover it is difficult, in the case of a structure, to decide which mechanism predominates. It is of course possible to describe the smooth transition in the uniaxial curves, for example by combining the ductile and brittle models. This approach will not be pursued here; rather, in the absence of large strains, it is common to ignore the ductile rupture mechanism and to concentrate on brittle rupture. This has led to the study of a new subject called *continuum damage mechanics*, being a system for the description of the deterioration of structures under load due to the growth of damage. One way to construct a mathematical model of this is to introduce the damage parameter as a state variable into the constitutive relations developed in the previous section, so that

$$\dot{\epsilon}_C = f(\alpha, \omega) \qquad \dot{\omega} = g(\sigma, \omega) \tag{2.8}$$

(indeed now ω need have no specific physical interpretation). No distinction is made between the initial stress σ_0 and the stress σ. For example, equations (2.8) may take the forms

$$\dot{\epsilon}_C = B\left(\frac{\sigma}{1-\omega}\right)^n \qquad \dot{\omega} = D\left(\frac{\sigma}{1-\omega}\right)^k \tag{2.9}$$

Such constitutive relations can be used in place of equation (2.1) if it is expected that tertiary creep is significant. They are assumed to hold true for variable loading, but can be made more realistic by introducing further state variables. We will leave the topic of creep rupture for the present, but return to it in a later chapter.

Further reading

Kennedy, A.J., *Processes of Creep and Fatigue in Metals,* Oliver & Boyd, 1962.
Garofalo, F., *Fundamentals of Creep and Creep Rupture in Metals*, Macmillan, 1965.
Gittus, J., *Creep, Viscoelasticity and Creep Fracture in Solids,* Applied Science Publ., 1975.
Krempl, E., *Cyclic creep – An Interpretive Literature Survey*, Bulletin No. 195, Welding Research Council, June 1974, pp. 63–123.
Hart, E.W., *et al.* Phenomenological theory : a guide to constitutive relations and fundamental deformation properties, in *Constitutive Equations in Plasticity*, Ed. A. Argon, MIT Press, 1975, Chap. 5, pp. 149–97.
Hart, E.W., Constitutive relations for the non-elastic deformation of metals, *Trans. ASME, J. Eng. Mat. Techn.,* 98, 193–202, 1976.
Miller, A.K., An inelastic constitutive model for monotonic, cyclic, and creep deformation, *Trans. ASME, J. Eng. Mat. Techn.,* 98, 97–105, 1976.
Ponter, A.R.S. and Leckie, F.A., Constitutive relationships for the time dependent deformation of metals, *Trans. ASME, J. Eng. Mat. Techn.,* 98, 47–51, 1976.
Robinson, D.N., *A Unified Creep–Plasticity Model for Structural Metals at High Temperatures,* Report ORNL/TM-5969, Oak Ridge Natl Lab., 1978.
Murakami, S. and Ohno, N., A constitutive equation of creep for high temperature component analysis, *Proc. 6th Int. Conf. on Structural Mechanics in Reactor Technology, Paris,* 1981, Paper L7/3.

Chapter 3

Simple component behaviour

The analysis of complex structures will normally require the uniaxial constitutive relations derived in the previous chapter to be extended to deal with multiaxial stress states. Nevertheless, some simple components – beams and bar structures – can be analysed using the uniaxial relationships. In this chapter the creep behaviour of a number of such components will be examined as a means of introducing the basic approaches to stress analysis for creep and of illustrating some simple approximations which can be made. Throughout, we will use the simplest constitutive relation in rate form for variable stress and power law creep, that is, generally

$$\dot{\epsilon} = \frac{\dot{\sigma}}{E} + g(t)\sigma^n$$

We will begin this study with a classic problem – the steady creep of a beam in bending. In *steady creep of a structure* the stresses and deformation rates are constant in time, so that the material is in secondary creep and the constitutive relation takes the simpler form

$$\dot{\epsilon} = B\sigma^n$$

3.1 Example: Steady creep of a beam in bending

Consider a beam of bisymmetrical cross-section, of area A, loaded by a constant bending moment (*Figure 3.1*). We assume pure bending with plane sections remaining plane during deformation: then only a longitudinal stress σ and longitudinal strain ϵ are of significance. Compatibility of the strain with a curvature κ is ensured if

$$\epsilon = \kappa y \tag{3.1}$$

and equilibrium between stress and applied moment dictates that

$$M = \int_A \sigma y \, dA \tag{3.2}$$

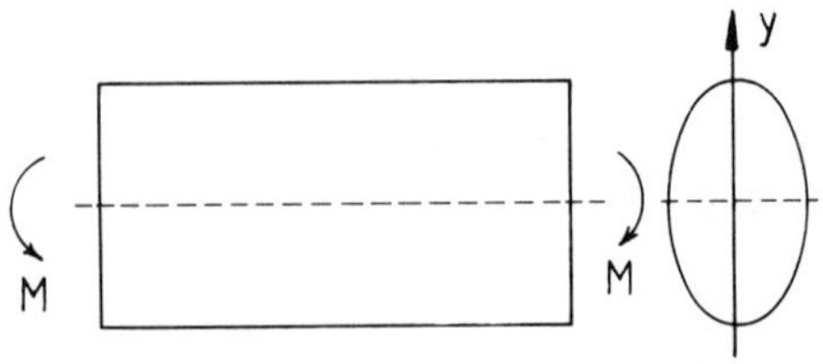

Figure 3.1 Geometry of beam in bending

For steady creep in a structure, the stress and rate of change of deformation are constant in time so that the constitutive relation reduces to

$$\dot{\epsilon} = B\sigma^n \tag{3.3}$$

The procedure used to resolve this problem is analogous to that in elasticity: first the constitutive relation (3.3) is inverted to give stress in terms of strain rate

$$\sigma = (\dot{\epsilon}/B)^{1/n}$$

which is then substituted into equilibrium (3.2)

$$M = \int_A \left(\frac{\dot{\epsilon}}{B}\right)^{1/n} y\,\mathrm{d}A$$

and using the strain–displacement relation (3.1) there results

$$\dot{\kappa} = B\left(\frac{M}{I_n}\right)^n \qquad I_n = \int_A y^{1+1/n}\mathrm{d}A \tag{3.4}$$

Substitution into (3.1) gives the strain rate, and substitution into the inverted constitutive relation gives the stress

$$\sigma = \frac{My^{1/n}}{I_n} \tag{3.5}$$

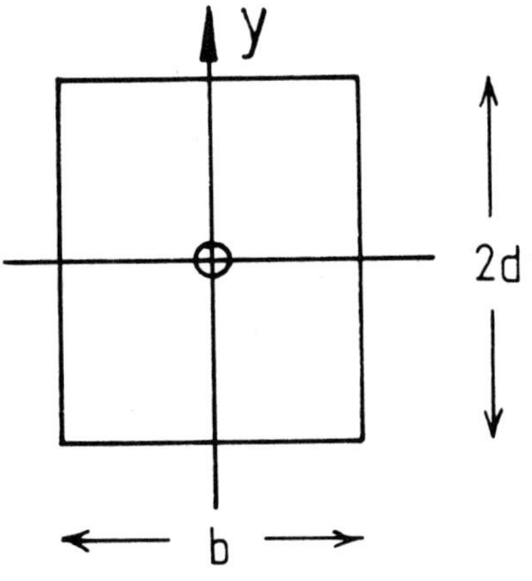

Figure 3.2 Rectangular cross-section beam

It is worth examining the stress distribution in more detail here. First suppose that the beam is of rectangular cross-section with breadth b and depth $2d$ (*Figure 3.2*). Then

$$I_n = \frac{2bd^{\,2+1/n}}{2 + 1/n}$$

The stress may be arranged in a non-dimensional form in terms of the maximum stress in a linear elastic beam of the same shape under the same load, i.e.

$$\frac{\sigma}{\hat{\sigma}_e} = \frac{2n+1}{3n}\left(\frac{y}{d}\right)^{1/n}$$

The maximum creep stress is evidently also at $y = d$ on the outside fibre. Therefore

$$\frac{\hat{\sigma}}{\hat{\sigma}_e} = \frac{2}{3} + \frac{1}{3} \cdot \frac{1}{n} \tag{3.6}$$

which is *linear* in the reciprocal of the exponent n. This has been shown to be a fairly general, but of course mostly approximate, feature of components evaluated with the power law and there is evidence of its validity for a wide range of components. This is rather useful. Finally, the stress distribution across the beam is shown in *Figure 3.3*.

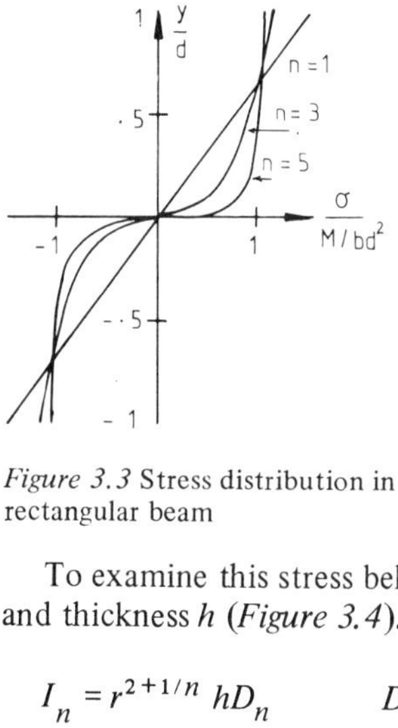

Figure 3.3 Stress distribution in rectangular beam

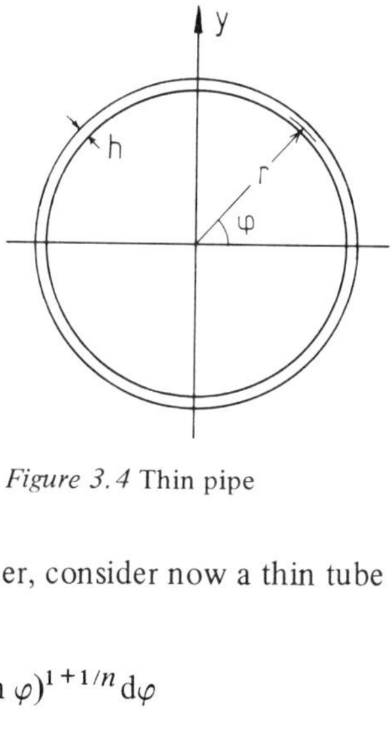

Figure 3.4 Thin pipe

To examine this stress behaviour further, consider now a thin tube of radius r and thickness h (*Figure 3.4*). Here

$$I_n = r^{2+1/n}\, hD_n \qquad D_n = \int_0^{2\pi} (\sin\varphi)^{1+1/n}\, d\varphi$$

and the peak creep stress in terms of the peak elastic stress is

$$\frac{\hat{\sigma}}{\hat{\sigma}_e} = \frac{\pi}{D_n} \tag{3.7}$$

where D_n can be found explicitly in terms of a Gamma function but is more readily obtained numerically. A few values are:

n	1	3	5	7	9	∞
D_n	π	3.643	3.776	3.836	3.864	4.0

If the peak creep stress (3.7) is plotted against the reciprocal of the exponent n, a straight line results (*Figure 3.5*). Of course, in this example the linearity is only approximate. This process can be applied in reverse and may be useful in design.

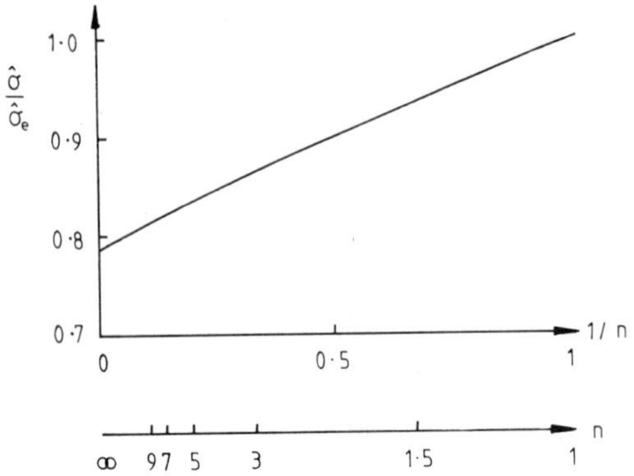

Figure 3.5 Variation of maximum stress in thin pipe with reciprocal of creep exponent

The case $n = 1$ is of course analogous to linear elasticity, but it is also noted that the case $n \to \infty$ is analogous to a rigid–plastic material. The latter is known as the *plastic analogy* and allows the creep stresses corresponding to $n \to \infty$ to be identified with the limit stresses of an equivalent rigid–plastic analysis. For the thin tube, the limit stress can be evaluated as

$$\hat{\sigma}_L = \frac{M}{4hr^2}$$

while the peak elastic stress is

$$\hat{\sigma}_e = \frac{M}{\pi hr^2}$$

If these are identified with the extremes of the creep behaviour for a power law for $n \to \infty$ and $n = 1$ respectively, and it is assumed that the peak stress is linear in the reciprocal of the exponent, then the creep stress for any other value of the exponent n is given approximately by

$$\hat{\sigma} \simeq \hat{\sigma}_\infty + \frac{1}{n}(\hat{\sigma}_1 - \hat{\sigma}_\infty)$$

or alternatively

$$\frac{\hat{\sigma}}{\hat{\sigma}_e} \simeq \frac{\pi}{4} + \frac{1}{n}\left(1 - \frac{\pi}{4}\right)$$

which can be compared numerically with equation (3.7). The agreement is excellent.

We conclude this study of the simple beam in bending by introducing the reader to one of the inherent problems in creep stress analysis. The problem arises since we cannot ordinarily determine either the functional form of the constitutive relation or values for the material parameters therein with any great precision for a real material. In our stress analysis we are usually forced to adopt an idealised mathematical model of the material behaviour. In elasticity, this uncertainty about the material behaviour is not critical since the response is linear. Unfortunately, the nonlinearity of the creep constitutive equation can amplify the effect of this uncertainty.

Let us examine the rectangular cross-section beam: the curvature rate is given by equation (3.4) as

$$\dot{\kappa} = \frac{B}{d}\left(1 + \frac{1}{2n}\right)^n \left(\frac{M}{bd^2}\right)^n$$

By way of illustration we take as a typical creep law

$B = 2.8 \times 10^{-26}$ $\qquad n = 4.7$

These values are not exact; for example, the exponent n may have been subject to rounding errors and lie in the range

$$4.6 \leqslant n \leqslant 4.8$$

For the particular case

$$\frac{M}{bd^2} = 10000$$

in appropriate units, the following range of values of the normalised curvature rate $\dot{\kappa}\, d$ is obtained:

n	4.6	4.7	4.8
$\dot{\kappa}d$	1.12E–7	2.80E–7	7.28E–7

It is obvious that the range of possible values of $\dot{\kappa}\, d$ for even a small range of the exponent n is large and clearly unacceptable. A little experimentation tells us that this situation has occurred due to the magnitude of the normalised grouping M/bd^2. Let us scale it and write

$$\frac{\dot{\kappa}}{\dot{\epsilon}_0} = \frac{1}{d}\left(1 + \frac{1}{2n}\right)^n \left(\frac{1}{\sigma_0}\,\frac{M}{bd^2}\right)^n = \delta(\sigma_0, n)$$

where σ_0 is arbitrary and $\dot{\epsilon}_0 = B\sigma_0^n$. For simplicity, let $\sigma_0 = \alpha M/bd^2$ where the scalar α is also arbitrary. After some experimentation we find that the least error occurs if the value of the scalar α is close to unity; *Figure 3.6* shows the variation in dδ with n for various values of α. Indeed, it can be seen that for $\alpha = 1$ it is virtually constant over the range in question!

Why this should happen can be explained by examining the whole range of the exponent n (*Figure 3.6*). It can be seen that

$\alpha > 1 \qquad \delta \to 0$

$\qquad\qquad\qquad\qquad n \to \infty$

$\alpha < 1 \qquad \delta \to \infty$

However if $\alpha = 1$ this grouping tends to a finite limit; in fact, it can be shown that

$$\delta \to \sqrt{e}/d$$

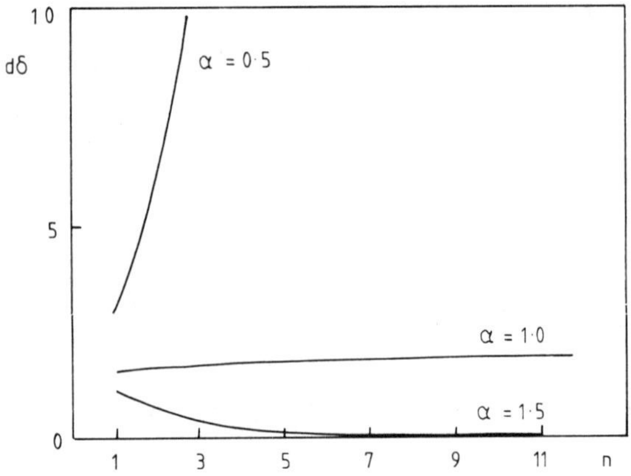

Figure 3.6 Variation of scaling factor with n for various values of α

Thus for sufficiently large n, say $n \geqslant 3$ if we choose the scalar $\alpha = 1$, we can write approximately

$$\dot{\kappa} \simeq \delta(\sigma_R, \infty) \times \dot{\epsilon}_R \tag{3.8}$$

where $\sigma_R = M/bd^2$

It should be pointed out that we have not quite eliminated our original difficulty since $\dot{\epsilon} = B\sigma_R^n$ is still sensitive to variations in the exponent n. The only way to overcome this remaining uncertainty is to get more information about the material behaviour. In this case, it may be asked which additional tests should be performed. If this is not feasible we would like to make the best use of the data that we have available, e.g. by making use of known scatter bands. The key to finding the most relevant test to perform or where to look in the available data can be found in relation (3.8) if we interpret $\dot{\epsilon}_R$ as the result of a uniaxial test held at the stress σ_R. We call this a *reference stress.* It has been chosen here such that the *scaling factor*

$$\delta(\sigma_0, n) = \dot{\kappa}/\dot{\epsilon}_0$$

is virtually independent of the material parameter n, i.e. $\delta(\sigma_R, n) \simeq \delta(\sigma_R, \infty)$.

We know that the scaling factor δ is in error by no more than 2% for $n \geqslant 3$, and to overcome the remaining uncertainty scatter bands on the reference stress test $\dot{\epsilon}_R$ may be obtained either from additional testing or from the original data. However, the result for $\dot{\kappa}$ has been desensitised to the value of n since equation (3.8) can now be used instead of equation (3.4).

3.2 Example: Steady creep of a non-uniformly heated structure

The study of space frameworks provides numerous examples wherein the uniaxial constitutive relations can be utilised. Consider the symmetric three-bar structure shown in *Figure 3.7* where bar 1 is at a temperature T_1 and bars 2 are at a temperature $T_2 < T_1$. A vertical mechanical load Q is also applied. If the bars have equal cross-sectional area A and stresses σ_1 and σ_2, then equilibrium requires that

$$\sigma_1 + 2\sigma_2 \cos \vartheta = Q/A$$

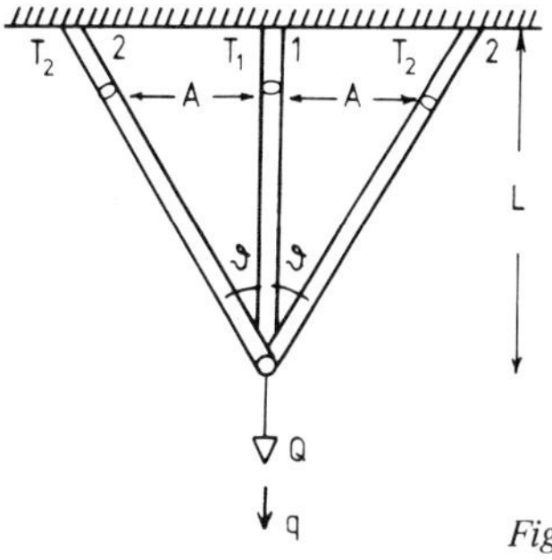

Figure 3.7 Geometry of three-bar structure

If the structure displaces a vertical amount q then the strains in each bar are given by

$$\epsilon_1 = q/L \qquad \epsilon_2 = (q\cos^2\vartheta)/L$$

The basic system of equations for this structure is completed by the constitutive relations which here take the form ($\gamma = \Delta H/R$)

$$\dot{\epsilon}_1 = C\exp(-\gamma/T_1)\sigma_1^n \qquad \dot{\epsilon}_2 = C\exp(-\gamma/T_2)\sigma_2^n$$

Following the procedure used in the previous example, the displacement rate is obtained as

$$\frac{\dot{q}}{L} = \frac{C(Q/A)^n\exp(-\gamma/T_1)}{[1 + 2(\cos\vartheta)^{1+2/n}e^{\beta}]^n} \tag{3.9}$$

while the stresses are

$$\begin{aligned}\sigma_1 &= \frac{Q/A}{1 + 2(\cos\vartheta)^{1+2/n}e^{\beta}}\\ \sigma_2 &= \frac{(Q/A)e^{\beta}\cos^{2/n}\vartheta}{1 + 2(\cos\vartheta)^{1+2/n}e^{\beta}}\end{aligned} \tag{3.10}$$

where

$$\beta = \frac{\gamma}{n}\left(\frac{1}{T_2} - \frac{1}{T_1}\right)$$

has some significance as a measure of the 'temperature sensitivity' of the material. Typically $0 \leqslant \beta \leqslant 2$ where $\beta = 0$ corresponds to the isothermal solution for the maximum temperature. One interesting feature is that if the stresses are plotted against $1/n$ for various values of β (*Figure 3.8*), then straight lines are again obtained. Let us now attempt to express this in the reference stress form by writing

$$\dot{q} = \delta(\sigma_0, n) \times \dot{\epsilon}_0$$

where now for an arbitrary σ_0

$$\dot{\epsilon}_0 = C\exp(-\gamma/T_1)\sigma_0^n$$

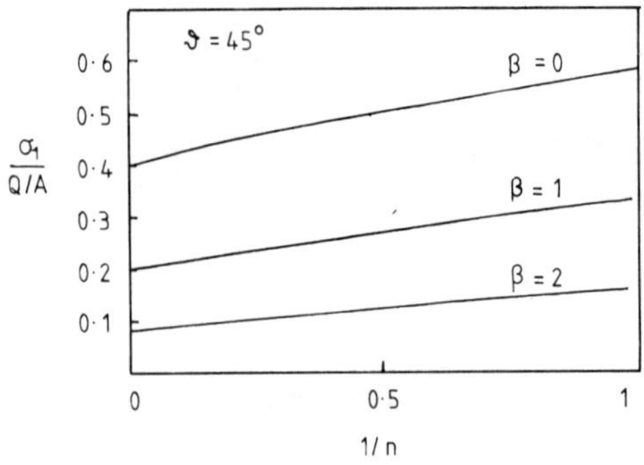

Figure 3.8 Variation of stress with reciprocal of creep exponent

corresponding to the maximum temperature. The scaling factor is

$$\frac{\delta(\sigma_0, n)}{L} = \frac{[(1/\sigma_0)(Q/A)]^n}{[1 + 2(\cos\vartheta)^{1+2/n} e^{\bar{\gamma}/n}]^n}$$

with

$$\bar{\gamma} = \gamma\left(\frac{1}{T_2} - \frac{1}{T_1}\right)$$

For convenience, we will write $\sigma_0 = \alpha(Q/A)$. Following the same procedure as before, we attempt to find α such that the scaling factor δ has a limiting value as $n \to \infty$. In fact, if we define

$$\alpha_R = \frac{1}{1 + 2\cos\vartheta}$$

for all $\bar{\gamma}$, then

$$\begin{array}{lll} \alpha > \alpha_R & \delta \to 0 & \\ & & n \to \infty \\ \alpha < \alpha_R & \delta \to \infty & \end{array}$$

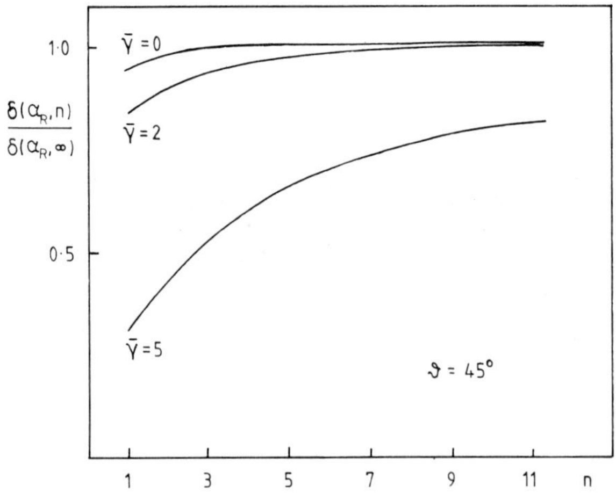

Figure 3.9 Variation of scaling factor with n

but if $\alpha = \alpha_R$, then

$$\delta(\alpha_R, n) \to (e^{\bar{\gamma}/2} \cos\vartheta)^{-4\cos\vartheta/(1+2\cos\vartheta)} \qquad n \to \infty$$

However, if we now plot the ratio $\delta(\alpha_R, n)/\delta(\alpha_R, \infty)$ (*Figure 3.9*), we see that for increasing values of the temperature parameter $\bar{\gamma}$ the limiting value of the scaling factor becomes a worse approximation for the whole range of the exponent n. We will find this generally true for non-uniformly heated structures later in Chapter 6.

3.3 Example: Forward creep of a hinged bar structure

In steady creep, the stresses remain constant, so that elastic strains may be neglected and the creep strain rate is constant. In general for a component under constant loading, there will be a phase of stress redistribution as the stresses change from their initial elastic values to their steady-state values. During this phase, elastic stresses must be included. We will demonstrate this for a simple bar structure.

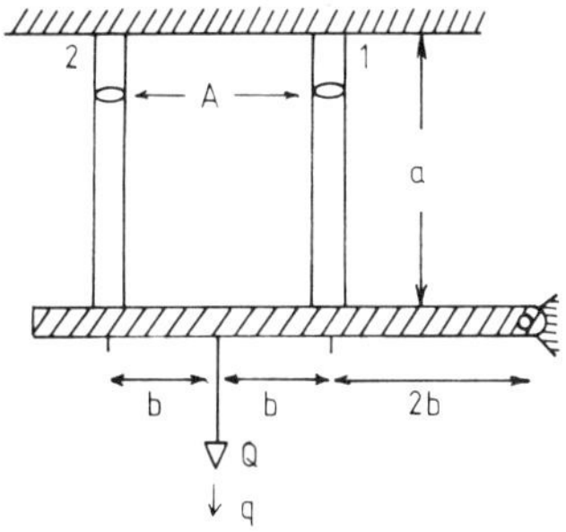

Figure 3.10 Geometry of hinged bar structure

Consider two bars of equal length a and cross-sectional area A which are attached to a rigid rod which has a hinge at one end (*Figure 3.10*). A load Q is applied as shown; then if σ_1 and σ_2 are the stresses in the bars, equilibrium requires that

$$\sigma_1 + 2\sigma_2 = 3Q/2A \qquad (3.11)$$

Compatibility of the system is ensured if there is no parting of the bars from the rigid rod. Since there is a small rotation about the hinge, the extension of the second bar is twice that of the first

$$\delta_2 = 2\delta_1$$

These are related to their respective strains by

$$\epsilon_1 = \delta_1/a \qquad \epsilon_2 = \delta_2/a$$

and if we define the displacement q at the point of application of the load Q, then the strain–displacement relations are

$$\epsilon_1 = \frac{2}{3}\, q/a \qquad \epsilon_2 = \frac{4}{3}\, q/a \qquad (3.12)$$

Finally the constitutive relations are assumed to have the form

$$\dot{\epsilon}_1 = \frac{\dot{\sigma}_1}{E} + g(t)\,\sigma_1^n \qquad \dot{\epsilon}_2 = \frac{\dot{\sigma}_2}{E} + g(t)\sigma_2^n \tag{3.13}$$

under isothermal conditions.

Initially the creep strains are zero, and on application of the load there is an instantaneous elastic response given by

$$q(0) = \frac{3}{10}\frac{a}{E}\frac{3Q}{2A}$$

$$\sigma_1(0) = \frac{1}{5}\frac{3Q}{2A} \tag{3.14}$$

$$\sigma_2(0) = \frac{2}{5}\frac{3Q}{2A}$$

We can derive suitable equations for the phase of stress redistribution as follows. The basic relations are written in rate form

$$\dot{\sigma}_1 + 2\dot{\sigma}_2 = 0$$

$$\dot{\epsilon}_1 = \frac{2}{3}\dot{q}/a\cdot \qquad \dot{\epsilon}_2 = \frac{4}{3}\dot{q}/a$$

$$\dot{\epsilon}_1 = (\dot{\sigma}_1/E) + \dot{\epsilon}_{C1} \quad \dot{\epsilon}_2 = (\dot{\sigma}_2/E) + \dot{\epsilon}_{C2}$$

where

$$\dot{\epsilon}_{C1} = g(t)\sigma_1^n \qquad \dot{\epsilon}_{C2} = g(t)\sigma_2^n$$

If we think of the creep strain rates $\dot{\epsilon}_{C1}$, $\dot{\epsilon}_{C2}$ as being given, then these equations are simply a linear elastic problem for the rates $\dot{\sigma}_1$, $\dot{\sigma}_2$ and $\dot{q}$. They can be solved to give

$$\dot{\sigma}_1 = E\left(\frac{2}{5}\dot{\epsilon}_{C2} - \frac{4}{5}\dot{\epsilon}_{C1}\right)$$

$$\dot{\sigma}_2 = E\left(\frac{2}{5}\dot{\epsilon}_{C1} - \frac{1}{5}\dot{\epsilon}_{C2}\right)$$

$$\dot{q} = \frac{3}{10}\left(\dot{\epsilon}_{C1} + 2\dot{\epsilon}_{C2}\right)a$$

which can be written as a coupled system of first-order differential equations, i.e.

$$\frac{d\sigma_1}{dt} = Eg(t)\left(\frac{2}{5}\sigma_2^n - \frac{4}{5}\sigma_1^n\right)$$

$$\frac{d\sigma_2}{dt} = Eg(t)\left(\frac{2}{5}\sigma_1^n - \frac{1}{5}\sigma_2^n\right) \tag{3.15}$$

$$\frac{dq}{dt} = \frac{3}{10} g(t)(\sigma_1^n + 2\sigma_2^n)a$$

forming an initial value problem for the unknowns σ_1, σ_2 and q together with the initial condition, equations (3.14)

These equations can be simplified if we introduce a time scale

$$\tau = E\sigma_0^{n-1} \int g(t)\,dt$$

and new variables

$$S_1 = \frac{\sigma_1}{\sigma_0} \qquad S_2 = \frac{\sigma_2}{\sigma_0} \qquad \Delta = \frac{Eq}{a\sigma_0} \qquad (\dot{\ }) = \frac{d(\ )}{d\tau}$$

where $\sigma_0 = 3Q/2A$. Then the equations become

$$\begin{aligned} \frac{dS_1}{d\tau} &= \left(\frac{2}{5} S_2^n - \frac{4}{5} S_1^n\right) \\ \frac{dS_2}{d\tau} &= \left(\frac{2}{5} S_1^n - \frac{1}{5} S_2^n\right) \\ \frac{d\Delta}{d\tau} &= \frac{3}{10}(S_1^n + 2S_2^n) \end{aligned} \tag{3.16}$$

which must be solved subject to the initial conditions

$$S_1(0) = \frac{1}{5} \qquad S_2(0) = \frac{2}{5} \qquad \Delta(0) = \frac{3}{10}$$

These equations cannot be solved in closed form and numerical techniques are necessary. The best way to do this is to use any standard routine for the solution of initial value problems which can be found in any computer scientific subroutine library. Alternatively, the simplest solution procedure is to use Euler's method. In this, the solution is obtained at a number of discrete times $\tau_0, \tau_1, \ldots, \tau_k, \tau_{k+1}$ such that

$$\tau_0 = 0 \qquad \tau_{k+1} - \tau_k = \Delta\tau$$

with $\Delta\tau$ prescribed. Then the solution at the $(k + 1)$th step is obtained from that at the kth step according to the formulae

$$\begin{aligned} S_1(\tau_{k+1}) &= S_1(\tau_k) + \Delta\tau \dot{S}_1(\tau_k) \\ S_2(\tau_{k+1}) &= S_2(\tau_k) + \Delta\tau \dot{S}_2(\tau_k) \\ \Delta(\tau_{k+1}) &= \Delta(\tau_k) + \Delta\tau \dot{\Delta}(\tau_k) \end{aligned}$$

starting from the initial conditions. In this way, the solution can be evaluated for a given value of the material parameter n. It is to be expected that as we decrease the time step $\Delta\tau$ the solution will tend to the exact solution. The time distributions of S_1 and S_2 are shown in *Figure 3.11* and that of Δ in *Figure 3.12*.

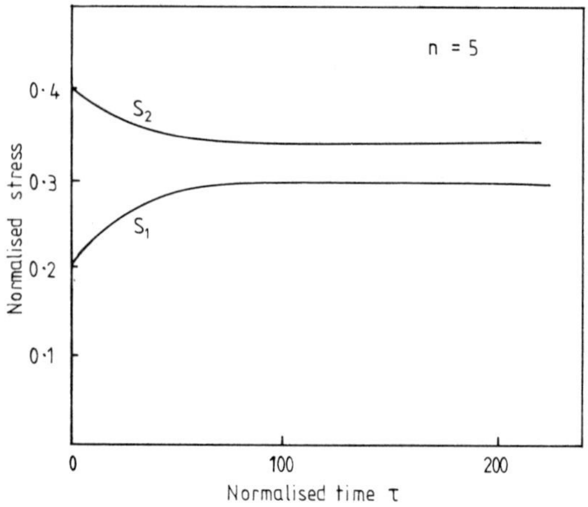

Figure 3.11 Redistribution of stress

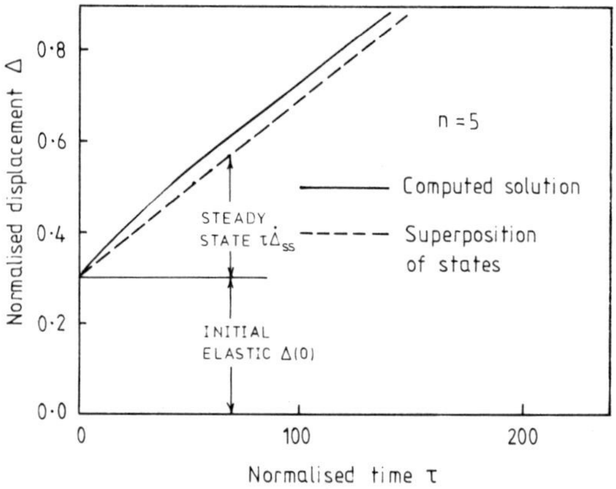

Figure 3.12 Evolution of vertical displacement

The response shown in *Figures 3.11* and *3.12* for a structural component under constant load typifies *forward creep*. The change in stress is due to stress redistribution within the component. What happens is that as time progresses the stress distribution, which at $\tau = 0$ is purely elastic, changes until it reaches a steady state. The higher elastic stresses tend to reduce and the lower elastic ones tend to increase due to creep. The process is one of *stress redistribution* and is termed *transient creep* (if primary creep is included in the constitutive relations, then transient creep includes this also). When the stresses become steady (constant for constant load), the strain and displacement rates are also steady. For this structure, the steady state can be evaluated as

$$\sigma_1^{ss} = \frac{3Q}{2A} \quad \frac{1}{1 + 2^{1+1/n}}$$

$$\sigma_2^{ss} = \frac{3Q}{2A} \quad \frac{2^{1/n}}{1 + 2^{1+1/n}}$$

where the superscript 'ss' denotes the steady state. It can be seen for this structure that the process of stress redistribution takes place in a relatively short time and that the amount of deformation accumulated during redistribution is quite small. Then, to a good approximation, the displacement can be evaluated by simply adding the initial elastic displacement to the amount accumulated during steady creep

$$\Delta(\tau) \simeq \Delta(0) + \tau \frac{d\Delta^{ss}}{d\tau} \tag{3.17}$$

We call this approximation *superposition of states* for brevity. *It is defined as the superposition of an elastic part, calculated as if there were no creep, and a pure creep part, calculated as if there were no elastic effects.* For constant loading this reduces to the approximation (3.17); however, we shall see later that it can also be used for other types of loading, in particular relaxation.

The above comments are generally valid, although we have used a simple time hardening model of creep. Specifically, we have removed the time function from the creep law

$$\frac{d\epsilon}{dt} = \frac{1}{E}\frac{d\sigma}{dt} + g(t)\sigma^n$$

by introducing another time scale

$$\tau = E\sigma_0^{n-1} \int g(t)\, dt$$

for some σ_0. The creep law then becomes

$$\frac{d\Sigma}{d\tau} = \frac{dS}{d\tau} + S^n$$

where there are new variables

$$S = \sigma/\sigma_0 \qquad \Sigma = E\epsilon/\sigma_0$$

It will be difficult to employ such a transformation if we use a more complex constitutive equation, for example an internal-variable model. However, the transformation is very attractive, and should always be applied for time hardening with a power law.

3.4 Example: Cyclic creep of a hinged bar structure

It is instructive to examine the behaviour of the hinged bar structure under cyclic loading. There is one difficulty – the use of the simple time-hardening constitutive relation

$$\dot{\epsilon} = \frac{\dot{\sigma}}{E} + g(t)\,\sigma^n$$

which, it is recalled, was developed from constant-load tests, is questionable. This difficulty can be avoided if it is presumed that the time function $g(t)$ is cyclic with a period equal to that of the loading of the structure. The form of $g(t)$ can then be deduced from cyclic uniaxial tests, but it is important to

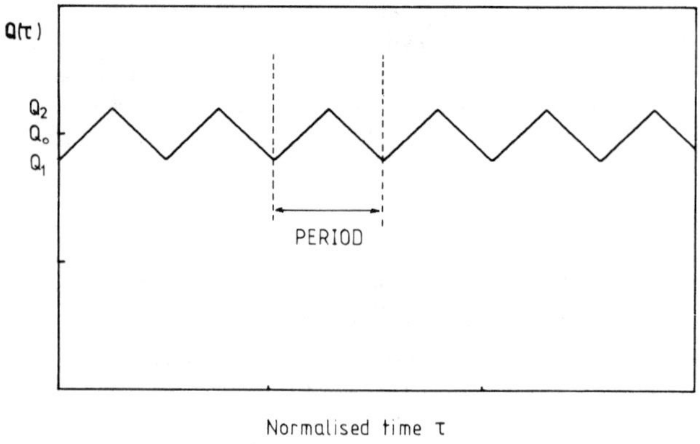

Figure 3.13 Cyclic load on hinged bar structure

appreciate that the resulting constitutive model is poor, although it is acceptable for engineering purposes.

Let us assume a cyclic load $Q(t)$ as prescribed in *Figure 3.13*. In this case the equations (3.15) for stress redistribution can be written as

$$\frac{d\sigma_1}{dt} = Eg(t)\left(\frac{2}{5}\sigma_2^n - \frac{4}{5}\sigma_1^n\right) + \dot{\sigma}_1^0$$

$$\frac{d\sigma_2}{dt} = Eg(t)\left(\frac{2}{5}\sigma_1^n - \frac{1}{5}\sigma_2^n\right) + \dot{\sigma}_2^0 \qquad (3.18)$$

where

$$\sigma_1^0(t) = \frac{1}{5}\,\frac{3Q(t)}{2A} \qquad \sigma_2^0(t) = \frac{2}{5}\,\frac{3Q(t)}{2A}$$

are the stresses obtained from the solution of an *equivalent elastic problem*. These stresses are, of course, cyclic. For convenience, we define the *residual stresses*

$$\rho_1(t) = \sigma_1(t) - \sigma_1^0(t)$$

$$\rho_2(t) = \sigma_2(t) - \sigma_2^0(t)$$

The principal advantage of doing this is to avoid problems in defining the stress rates $\dot{\sigma}_1^0$ and $\dot{\sigma}_2^0$ if there is a change in direction of the loads.

The two differential equations (3.18), rewritten in terms of the residual stresses, can now be solved numerically in an identical manner to that of the previous section. In *Figure 3.14* the variation of $\rho_1(t)$ with the normalised time scale

$$\tau = E\sigma_0{}^{n-1}\int g(t)dt, \; \sigma_0 = \frac{3Q_0}{2A}$$

is shown. It can be seen that a cyclic residual stress state, with a period equal to that of the applied load, is eventually reached. This will be found to be generally true for structures under cyclic load, even if a better material model is used.

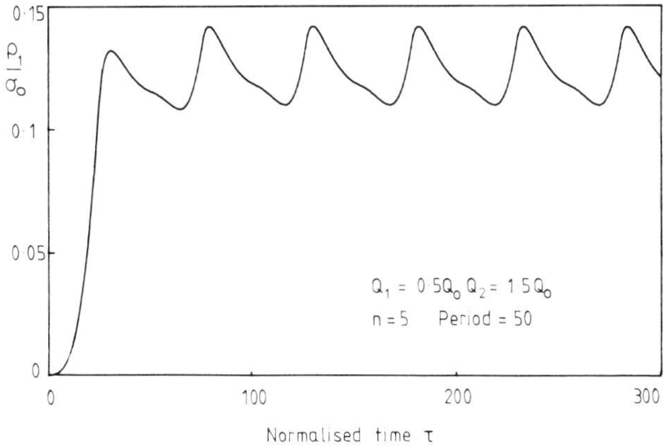

Figure 3.14 Variation of residual stress with time

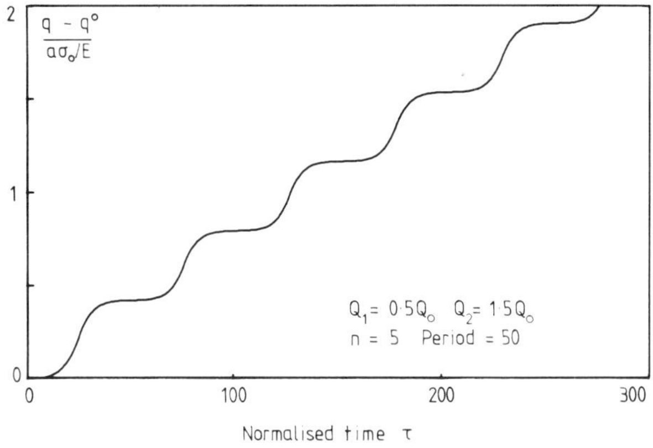

Figure 3.15 Variation of residual displacement with time

This asymptotic stress state is called the *cyclic stationary state*. In *Figure 3.15* the residual vertical deflection $q(t) - q^0(t)$ is shown where q^0 is the equivalent elastic solution; it is to be noted that this does *not* reach a cyclic state.

3.5 Example: Relaxation of a beam in bending

There is a tendency in thinking about creep to consider constant-load, or prescribed load, situations, where the deformations have to be estimated with time, exclusively. However, there is another important problem known as *relaxation* where the displacement is fixed and creep causes a reduction of the load required to maintain that displacement. For this situation, stress redistribution again takes place as the load relaxes. It is instructive first to examine what happens in the uniaxial case. Suppose that the specimen is of length l and cross-sectional area A subject to a constant displacement δ. The strain is given as

$$\epsilon = \delta / l$$

and the material is assumed to obey a time hardening power law of creep

$$\frac{d\epsilon}{dt} = \frac{1}{E}\frac{d\sigma}{dt} + g(t)\,\sigma^n$$

Since δ is assumed to be constant, $\dot{\delta} = 0$, and we have a single equation for the stress which can be solved in terms of the initial condition $\sigma = \sigma_0$, the stress on loading.

On rearranging

$$\int \frac{d\sigma}{\sigma^n} = -E \int g(t)\,dt$$

which yields after integration

$$\frac{\sigma(t)}{\sigma_0} = \left(1 + (n-1)\,E\sigma_0^{n-1} \int g(t)dt\right)^{-1/(n-1)} \tag{3.19}$$

Thus the load required to maintain this displacement

$$\sigma(t) = P(t)/A$$

can be obtained. According to equation (3.19) the load reduces with time, eventually reaching zero.

We turn now to the problem of a beam in bending, examined in steady creep in Section 3.1, *Figure 3.1*..It is supposed that the beam is bent under the action of a bending moment M_0 to a curvature κ_0 which is then fixed. If the material subsequently creeps according to a relation of the form

$$\dot{\epsilon} = \frac{\dot{\sigma}}{E} + g(t)\,\sigma^n$$

the problem is to determine the subsequent relaxation of the moment due to creep.

To solve this, we note that, since the curvature is held constant, the longitudinal strain, from equation (3.1), is also constant in time. Thus the longitudinal stress σ is essentially similar to the uniaxial condition, and the solution is the same, equation (3.18),

$$\frac{\sigma(t)}{\sigma(0)} = \left(1 + (n-1)E \int g(t)\,dt\,\sigma(0)^{n-1}\right)^{-1/(n-1)}$$

where the initial stress is

$$\sigma(0) = M_0 y / I$$

Then from equilibrium, equation (3.2), the moment can be evaluated at any time by

$$\frac{M(t)}{M_0} = \frac{1}{I}\int_A y^2 \left[1 + (n-1)E \int g(t)\,dt \left(\frac{M_0 y}{I}\right)^{n-1}\right]^{-1/(n-1)} dA \tag{3.20}$$

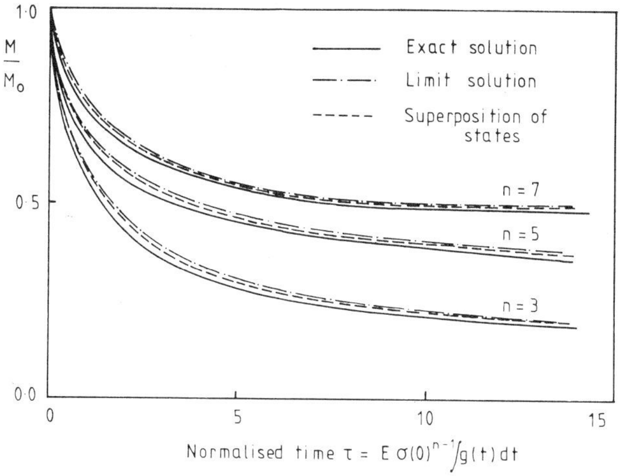

Figure 3.16 Relaxation of a rectangular beam in bending: comparison of exact and approximate solutions

The integral in equation (3.20) should be evaluated numerically, for example using Simpson's rule. Solution curves for a rectangular cross-section are shown in *Figure 3.16*. Inspection of these shows that the curves are similar to the uniaxial curves, equation (3.19). Let us write

$$\frac{M(t)}{M_0} = [1 + (n-1)\,k(t^*)t^*\sigma_0^{n-1}]^{-1/(n-1)}$$

for some σ_0, where for convenience we have defined

$$t^* = E\int g(t)\,dt$$

By definition

$$k(t^*) = \frac{1}{(n-1)\sigma_0^{n-1}}\left[\left(\frac{I/(t^*)^{1/(n-1)}}{\int_A y^2[1+(n-1)t^*(M_0 y/I^{n-1}]^{-1/(n-1)}dA}\right)^{n-1} - \frac{1}{t^*}\right]$$

and if we take the limit as $t^* \to \infty$, then

$$k(t^*) \to \left(\frac{M_0}{\sigma_0 \int_A |y| dA}\right)^{n-1}$$

Therefore the relaxation curve for the moment is asymptotically equivalent to a uniaxial test, equation (3.19), if we choose σ_0 such that $k(t^*) \to 1$; that is, if

$$\sigma_0 = \frac{M_0}{\int_A |y|\,dA}$$

Thus

$$\frac{M(t)}{M_0} \to \left(1 + (n-1)\,\sigma_0^{n-1}\,E\int g(t)\,dt\right)^{-1/(n-1)} \qquad (3.21)$$

which is also shown in *Figure 3.16*, for a rectangular cross-section.

Finally we employ the approximation of superposition of states – according to which the curvature is supposed to be composed of an elastic part, found from an elastic analysis, and a pure creep part, found from a creep analysis ignoring the elastic strains, assuming a history $M(t)$ of loading. The elastic solution is readily obtained as

$$\kappa_E(t) = M(t)/EI$$

and the pure creep solution, analogous to the steady-state solution of Section 3.1, is

$$\dot{\kappa}_C(t) = g(t)\left(\frac{M(t)}{I_n}\right)^n$$

Hence the total curvature is given by

$$\kappa_0 \simeq \kappa_E + \kappa_C \tag{3.22}$$

Since $\dot{\kappa}_0 = 0$, an equation for the unknown bending moment can be found as

$$\frac{1}{EI}\frac{dM}{dt} + g(t)\left(\frac{M}{I_n}\right)^n = 0$$

with the initial condition

$$M(0) = 0$$

This can easily be solved as in the uniaxial case to yield the approximation

$$\frac{M(t)}{M_0} \simeq \left(1 + (n-1)E\int g(t)\,dt\,M_0^{n-1}\frac{I}{I_n^n}\right)^{-1/(n-1)} \tag{3.23}$$

which is in the form of a uniaxial test, equation (3.19), on identifying

$$\sigma_0 = M_0\left(\frac{I}{I_n^n}\right)^{1/(n-1)}$$

The superposition of states approximation, equation (3.23), is shown for the rectangular cross-section also in *Figure 3.16*; in this case

$$\sigma_0 = \alpha(n)\,M/bd^2$$

where

$$\alpha(n) = \left(\frac{2}{3}\left(1 + \frac{1}{2n}\right)^n\right)^{1/(n-1)}$$

It is interesting to note that $\alpha(n)$ is virtually constant.

Further reading

Hult, J., *Creep in Engineering Structures*, Blaisdell, 1966.
Kachanov, L.M., *Theory of Creep*, 1960 (Trans. National Lending Library for Science and Technology, 1967).

Chapter 4

Creep under multiaxial states of stress

In this chapter we will briefly develop the constitutive equations for material creep under multiaxial states of stress. In practice, it is found that the multiaxial characteristics of creep are very similar to those of classical plasticity on which there is a considerable literature. The use of these constitutive equations in stress analysis will then be illustrated by means of a number of problems in steady creep for which analytical solutions are available.

4.1 A convenient multiaxial constitutive relation

The general state of stress at a point in a solid material element is prescribed by nine stress components. These are depicted in *Figure 4.1* using right-handed Cartesian axes (x, y, z) for directional reference. A plane area of the material

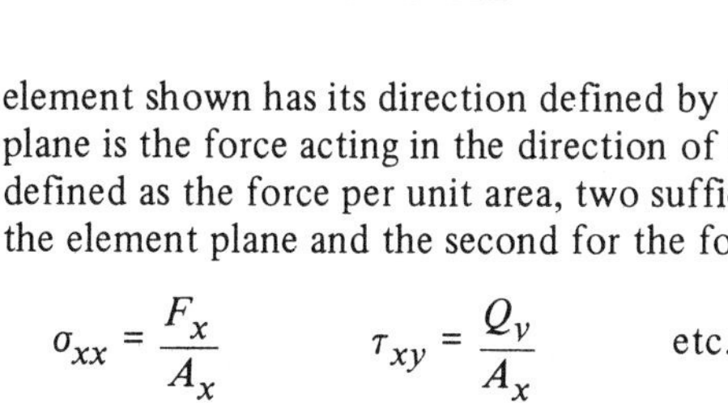

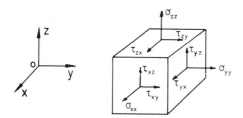

Figure 4.1 Components of stress

element shown has its direction defined by its outward normal; a force on this plane is the force acting in the direction of this outward normal. Since stress is defined as the force per unit area, two suffices are used, the first for the area of the element plane and the second for the force. Thus

$$\sigma_{xx} = \frac{F_x}{A_x} \qquad \tau_{xy} = \frac{Q_y}{A_x} \qquad \text{etc.}$$

where A_x is the area of the x-plane, F_x is the normal force in the x-direction, and Q_y is the shear force in the y-direction. The stress is represented algebraically by the *stress tensor*

$$\sigma_{ij} = \begin{bmatrix} \sigma_{xx} & \tau_{xy} & \tau_{xz} \\ \tau_{yx} & \sigma_{yy} & \tau_{yz} \\ \tau_{zx} & \tau_{zy} & \sigma_{zz} \end{bmatrix}$$

where normally $\tau_{xy} = \tau_{yx}, \ldots$, so that only six quantities are needed to specify the state of stress.

In any three-dimensional state of stress, there is always a particular set of orthogonal (mutually perpendicular) directions for which the stresses are entirely normal. These are the *principal directions* specified by the *principal axes* (1, 2, 3). The planes whose normals are in the principal directions are the *principal planes* and the stresses on them are the *principal (normal) stresses* σ_1, σ_2 and σ_3 annotated such that $\sigma_1 > \sigma_2 > \sigma_3$. The *principal shear stresses* act in the sections at 45° to the principal planes

$$\tau_1 = \frac{1}{2}(\sigma_2 - \sigma_3) \qquad \tau_2 = \frac{1}{2}(\sigma_3 - \sigma_1) \qquad \tau_3 = \frac{1}{2}(\sigma_1 - \sigma_2)$$

In a similar manner, there is introduced the infinitesimal *strain tensor* analogous to the stress tensor with lineal strain ϵ_{xx} analogous to the normal stress σ_{xx} and semi-shear strain $\gamma_{xy}/2$ analogous to shear stress τ_{xy}:

$$\epsilon_{ij} = \begin{bmatrix} \epsilon_{xx} & \gamma_{xy}/2 & \gamma_{xz}/2 \\ \gamma_{yx}/2 & \epsilon_{yy} & \gamma_{yz}/2 \\ \gamma_{zx}/2 & \gamma_{zy}/2 & \epsilon_{zz} \end{bmatrix}$$

and the principal directions of strain (which are not *a priori* the same as those for stress) together with the principal strains ϵ_1, ϵ_2 and ϵ_3 and *principal shears*

$$\gamma_1 = \epsilon_2 - \epsilon_3 \qquad \gamma_2 = \epsilon_3 - \epsilon_1 \qquad \gamma_3 = \epsilon_1 - \epsilon_2$$

In addition to small strains, it is assumed that the deformations are slow (or quasistatic) so that the strain-rate tensor is given by

$$\dot{\epsilon}_{ij} = \frac{d}{dt}\epsilon_{ij} = \frac{\partial}{\partial t}\epsilon_{ij} + \frac{\partial \epsilon_{ij}}{\partial \sigma_{kl}}\frac{\partial \sigma_{kl}}{\partial t} \simeq \frac{\partial}{\partial t}\epsilon_{ij}$$

In order to develop multiaxial constitutive relations, some experimental evidence is obviously required. In the following development, only creep secondary strains are considered, ignoring for the time being the contribution of elastic strains. Experimentally, it is found that materials react in different ways to normal and shear stresses, characterised by the *hydrostatic pressure*

$$\sigma_v = \frac{1}{3}(\sigma_1 + \sigma_2 + \sigma_3)$$

being the mean value of the principal stresses and the *effective (shear) stress*

$$\bar{\sigma} = \sqrt{2}\,(\tau_1^2 + \tau_2^2 + \tau_3^2)^{1/2}$$

These stresses are invariant to the choice of the coordinate system and are related to the stresses on the so-called *octahedral planes*. These are planes whose normals are equally inclined to the three principal directions (together they form an octahedron whose apexes lie at the centres of the faces of the principal cube). The hydrostatic pressure σ_v is the normal stress on the octahedral planes; the effective stress $\bar{\sigma}$ is proportional to the shear stress on the octahedral planes but normalised so that for uniaxial behaviour $\sigma_2 = \sigma_3 = 0$, $\sigma_1 = \bar{\sigma}$. In fact, if τ_{oct} is the shear stress on the octahedral plane, then

$$\tau_{oct} = \sqrt{2}\,\bar{\sigma}/3$$

In a similar manner the *octahedral (lineal) strain*

$$\epsilon_v = \frac{1}{3}(\epsilon_1 + \epsilon_2 + \epsilon_3)$$

and the *effective (shear) strain rate*

$$\bar{\dot{\epsilon}} = \sqrt{2}\,(\dot{\gamma}_1^2 + \dot{\gamma}_2^2 + \dot{\gamma}_3^2)^{1/2}/3$$

are introduced. It should be noted that $\dot{\bar{\epsilon}} \neq \bar{\dot{\epsilon}}$.

Experimental studies of multiaxial creep behaviour are commonly concerned with the stretching and shearing of plane sheets, tubes under internal pressure, bars under bending and axial tension or twisting and other simple components. It turns out that, for *isotropic* and *homogeneous* materials, creep is essentially a shear-dominated process; the following observations are made:

(1) Changes in the hydrostatic pressure have no influence on the creep behaviour. This is related to the observation that creep deformations appear as distortions – there is a change in shape without a change in volume. (Unlike for elastic behaviour where there is a change in volume.) The rate of change of volumetric strain (proportional to the octahedral strain) is then zero,

$$\dot{\epsilon}_1 + \dot{\epsilon}_2 + \dot{\epsilon}_3 = 0$$

(2) The principal shear strain rates are proportional to the principal shear stresses

$$\frac{\dot{\gamma}_1}{\tau_1} = \frac{\dot{\gamma}_2}{\tau_2} = \frac{\dot{\gamma}_3}{\tau_3} = 2\psi \qquad (4.1)$$

where ψ is a constant (but dependent on position in the stressed solid).

Combining these two results leads to the relations

$$\begin{aligned} \dot{\epsilon}_1 &= \psi(\sigma_1 - \sigma_v) \\ \dot{\epsilon}_2 &= \psi(\sigma_2 - \sigma_v) \\ \dot{\epsilon}_3 &= \psi(\sigma_3 - \sigma_v) \end{aligned} \qquad (4.2)$$

It still remains to discover the form of the constant ψ. Mutually subtracting the relations (4.2), squaring and subsequently adding, there results

$$\psi = 3\bar{\dot{\epsilon}}/2\bar{\sigma}$$

As a final experimental observation it is found that:

(3) The effective strain rate is related to the effective stress in the same way as the uniaxial relation, e.g.

$$\dot{\bar{\epsilon}} = g(t)f(\bar{\sigma}) \tag{4.3}$$

Therefore

$$\begin{aligned}\dot{\epsilon}_1 &= \frac{3}{2}\, g(t)\, \frac{f(\bar{\sigma})}{\bar{\sigma}}\, (\sigma_1 - \sigma_v)\\ \dot{\epsilon}_2 &= \frac{3}{2}\, g(t)\, \frac{f(\bar{\sigma})}{\bar{\sigma}}\, (\sigma_2 - \sigma_v)\\ \dot{\epsilon}_3 &= \frac{3}{2}\, g(t)\, \frac{f(\bar{\sigma})}{\bar{\sigma}}\, (\sigma_3 - \sigma_v)\end{aligned} \tag{4.4}$$

For the Norton power law $f(\bar{\sigma}) = \bar{\sigma}^n$. The form of relations (4.4) depends on the assumption of time hardening. For a strain hardening material

$$\dot{\bar{\epsilon}} = mB^{1/m} \frac{\bar{\sigma}^{n/m}}{\bar{\epsilon}^{(1-m)/m}} \qquad \psi = \frac{3}{2} mB^{1/m} \bar{\sigma}^{(n-m)/m} \bar{\epsilon}^{(1-1/m)}$$

Before proceeding, two points should be emphasised. First, it is apparent that the creep strain rates are uninfluenced by the superposition of a hydrostatic pressure p and are dependent only on changes in the *stress deviators* s_1, s_2 and s_3 defined by

$$s_i = \sigma_i - \sigma_v \qquad i = 1, 2, 3$$

for if σ_i is replaced by $\sigma_i' = \sigma_i + p$, $i = 1, 2, 3$, then noting that $\bar{\sigma} = \sqrt{\frac{3}{2}}(s_1^2 + s_2^2 + s_3^2)^{1/2}$ it follows that

$$\bar{\sigma}' = \bar{\sigma} \qquad s_i' = s_i \qquad i = 1, 2, 3$$

This is important if it is desired to invert the relations (4.4) in order to express stress in terms of strain rate; from (4.2)

$$s_i = \frac{1}{\psi}\, \dot{\epsilon}_i = \frac{2}{3} \frac{\bar{\sigma}}{\dot{\bar{\epsilon}}}\, \dot{\epsilon}_i \qquad i = 1, 2, 3$$

and assuming the constitutive relation (4.4) is invertible

$$\bar{\sigma} = f^{-1}\left(\frac{\dot{\bar{\epsilon}}}{g(t)}\right)$$

there results

$$s_i = \frac{2}{3}\left(\frac{f^{-1}(\dot{\bar{\epsilon}}/g(t))}{\dot{\bar{\epsilon}}}\right) \dot{\epsilon}_i \qquad i = 1, 2, 3 \tag{4.5}$$

In general, the state of stress σ_1, σ_2 and σ_3 may not be determined from (4.5), since $s_1 + s_2 + s_3 = 0$ and is therefore only known to within an arbitrary superimposed hydrostatic pressure. Nevertheless for the special case of plane stress $\sigma_3 = 0$ it can be shown that

$$\sigma_1 = \frac{4}{3}\left(\frac{f^{-1}(\dot{\bar{\epsilon}}/g(t))}{\dot{\bar{\epsilon}}}\right)(\dot{\epsilon}_1 + \frac{1}{2}\dot{\epsilon}_2)$$

$$\sigma_2 = \frac{4}{3}\left(\frac{f^{-1}(\dot{\bar{\epsilon}}/g(t))}{\dot{\bar{\epsilon}}}\right)(\dot{\epsilon}_2 + \frac{1}{2}\dot{\epsilon}_1) \tag{4.6}$$

Again for the power law

$$f^{-1}(\dot{\bar{\epsilon}}/g(t)) = (\dot{\bar{\epsilon}}/g(t))^{1/n}$$

Secondly, it should be pointed out that this is only one particular multiaxial form dependent on the assumptions of relation (4.3). Others are possible which, instead of the effective stress and strain rate, use the maximum shear stress $\tau_{max} = \tau_2$ (since $\sigma_1 > \sigma_2 > \sigma_3$) and corresponding maximum shear strain rate $\dot{\gamma}_2$. However, the form introduced here is most commonly used in practice.

Finally, if elastic strains are included, then as before the total strain is decomposed into an elastic part and a creep part

$$\epsilon_i = \epsilon_i^E + \epsilon_i^C \qquad i = 1, 2, 3 \tag{4.7}$$

and the preceding development is applied to the creep strain rates from (4.4)

$$\dot{\epsilon}_1^C = \frac{3}{2} g(t) \frac{f(\bar{\sigma})}{\bar{\sigma}} (\sigma_1 - \sigma_v)$$

$$\dot{\epsilon}_2^C = \frac{3}{2} g(t) \frac{f(\bar{\sigma})}{\bar{\sigma}} (\sigma_2 - \sigma_v) \tag{4.8}$$

$$\dot{\epsilon}_3^C = \frac{3}{2} g(t) \frac{f(\bar{\sigma})}{\bar{\sigma}} (\sigma_3 - \sigma_v)$$

The elastic strains ϵ_i^E are related to stress in the usual manner

$$E\epsilon_1^E = \sigma_1 - \nu(\sigma_2 + \sigma_3)$$

$$E\epsilon_2^E = \sigma_2 - \nu(\sigma_3 + \sigma_1) \tag{4.9}$$

$$E\epsilon_3^E = \sigma_3 - \nu(\sigma_1 + \sigma_2)$$

where E is Young's modulus and ν is Poisson's ratio.

4.2 Further generalisations

We will use the constitutive relations (4.7)–(4.9) almost exclusively. If a more complex model of material behaviour is required, for example the state-variable theory introduced in Section 2.2, then a suitable multiaxial generalisation should be given. In a multiaxial state of stress, the 'rest stress' is also orientated and has three components α_1, α_2 and α_3 in the principal directions while the 'drag stress'

R remains a scalar quantity. The constitutive equations (2.4) and (2.5) then take the forms (cf. equation (4.2))

$$\begin{aligned} \dot{\epsilon}_i &= \psi(s_i - a_i) \qquad i = 1, 2, 3 \\ \dot{\alpha}_i &= f_1 \dot{\epsilon}_i^C - f_2 \alpha_i \\ \dot{R} &= g_1 \dot{\bar{\epsilon}}^C - g_2 (R - R_0) \end{aligned} \tag{4.10}$$

where $\alpha_i, i = 1, 2, 3$, is the deviatoric rest stress

$$\alpha_i = \alpha_i - \alpha_v \qquad i = 1, 2, 3$$

$$\alpha_v = \frac{1}{3}(\alpha_1 + \alpha_2 + \alpha_3)$$

The scalar quantities ψ, f_1, f_2, g_1 and g_2 are functions of the effective stress $\bar{\sigma}$, the effective rest stress $\bar{\alpha}$, the drag stress R and, in general, the effective stress difference between the actual stress and the rest stress, denoted here by

$$\bar{\sigma}_\alpha = \sqrt{\left\{\frac{3}{2}\left[(s_1 - a_1)^2 + (s_2 - a_2)^2 + (s_3 - a_3)^2\right]\right\}}$$

The forms of these functions are determined from the appropriate uniaxial theory.

For example, for Hart's theory, which we will use later, the equations (4.10) take the special forms

$$\dot{\epsilon}_i^C = \frac{3}{2} \frac{\dot{\bar{\epsilon}}^C}{\bar{\sigma}_\alpha} (s_i - a_i)$$

$$\dot{a}_i = \frac{2}{3} c_2 \left[\dot{\epsilon}_i^C - \frac{3}{2} f_2(\bar{\alpha}, \sigma^*) a_i\right]$$

where from the uniaxial theory

$$\dot{\bar{\epsilon}}^C = c_1 (\bar{\sigma}_\alpha)^m \qquad f_2 = \frac{\dot{\epsilon}^*}{\bar{\alpha}} \left[\ln(\sigma^*/\bar{\alpha})\right]^{-1/\lambda}$$

and as before

$$\dot{\epsilon}^* = c_3(\sigma^*)^n \exp(-\Delta H/RT) \qquad \dot{\sigma}^* = f_2(\bar{\alpha}, \sigma^*)\sigma^* \, \bar{\alpha}\Gamma(\sigma^*, \bar{\alpha})$$

4.3 Example: Steady creep of a thick cylinder

The first problem which will be considered here is that of a long, uniformly heated, thick cylinder of inner radius a, outer radius b under an internal pressure p as shown in *Figure 4.2*. For convenience, we will assume that the length of the cylinder is *fixed* so that conditions of *plane strain* are present. (In fact, for an incompressible material, it can be argued that the same solution holds if the ends are free.) The geometry of the cylinder is best described by the cylindrical polar coordinate system (r, ϑ, z) (*Figure 4.2*). Owing to the nature of the loading, it can be assumed that a state of rotationally symmetric plane strain prevails.

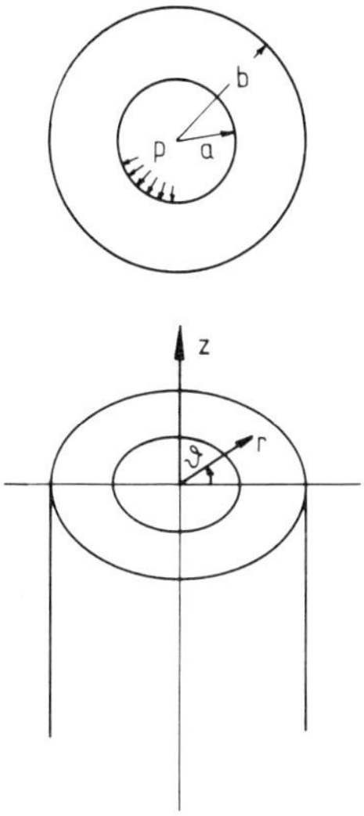

Figure 4.2 Geometry of thick cylinder

This being so, there is no displacement in the z-direction or in the ϑ-direction, whilst the radial displacement u is a function of r alone. Additionally, the coordinate system can be taken as the principal strain and principal stress axes. Let $(\epsilon_r, \epsilon_\vartheta, \epsilon_z)$ be the principal strains and $(\sigma_r, \sigma_\vartheta, \sigma_z)$ be the principal stresses. For plane strain, assuming that the length of the cylinder remains fixed, $\epsilon_z = 0$.

The strain–displacement relations are then obtained from first principles for polar coordinates as

$$\epsilon_\vartheta = u/r \qquad \epsilon_r = \mathrm{d}u/\mathrm{d}r \tag{4.11}$$

together with the compatibility relation

$$\frac{\mathrm{d}\epsilon_\vartheta}{\mathrm{d}r} = \frac{1}{r}(\epsilon_r - \epsilon_\vartheta) \tag{4.12}$$

The equilibrium relations are also given by

$$\frac{\mathrm{d}\sigma_r}{\mathrm{d}r} = \frac{1}{r}(\sigma_\vartheta - \sigma_r) \tag{4.13}$$

with solution

$$\sigma_r = \frac{1}{r}\frac{\mathrm{d}\phi}{\mathrm{d}r} \qquad \sigma_\vartheta = \frac{\mathrm{d}^2\phi}{\mathrm{d}r^2}$$

where ϕ is the (Airy) stress function. These must satisfy the boundary conditions

$$\sigma_r(a) = -p \qquad \sigma_r(b) = 0 \tag{4.14}$$

These basic equations are completed by the steady-state creep constitutive relations, i.e. stresses and deformation rates constant (so that elastic strains can be ignored). Because of plane strain $\epsilon_z = 0$ and consequently $\sigma_z = \sigma_v$ or

$$\sigma_z = \frac{1}{2}(\sigma_r + \sigma_\vartheta)$$

so that the constitutive relations are

$$\dot{\epsilon}_r = \frac{3}{2}\frac{f(\bar{\sigma})}{\bar{\sigma}}\frac{1}{2}(\sigma_r - \sigma_\vartheta)$$

$$\dot{\epsilon}_\vartheta = \frac{3}{2}\frac{f(\bar{\sigma})}{\bar{\sigma}}\frac{1}{2}(\sigma_\vartheta - \sigma_r) \qquad (4.15)$$

$$\dot{\epsilon}_z = 0$$

where

$$\bar{\sigma}^2 = \frac{3}{4}(\sigma_\vartheta - \sigma_r)^2$$

In principle, the problem may now be solved using well known techniques from the elastic analysis of plane strain problems to derive either an equation for the displacement rates or for the stress function. As shall be seen, this invariably leads to complex nonlinear differential equations. However, owing to the assumption of incompressibility for plane strain the problem becomes rather simple

$$\dot{\epsilon}_r = -\dot{\epsilon}_\vartheta$$

so that the compatibility relation can be reduced to

$$\frac{d\dot{\epsilon}_\vartheta}{dr} = -\frac{2}{r}\dot{\epsilon}_\vartheta$$

with solution

$$\dot{\epsilon}_\vartheta = A/r^2 \qquad (4.16)$$

where A is a constant of integration. From the second of equations (4.15)

$$f(\bar{\sigma}) = (2A/\surd 3)/r^2$$

and so

$$\bar{\sigma} = \surd 3\,(\sigma_\vartheta - \sigma_r)/2 = f^{-1}((2A/\surd 3)/r^2)$$

Substitution into equation (4.13) then yields

$$\frac{d\sigma_r}{dr} = \frac{1}{r}\frac{2}{\surd 3}f^{-1}\left(\frac{2A/\surd 3}{r^2}\right)$$

which can be integrated along with the boundary condition (4.14) at $r = a$ to give the solution

$$\sigma_r = -p + \frac{2}{\surd 3}\int_a^r \frac{f^{-1}((2A/\surd 3)/r^2)}{r}\,dr \qquad (4.17)$$

$$\sigma_\vartheta = -p + \frac{2}{\surd 3}\int_a^r \frac{f^{-1}((2A/\surd 3)/r^2)}{r}\,dr + \frac{2}{\surd 3}f^{-1}\left(\frac{2A/\surd 3}{r^2}\right) \qquad (4.18)$$

Therefore the stresses are known if the uniaxial constitutive function f is prescribed; the constant A is then determined from the boundary condition at $r = b$.

For the power law $f = B\sigma^n$, $f^{-1} = (\dot{\epsilon}/B)^{1/n}$

$$\sigma_r = -p + \frac{2}{\sqrt{3}} \int_a^r \left(\frac{(2A/\sqrt{3})/B}{r^2}\right)^{1/n} \frac{dr}{r} \tag{4.19}$$

and after integration and some rearrangement with application of the boundary condition there results

$$\sigma_r = -p\left(\frac{(b/r)^{2/n} - 1}{(b/a)^{2/n} - 1}\right)$$
$$\sigma_\vartheta = p\left(\frac{1 - (1 - 2/n)(b/r)^{2/n}}{(b/a)^{2/n} - 1}\right) \tag{4.20}$$

Also from equations (4.11) and (4.17) the displacement rate is given as

$$\dot{u} = B\left(\frac{\sqrt{3}}{2}\right)^{n+1} \frac{b^2}{(b/a)^{2/n} - 1} \left(\frac{2}{n}\right)^n \frac{p^n}{r} \tag{4.21}$$

One interesting aspect of this solution is that the stress distribution essentially changes for varying n. For $n = 2$, σ_ϑ = constant; for $n < 2$, σ_ϑ diminishes with r; and for $n > 2$, σ_ϑ increases with r reaching a maximum at the outer surface. Typical stress distributions are shown in *Figure 4.3*. It should also be noted that the variations of $\sigma_\vartheta|_{r=a}$ with $1/n$ is approximately linear (*Figure 4.4*). The limit solution as $n \to \infty$ is given by

$$\sigma_r = -p \frac{\ln(b/r)}{\ln(b/a)} \qquad \sigma_\vartheta = p \frac{1 - \ln(b/r)}{\ln(b/a)}$$

on noting the relation

$$x^{1/n} = 1 + (1/n)\ln x + o\,(1/n^2)$$

Thus, for a perfectly plastic material with yield stress σ_y the limit load can be

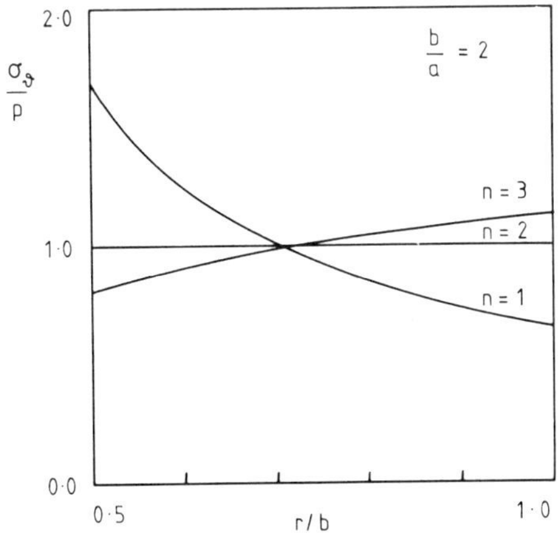

Figure 4.3 Circumferential stress distribution

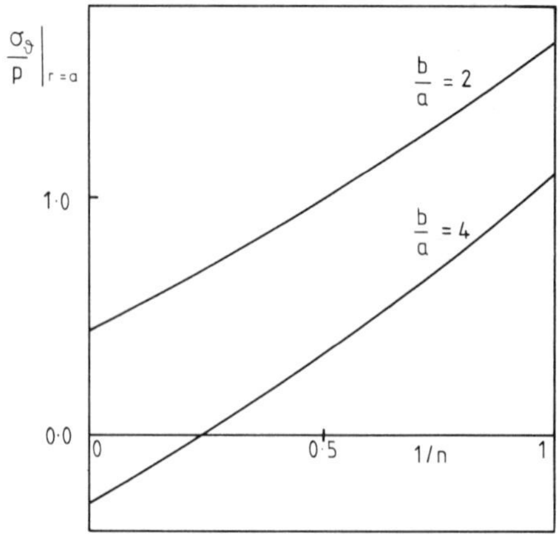

Figure 4.4 Variation of circumferential stress at inner surface with $1/n$

evaluated from $\sigma_\vartheta - \sigma_r = \sigma_y = p/\ln(b/a)$; hence $p_L = \sigma_y \ln(b/a)$. Finally, since σ_z is the average of the other two stresses, it is no longer constant across the thickness as in the linear elastic case. Nevertheless, it can be shown that the integrated effect of this longitudinal stress through the thickness is statically equivalent to the force of the internal pressure acting on the end of the cylinder

$$2\pi \int_a^b \sigma_z r \,\mathrm{d}r = \pi a^2 p$$

as it should be.

In practice, many thick-walled pressure vessels are accompanied by a radial heat flux caused by a temperature difference between the walls. If T_a and T_b are the temperatures on the walls, then from basic heat transfer

$$T = T_a + (T_b - T_a)\,\frac{\ln(r/a)}{\ln(b/a)} \tag{4.22}$$

It is possible to obtain a closed-form solution for this problem for the power law, which is written in the form

$$f(\sigma) = C\,\mathrm{e}^{-\gamma/T}\sigma^n$$

to allow for the material response to varying temperature. Using equation (4.22), equation (4.19) can be rewritten, after a little manipulation, as

$$\sigma_r = -p + \frac{2}{\sqrt{3}}\int_a^r \left(\frac{(2A/\sqrt{3})/B_r}{r^2}\right)^{1/n_T} \frac{\mathrm{d}r}{r}$$

where

$$n_T = \left(\frac{1}{n} + \frac{\beta}{2\ln(b/a)}\right)^{-1} \qquad \beta = \frac{\gamma}{n}\,\frac{(T_b - T_a)}{T_b^2}$$

$$B_T = C\exp^{(-\gamma/T_b)}\exp\left[-\gamma\left(\frac{T_b - T_a}{T_b^2}\right)\right]$$

and we have used the approximation

$$\gamma\left(\frac{1}{T_a}-\frac{1}{T_b}\right) \simeq \gamma\left(\frac{T_b-T_a}{T_b^2}\right)$$

Therefore, the previous solution (4.20) remains valid with these new values of

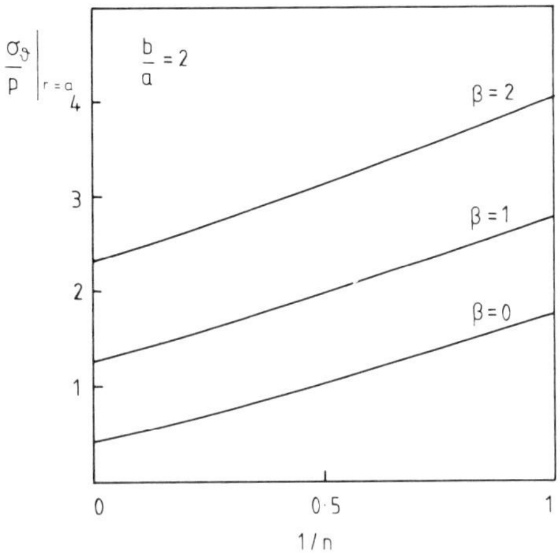

Figure 4.5 Variation of circumferential stress at inner surface with $1/n$: non-isothermal case

B_T and n_T. The variation of $\sigma_\vartheta|_{r=a}$ with $1/n$ is given in *Figure 4.5* and is again seen to be approximately linear for given values of β.

4.4 Example: Steady creep of a holed plate under uniform tension

The previous example dealt with the problem of axisymmetric plane strain which is one of the few problems in creep analysis to admit a closed-form solution. When the corresponding problem of plane stress is encountered, the solution is, perhaps surprisingly, not so straightforward. Here we will examine a special plane stress

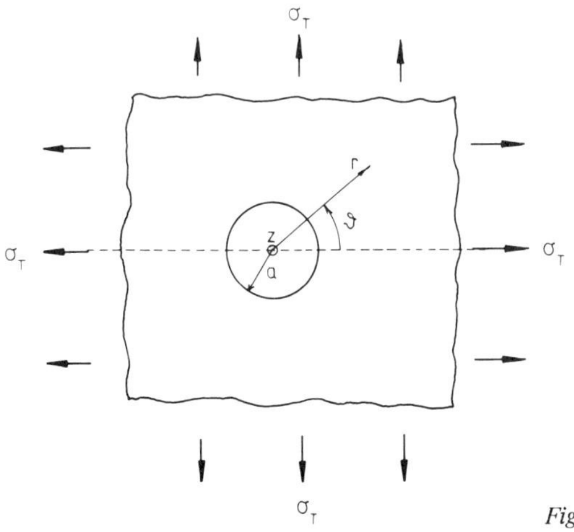

Figure 4.6 Geometry of holed plate

problem which does admit a closed-form solution. It is that of an infinite plate, subject to a uniform tension, weakened by a circular hole. Ultimately, we will be interested in the stress concentration induced by this hole.

Thus consider an infinite plate containing a hole of radius a. The geometry is again best described by the polar coordinate system (r, ϑ, z) as shown in *Figure 4.6.* Owing to the loading, it is safe to assume a state of *axisymmetric plane stress* with the coordinate system being the principal axes of stress $(\sigma_r, \sigma_\vartheta, \sigma_z)$ and strain $(\epsilon_r, \epsilon_\vartheta, \epsilon_z)$. For plane stress σ_r and σ_ϑ are functions of r alone, while $\sigma_z = 0$. The same field equations as the previous example hold; namely, strain–displacement and compatibility

$$\epsilon_\vartheta = u/r \qquad \epsilon_r = \mathrm{d}u/\mathrm{d}r$$
$$\frac{\mathrm{d}\epsilon_\vartheta}{\mathrm{d}r} = \frac{1}{r}(\epsilon_r - \epsilon_\vartheta) \tag{4.23}$$

and equilibrium, assuming the plate is of constant thickness,

$$\frac{\mathrm{d}\sigma_r}{\mathrm{d}r} = \frac{1}{r}(\sigma_\vartheta - \sigma_r) \tag{4.24}$$

with boundary conditions replaced by

$$\sigma_r(a) = 0$$
$$\sigma_r = \sigma_\vartheta = \sigma_\mathrm{T} \qquad r \to \infty \tag{4.25}$$

where σ_T is the applied tension.

The constitutive relations for plane stress are

$$\dot{\epsilon}_r = \frac{f(\bar{\sigma})}{\bar{\sigma}}\left(\sigma_r - \frac{1}{2}\sigma_\vartheta\right)$$
$$\dot{\epsilon}_\vartheta = \frac{f(\bar{\sigma})}{\bar{\sigma}}\left(\sigma_\vartheta - \frac{1}{2}\sigma_r\right)$$
$$\dot{\epsilon}_z = -\frac{1}{2}\frac{f(\bar{\sigma})}{\bar{\sigma}}(\sigma_r + \sigma_\vartheta) \tag{4.26}$$
$$\bar{\sigma}^2 = (\sigma_r^2 - \sigma_r\sigma_\vartheta + \sigma_\vartheta^2)$$

It should become apparent that the rather fortunate solution procedure used in the previous plane strain problem cannot be followed here. The well known elastic procedures would lead to complicated nonlinear differential equations. However, it has been found that for plane stress problems, a simple transformation can clarify the equations.

The principal stresses can be expressed in the form

$$\sigma_r = \frac{2}{\sqrt{3}}\bar{\sigma}\sin(\varphi - \pi/6) \qquad \sigma_\vartheta = \frac{2}{\sqrt{3}}\bar{\sigma}\sin(\varphi + \pi/6)$$

where the angle $\varphi - \pi/6$ is called the *angle of similarity* of the stress distribution.

The constitutive relations then become

$$\dot{\epsilon}_r = -\frac{3}{2} f(\bar{\sigma}) \cos(\varphi + \pi/6) \qquad \dot{\epsilon}_\vartheta = \frac{3}{2} f(\bar{\sigma}) \cos(\varphi - \pi/6)$$

Then introducing new variables

$$S = \ln(\bar{\sigma}/\sigma_0) \qquad E = \ln(f(\bar{\sigma})/\dot{\epsilon}_0) \qquad R = \ln(r/a)$$

where σ_0 and $\dot{\epsilon}_0$ are for the present arbitrary, substitution into equations (4.23) and (4.24) gives, after some rearrangement,

$$\frac{\mathrm{d}\varphi}{\mathrm{d}R} \cos(\varphi - \pi/6) + \frac{\mathrm{d}S}{\mathrm{d}R} \sin(\varphi - \pi/6) - \cos\varphi = 0$$

$$\frac{\mathrm{d}\varphi}{\mathrm{d}R} \sin(\varphi - \pi/6) - \frac{\mathrm{d}E}{\mathrm{d}R} \cos(\varphi - \pi/6) - \sqrt{3}\cos\varphi = 0$$

Eliminating $\cos\varphi$ by multiplying the first equation by $\sqrt{3}$ and subtracting the second, and then eliminating $\mathrm{d}R$, yields the single equation

$$2\cos\varphi \frac{\mathrm{d}\varphi}{\mathrm{d}S} + \sqrt{3}\sin(\varphi - \pi/6) + \frac{\mathrm{d}E}{\mathrm{d}S}\cos(\varphi - \pi/6) = 0 \tag{4.27}$$

In general, E is a function of S and this may be solved subject to the boundary conditions

$$\begin{array}{llll} r = a & R = 0 & \varphi = \pi/6 & S = \ln(\bar{\sigma}/\sigma_0) = \ln[\sigma_\vartheta(a)/\sigma_0] = S_a \\ r \to \infty & R \to \infty & \varphi = \pi/2 & S = \ln(\bar{\sigma}/\sigma_0) = \ln(\sigma_T/\sigma_0) \end{array} \tag{4.28}$$

where S_a is an unknown quantity to be determined.

For the special case of the power law $f = B\sigma^n$, on choosing $\dot{\epsilon}_0 = B\sigma_0^n$ it is seen that

$$\mathrm{d}E/\mathrm{d}S = n$$

Finally, on choosing $\sigma_0 = \sigma_T$, equation (4.27) can be integrated as

$$S_a = -\int_{\pi/6}^{\pi/2} \frac{2\cos\varphi\,\mathrm{d}\varphi}{\sqrt{3}\sin(\varphi - \pi/6) + n\cos(\varphi - \pi/6)} \tag{4.29}$$

The latter integral can be evaluated on employing the transformation

$$\varphi = \psi + \pi/6 - \eta$$

where η is defined by

$$\frac{\sqrt{3}}{\sqrt{(3 + n^2)}} = \cos\eta \qquad \frac{n}{\sqrt{(3 + n^2)}} = \sin\eta \tag{4.30}$$

and so

$$S_a = -\frac{2}{\sqrt{(3 + n^2)}} \int_{\eta + \pi/3}^{\eta} \frac{\cos(\psi + \pi/6 - \eta)}{\sin\psi}\,\mathrm{d}\psi$$

On expanding the numerator, the integral can be evaluated as

$$S_a = \frac{2}{\sqrt{(3+n^2)}}\left[\cos\left(\frac{\pi}{6-\eta}\right)\ln\left(\frac{\sin(\pi/3+\eta)}{\sin\eta}\right) - \frac{\pi}{3}\sin\left(\frac{\pi}{6}-\eta\right)\right] \quad (4.31)$$

and η can be eliminated using equations (4.30), $\eta = \tan^{-1}(n/\sqrt{3})$ to give S_a as a function of n.

The *stress concentration factor* for the hole is defined by

$$K = \frac{\sigma_\vartheta(a)}{\sigma_T} = \exp S_a$$

so that immediately from equation (4.31), after rearrangement,

$$K = \left(\frac{n+3}{2n}\right)^{(n+3)/(n^2+3)} \exp\left[\frac{\pi}{\sqrt{3}}\left(\frac{n-1}{n^2+3}\right)\right] \quad (4.32)$$

It is left to the reader to verify that to a good approximation

$$K \simeq 1 + 1/n$$

4.5 Example: Steady creep in membrane shells

As a final example of a class of problems which can readily be solved in closed form, we examine creep deformations in membrane shells. Without going into too much detail, we briefly summarise the notion of a shell. By a *shell* is commonly meant a curved, thin-walled structure – its thickness h is small compared to its other dimensions. The curved surface which lies mid-way between the shell surfaces is called the mid-surface. In the stress analysis of shells, the deformation is described in terms of the deformation of the mid-surface and the stresses are described by stress resultants or forces per unit length of section through the thickness.

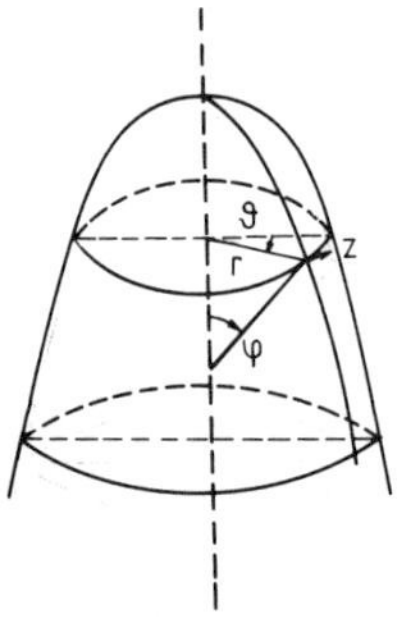

Figure 4.7 Geometry of thin shell mid-surface

In the following, we restrict our attention to shells of rotational symmetry under symmetrical loading. The mid-surface of such a shell can be described geometrically by the coordinate system (φ, ϑ, z) (*Figure 4.7*), where φ is the angle describing the meridional curve which is rotated about the axis of revolution, and z is distance through the thickness. Since the loading is presumed to be symmetric, this coordinate system may be taken as the principal axes for stress and strain. The stress and strain depend on φ alone.

The following assumptions are consistent with the thin-shell approximation:

(a) the stress σ_z is negligible;
(b) a line originally normal to the shell mid-surface will remain normal during deformation.

Then if $\in_\varphi$, $\in_\vartheta$ are the (stretching) strains in the mid-surface and κ_φ, κ_ϑ are the curvatures of the mid-surface, the principal strains are

$$\epsilon_\varphi = \in_\varphi + z\kappa_\varphi \qquad \epsilon_\vartheta = \in_\vartheta + z\kappa_\vartheta$$

and the stress and moment resultants are defined by

$$N_\vartheta = \int_{-h/2}^{h/2} \sigma_\vartheta \, dz \qquad M_\vartheta = \int_{-h/2}^{h/2} \sigma_\vartheta z \, dz$$

$$N_\varphi = \int_{-h/2}^{h/2} \sigma_\varphi \, dz \qquad M_\varphi = \int_{-h/2}^{h/2} \sigma_\varphi z \, dz$$

The strain quantities ($\in_\varphi$, $\in_\vartheta$, κ_φ, κ_ϑ) are sometimes called the *generalised strains* whilst the stress resultants (N_φ, N_ϑ, M_φ, M_ϑ) are called the *generalised stresses.* In shell analysis, they take the place of the classical strain and stress measures.

In membrane shells the state of stress is largely a membrane state because we set

$$M_\varphi = M_\vartheta = 0 \qquad \kappa_\varphi = \kappa_\vartheta = 0$$

so that the strains, and consequently the stresses, are constant through the thickness

$$N_\varphi = \sigma_\varphi h \qquad N_\vartheta = \sigma_\vartheta h$$

$$\in_\varphi = \epsilon_\varphi \qquad \in_\vartheta = \epsilon_\vartheta$$

The equations of equilibrium for a membrane shell can be obtained as

$$rN_\varphi - N_\vartheta r_\varphi \cos\varphi + q_\varphi r r_\varphi = 0$$
$$\frac{N_\varphi}{r_\varphi} + \frac{N_\vartheta}{r_\vartheta} + q_\mathrm{n} = 0 \tag{4.33}$$

where r_φ, r_ϑ are the principal radii of curvature of the mid-surface

$$r_\vartheta = \frac{r}{\sin\varphi} \qquad r_\varphi = \frac{1}{\cos\varphi}\frac{\mathrm{d}r}{\mathrm{d}\varphi}$$

where r is shown in *Figure 4.7.* The applied forces are q_φ in the meridional direction and q_n in the normal direction per unit area. On eliminating N_ϑ and integrating equation (4.33), there results

$$N_\varphi = \frac{1}{r\sin^2\varphi}\left(c_1 - \int (q_\varphi r_\varphi r_\vartheta \sin^2\varphi + q_\mathrm{n} r_\varphi r_\vartheta \sin\varphi\cos\varphi)\mathrm{d}\varphi\right) \tag{4.34}$$

where c_1 is a constant of integration. Thus in principle we have determined the stresses for the problem from the loading alone, viz. the problem is *statically determinate.*

The deformations can then be evaluated if we can integrate the strain–displacement relations for a membrane shell

$$\begin{aligned} \epsilon_\varphi &= \frac{1}{r_\varphi}\left(\frac{du}{d\varphi} + w\right) \\ \epsilon_\vartheta &= \frac{1}{r}(u\cos\varphi + w\sin\varphi) \end{aligned} \tag{4.35}$$

where u is the meridional displacement and w the normal displacement of the mid-surface. In fact, eliminating w and integrating leads to

$$u = \sin\varphi\left(\int (r_\varphi \epsilon_\varphi \sin\varphi - r\epsilon_\vartheta)\frac{d\varphi}{\sin^2\varphi} + c_2\right) \tag{4.36}$$

where c_2 is another constant of integration. The constants c_1 and c_2 are determined from the boundary conditions.

The solution is complete if the constitutive relations are given from equations (4.26) as

$$\begin{aligned} \dot{\epsilon}_\varphi &= \frac{f(\bar{N}/h)}{\bar{N}}\left(N_\varphi - \frac{1}{2}N_\vartheta\right) \\ \dot{\epsilon}_\vartheta &= \frac{f(\bar{N}/h)}{\bar{N}}\left(N_\vartheta - \frac{1}{2}N_\varphi\right) \\ \dot{\epsilon}_z &= -\frac{f(\bar{N}/h)}{\bar{N}}\,\frac{(N_\varphi + N_\vartheta)}{2} \\ \bar{N}^2 &= N_\varphi^2 - N_\varphi N_\vartheta + N_\vartheta^2 \end{aligned} \tag{4.37}$$

The last relation which gives $\dot{\epsilon}_z \neq 0$ is largely a consequence of the membrane assumption; in shell theory as a deformation assumption we should have $\epsilon_z = 0$. The relations given by equations (4.37) may be called *generalised constitutive relations*, since they relate the generalised stresses and strain rates for a shell. In the stress analysis of (membrane) shells they replace the actual constitutive relations – we shall meet these again later.

As an example, we consider the problem of the mating of a pressurised cylinder with a pressurised hemisphere – to form a head – of the same radius a subject to the same pressure p and manufactured from the same material. Their respective thicknesses may be different, say h_{cyl}, h_{cap}. According to the loading

$$q_n = -p \qquad q_\varphi = 0$$

For the cylinder

$$r_\vartheta = a \qquad r_\varphi \to \infty \qquad N_\varphi = pa/2 \qquad N_\vartheta = pa \qquad \bar{N} = \sqrt{3}pa/2$$

and for the sphere

$$r_\vartheta = r_\varphi = a \qquad N_\vartheta = N_\varphi = pa/2 \qquad \bar{N} = pa/2$$

from equation (4.34) where the constant of integration $c_1 = 0$ satisfies the condition that N_φ is finite as $\varphi \to 0$.

From equation (4.35) for the cylinder, $r_\varphi d\varphi \to dx$, $\varphi \to \pi/2$, so that

$$\epsilon_x = du/dx \qquad \epsilon_\vartheta = w/a$$

and

$$\dot{w}_{cyl} = a\dot{\epsilon}_\vartheta = \frac{2}{\sqrt{3}} f\left(\frac{pa}{h_{cyl}}\frac{\sqrt{3}}{2}\right)\frac{3}{4}$$

For the cap, $r = a \sin\varphi$ and equations (4.35) become

$$\epsilon_\varphi = \frac{1}{a}\left(\frac{du}{d\varphi} + w\right) \qquad \epsilon_\vartheta = \frac{u\cos\varphi + w\sin\varphi}{a\sin\varphi}$$

and at the base

$$\dot{w}_{cap} = a\dot{\epsilon}_\vartheta\Big|_{\pi/2} = 2f\left(\frac{pa}{h_{cap}}\frac{1}{2}\right)\frac{1}{4}$$

Hence $\dot{w}_{cyl} = \dot{w}_{cap}$ provided

$$\frac{\sqrt{3}f}{2}\left(\frac{pa}{h_{cyl}}\frac{\sqrt{3}}{2}\right) = \frac{1}{2}f\left(\frac{pa}{h_{cap}}\frac{1}{2}\right) \qquad (4.38)$$

For the power law

$$\frac{h_{cap}}{h_{cyl}} = 3^{-(n+1)/2n}$$

In general, the mating of two shells causes local bending stresses near the joint and membrane theory is insufficient.

4.6 Example: Steady creep of thin plates in bending

The concept of a thin shell introduced in the previous section can be specialised to deal with thin *plates*. As before, the geometry of the plate is defined by its mid-surface, which here is a flat plane. The thickness h is assumed to be uniform and small compared to the least dimensions of the mid-surface. If we introduce two principal directions on the mid-surface and principal stresses (σ_1, σ_2) together with principal strains (ϵ_1, ϵ_2), then the stress resultants and moments associated with the mid-surface are defined by

$$N_1 = \int_{-h/2}^{h/2} \sigma_1 \, dz \qquad M_1 = \int_{-h/2}^{h/2} \sigma_1 z \, dz \qquad \text{etc.} \qquad (4.39)$$

where z is measured through the thickness normal to the mid-surface, while the mid-surface strains and curvatures are

$$\epsilon_1 = \varepsilon_1 + z\kappa_1 \qquad \epsilon_2 = \varepsilon_2 + z\kappa_2$$

assuming the strains are linear through the thickness.

As a special case we consider here the important problem of the bending of a thin plate out of its plane, being a problem in plane stress. Thus there are no forces in the plane of the plate, or any edge loads. Under this assumption

$$N_1 = N_2 = 0 \qquad \epsilon_1 = \epsilon_2 = 0 \tag{4.40}$$

With this simplification we can obtain a generalised constitutive relation between the mid-surface moments M_1, M_2 and curvature rates $\dot{\kappa}_1$, $\dot{\kappa}_2$ for steady creep according to a power law.

For plane stress, from equation (4.6) with $\dot{\epsilon} = B\sigma^n$

$$\sigma_1 = \frac{4}{3}\left(\frac{1}{B}\right)^{1/n} (\bar{\dot{\epsilon}})^{1/n-1} (\dot{\epsilon}_1 + \frac{1}{2}\dot{\epsilon}_2)$$

$$\sigma_2 = \frac{4}{3}\left(\frac{1}{B}\right)^{1/n} (\bar{\dot{\epsilon}})^{1/n-1} (\dot{\epsilon}_2 + \frac{1}{2}\dot{\epsilon}_1)$$

But from equation (4.40)

$$\dot{\epsilon}_1 = z\dot{\kappa}_1 \qquad \dot{\epsilon}_2 = z\dot{\kappa}_2$$

so that substituting into equation (4.39)

$$M_1 = D_n \frac{4}{3}\left(\frac{1}{B}\right)^{1/n} (\bar{\dot{\kappa}})^{1/n-1} (\dot{\kappa}_1 + \frac{1}{2}\dot{\kappa}_2)$$

$$M_2 = D_n \frac{4}{3}\left(\frac{1}{B}\right)^{1/n} (\bar{\dot{\kappa}})^{1/n-1} (\dot{\kappa}_2 + \frac{1}{2}\dot{\kappa}_1) \tag{4.41}$$

$$\bar{\dot{\kappa}}^2 = \frac{4}{3}(\dot{\kappa}_1^2 + \dot{\kappa}_1\dot{\kappa}_2 + \dot{\kappa}_2^2)$$

where

$$D_n = \int_{-h/2}^{h/2} z^{1+1/n}\, dz = \frac{2n}{2n+1}(h/2)^{2+1/n}$$

is the *flexural rigidity* for a power law.

Inverting equations (4.41) we find

$$\dot{\kappa}_1 = \frac{B}{D_n^n} \bar{M}^{n-1} (M_1 - \frac{1}{2}M_2)$$

$$\dot{\kappa}_2 = \frac{B}{D_n^n} \bar{M}^{n-1} (M_2 - \frac{1}{2}M_1) \tag{4.42}$$

$$\bar{M}^2 = M_1^2 - M_1 M_2 + M_2^2$$

being the required generalised constitutive relation.

In general, strain–displacement and equilibrium equations can be established and, with suitable boundary conditions, our problem is defined. For example, if we restrict our attention to axially symmetric plates under axially symmetric bending, then the polar coordinates (r, φ) can be taken as principal direction for the mid-surface, and all quantities are functions of the radius r alone.

If w is the out-of-plane deflection, then the strain–displacement relations are

$$\kappa_r = -\frac{d^2 w}{dr^2} \qquad \kappa_\varphi = -\frac{1}{r}\frac{dw}{dr} \tag{4.43}$$

with compatibility

$$\frac{d}{dr}(r\kappa_\varphi) = \kappa_r$$

The equilibrium equations are

$$-\frac{d}{dr}(rQ_r) + pr = 0$$

$$Q_r = -\frac{1}{r}\left(\frac{d}{dr}(rM_r) - M_\varphi\right) \tag{4.44}$$

where Q_r is the shearing force and p is a distributed pressure over the surface of the plate.

A differential equation for the displacement rate $\dot{w}$ may be obtained on substituting equation (4.41) into equilibrium equations (4.44) and using the rate form of strain displacement, equation (4.43); then

$$\frac{4}{3}\left(\frac{1}{B}\right)^{1/n} D_n \frac{d}{dr}\left\{\frac{d}{dr}\left[F_n(\dot{w})\left(r^2\frac{d^2\dot{w}}{dr^2} + \frac{1}{2}\frac{d\dot{w}}{dr}\right)\right] - F_n(\dot{w})\left(\frac{d\dot{w}}{dr} + \frac{r}{2}\frac{d^2\dot{w}}{dr^2}\right)\right\} = pr$$

$$F_n(\dot{w}) = \left\{\frac{4}{3}\left[\left(\frac{d^2\dot{w}}{dr^2}\right)^2 + \frac{1}{r}\left(\frac{d^2\dot{w}}{dr^2}\right)\left(\frac{d\dot{w}}{dr}\right) + \frac{1}{r^2}\left(\frac{d\dot{w}}{dr}\right)^2\right]\right\}^{(1/n-1)/2} \tag{4.45}$$

which has to be solved with suitable boundary conditions, e.g.

Encastre:	$\dot{w} = 0$	$d\dot{w}/dr = 0$	
Simply supported:	$\dot{w} = 0$	$M_r = 0$	(4.46)
Free:	$Q_r = 0$	$M_r = 0$	

We are faced here with a rather complex nonlinear differential equation, which is typical of the type we shall encounter in steady creep analysis, and which has *no* simple solution of the type we have seen so far. The problem must be solved using an approximate numerical technique.

As a rather elementary, yet technically important, problem, we consider the bending of an annular plate of inner radius a and outer radius b (*Figure 4.8*), such that their ratio a/b is close to unity. Then with the types of loading we have proposed, it can be assumed that the cross-section of the ring does not deform. For example, with the outer edge simply supported and the inner edge free, with a uniform pressure distribution, there is only a bodily rotation about the support (*Figure 4.8*).

If ϑ is the angle of this rotation, then $w \simeq \vartheta(b - r)$, i.e.

$$\kappa_\varphi \simeq \vartheta/r \qquad \kappa_r \simeq 0$$

and from the generalised constitutive relation, equation (4.41),

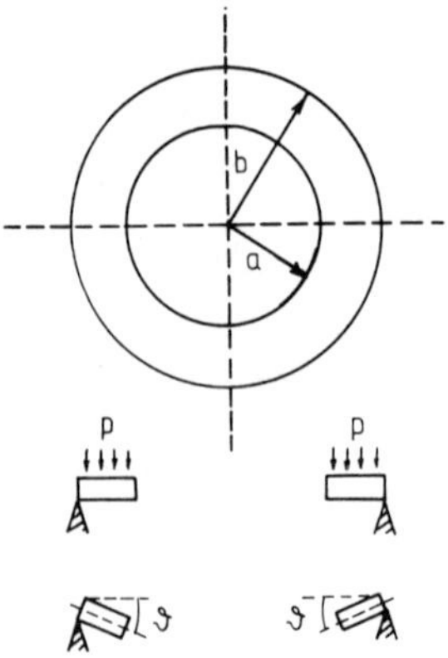

Figure 4.8 Geometry of annular plate

$$M_\varphi \simeq D_n \left(\frac{4}{3}\ \frac{1}{B}\right)^{1/n} \left(\frac{\dot{\vartheta}}{r}\right)^{1/n} \tag{4.47}$$

This must be balanced by the bending moment M causing the rotation

$$\int_a^b M_\varphi \ \mathrm{d}r = M \tag{4.48}$$

From simple statics, considering the equilibrium of half the ring, this moment is given by

$$M = \frac{1}{6}\ p\,(b^3 + 2a^3 - 3a^2 b)$$

Thus substituting equation (4.47) into (4.48), the following solution for $\dot{\vartheta}$ is obtained:

$$\dot{\vartheta} = \frac{3}{4}B \left(\frac{M}{D_n}\right)^n \left(\frac{1 - 1/n}{b^{1-1/n} - a^{1-1/n}}\right)^n \tag{4.49}$$

Consequently, at the inner free edge

$$\dot{w}_{\max} \simeq \dot{\vartheta}(b - a) \tag{4.50}$$

We will compare this simple solution with a more accurate one later.

4.7 The general boundary value problem for steady creep

Consider a solid body occupying a volume V bounded by a surface S (*Figure 4.9*). Points in the body are described by rectangular Cartesian axes (x_1, x_2, x_3) or x_i, $i = 1, 2, 3$.

In Section 4.1 we introduced the symmetric *stress tensor* σ_{ij} expressed alternatively here as

$$\sigma_{ij} = \begin{bmatrix} \sigma_{11} & \sigma_{12} & \sigma_{13} \\ \sigma_{21} & \sigma_{22} & \sigma_{23} \\ \sigma_{31} & \sigma_{32} & \sigma_{33} \end{bmatrix}$$

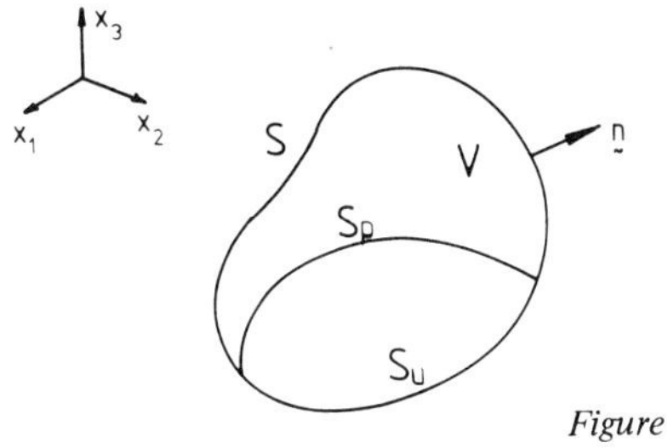

Figure 4.9 Geometry of a continuous body

We adopt the usual notation: the indicial convention implies that whenever a subscript occurs unrepeated it should take values 1, 2, 3; the summation convention implies that whenever a subscript occurs twice a summation with respect to that subscript should be taken over the values 1, 2, 3, e.g.

$$\sigma_{kk} = \sum_{k=1}^{3} \sigma_{kk} = \sigma_{11} + \sigma_{22} + \sigma_{33}$$

$$\sigma_{ij}\sigma_{ij} = \sum_{i=1}^{3}\sum_{j=1}^{3} \sigma_{ij}\,\sigma_{ij} = \sigma_{11}\,\sigma_{11} + \sigma_{12}\sigma_{12} + \sigma_{13}\sigma_{13} + \sigma_{21}\sigma_{21} + \sigma_{22}\sigma_{22} + \sigma_{23}\sigma_{23} + \sigma_{31}\sigma_{31} + \sigma_{32}\sigma_{32} + \sigma_{33}\sigma_{33}$$

and we define the *unit tensor* (or Kronecker delta) by

$$\delta_{ij} = \begin{cases} 1 & i = j \\ 0 & i \neq j \end{cases}$$

$$= \begin{bmatrix} 1 & 0 & 0 \\ 0 & 1 & 0 \\ 0 & 0 & 1 \end{bmatrix}$$

The *stress deviator* is then

$$s_{ij} = \sigma_{ij} - \frac{1}{3}\sigma_{kk}\delta_{ij}$$

The *infinitesimal strain tensor* ϵ_{ij} is derived from a displacement field u_i as

$$\epsilon_{ij} = \frac{1}{2}\left(\frac{\partial u_i}{\partial x_j} + \frac{\partial u_j}{\partial x_i}\right) = \frac{1}{2}(u_{i,j} + u_{j,i}) \qquad (4.51)$$

and in a similar manner we define the *strain-rate tensor*

$$\dot{\epsilon}_{ij} = \frac{1}{2}\left(\frac{\partial \dot{u}_i}{\partial x_j} + \frac{\partial \dot{u}_j}{\partial x_i}\right) = \frac{1}{2}(\dot{u}_{i,j} + \dot{u}_{j,i}) \qquad (4.52)$$

If the body suffers body forces $\bar{b}_i$, then the stress tensor must satisfy the equations of equilibrium

$$\frac{\partial \sigma_{ij}}{\partial x_i} + \bar{b}_j = \sigma_{ij,i} + \bar{b}_j = 0 \qquad (4.53)$$

If n_i are the direction cosines of the outer normal of the surface S (*Figure 4.9*), the surface tractions will be

$$p_i = \sigma_{ij} n_j$$

It is assumed that over part of the surface S_p, surface tractions are given, while over the remainder S_u, displacements are imposed; the boundary conditions are then

$$\begin{aligned} \dot{u}_i &= \bar{u}_i \qquad \text{on } S_u \\ p_i &= \bar{p}_i \qquad \text{on } S_p \end{aligned} \tag{4.54}$$

The constitutive relations for steady creep can be derived if we define the effective stress

$$\bar{\sigma}^2 = \frac{3}{2} s_{ij} s_{ij}$$

and the effective strain rate

$$\dot{\bar{\epsilon}}^2 = \frac{2}{3} \dot{\epsilon}_{ij} \dot{\epsilon}_{ij}$$

noting that incompressibility implies $\dot{\epsilon}_{kk} = 0$.

The constitutive relations for steady creep, equations (4.4) and (4.5), are

$$\begin{aligned} \dot{\epsilon}_{ij} &= \frac{3}{2} \frac{f(\bar{\sigma})}{\bar{\sigma}} s_{ij} \\ s_{ij} &= \frac{2}{3} \frac{f^{-1}(\dot{\bar{\epsilon}})}{\dot{\bar{\epsilon}}} \dot{\epsilon}_{ij} \end{aligned} \tag{4.55}$$

4.8 Example: Steady creep of a bar in torsion

Consider a cylindrical bar with an axis z whose ends lie at $z = 0$ and $z = l$. The lower end is fixed. The cross-section is described by a closed curve C (*Figure 4.10*). A set of Cartesian coordinates (x, y, z) is used to describe the geometry.

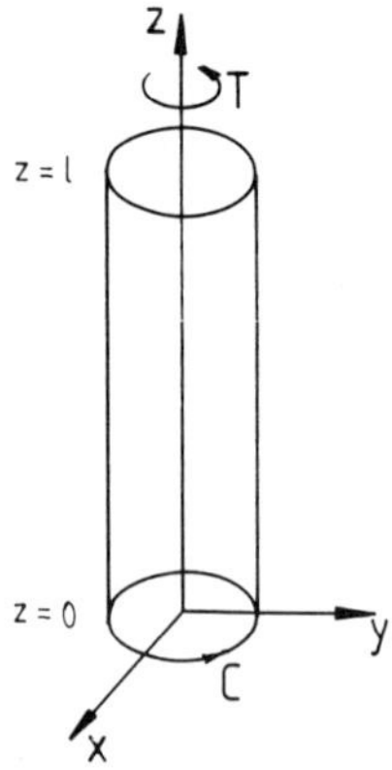

Figure 4.10 Geometry of circular bar

The displacement components in the (x, y, z) directions are written as u, v, w. According to the Saint Venant theory of torsion, as the bar twists the plane cross-sections are warped but the projection on the x, y-plane rotates as a rigid body. Then

$$u = -\vartheta z y \qquad v = \vartheta z x \qquad w = \vartheta \varphi(x, y)$$

where φ is called the 'warping function' and ϑ is the angle of twist per unit length of the bar.

The strain–displacement relations, equation (4.51), reduce to

$$\epsilon_{xx} = \epsilon_{yy} = \epsilon_{zz} = 0$$

$$\epsilon_{xy} = 0 \qquad \epsilon_{xz} = \frac{\vartheta}{2}\left(\frac{\partial\varphi}{\partial x} - y\right) \qquad \epsilon_{yz} = \frac{\vartheta}{2}\left(\frac{\partial\varphi}{\partial y} + x\right)$$

and we can derive a single equation of compatibility

$$\frac{\partial}{\partial y}\epsilon_{xz} - \frac{\partial}{\partial x}\epsilon_{yz} = -\vartheta$$

We adopt Prandtl's approach and assume that only σ_{xz} and σ_{yz} differ from zero, so that the equations of equilibrium, equation (4.53), are

$$\frac{\partial}{\partial x}\sigma_{xz} + \frac{\partial}{\partial y}\sigma_{yz} = 0$$

which is identically satisfied if σ_{xz} and σ_{yz} are derived from a stress function $\psi(x, y)$ such that

$$\sigma_{xz} = \frac{\partial\psi}{\partial y} \qquad \sigma_{yz} = -\frac{\partial\psi}{\partial x}$$

These must be in equilibrium with the applied torque T

$$T = \iint_C (x\sigma_{yz} - y\sigma_{xz})\, dx\, dy = 2\iint_C \psi\, dx\, dy$$

assuming the boundary condition

$$\psi = 0 \qquad \text{on C}$$

A nonlinear differential equation for ψ may be obtained on substituting the creep law, equation (4.54), into the compatibility relation, nothing that

$$s_{yz} = \sigma_{xz} \qquad s_{yz} = \sigma_{yz} \qquad \bar{\sigma}^2 = 3\left[\left(\frac{\partial\psi}{\partial y}\right)^2 + \left(\frac{\partial\psi}{\partial x}\right)^2\right]$$

Thus

$$\frac{\partial}{\partial y}\left(\frac{f(\bar{\sigma})}{\bar{\sigma}}\frac{\partial\psi}{\partial y}\right) + \frac{\partial}{\partial x}\left(\frac{f(\bar{\sigma})}{\bar{\sigma}}\frac{\partial\psi}{\partial x}\right) = -\frac{2}{3}\dot{\vartheta} \tag{4.56}$$

$$\psi = 0 \qquad \text{on C}$$

For a circular bar of radius a we may assume $\varphi(x, y) = 0$, and then

$$\epsilon_{xz} = -\vartheta y/2 \qquad \epsilon_{yz} = \vartheta x/2$$

We seek stress as a function of $r = \sqrt{(x^2 + y^2)}$; then

$$\frac{\partial\psi}{\partial x} = \psi' \frac{x}{r} \qquad \frac{\partial\psi}{\partial y} = \psi' \frac{y}{r} \qquad \overline{\sigma} = -\sqrt{3}\,\psi' \qquad (\;)' = \frac{\partial(\;)}{\partial r}$$

and equation (4.56) becomes

$$\frac{1}{r}\frac{\mathrm{d}}{\mathrm{d}r}\left[rf(\sqrt{3}\,\psi')\right] = 2\sqrt{3}\dot{\vartheta}$$

But we have $\psi' = 0$, $r = 0$, so that

$$\psi' = \frac{1}{\sqrt{3}} f^{-1}(\sqrt{3}\dot{\vartheta}r)$$

The torsional moment is

$$T = 2\iint_C \psi \,\mathrm{d}s = 4\pi\int_0^a \psi r \,\mathrm{d}r$$

and after integration by parts, bearing in mind that $\psi(a) = 0$,

$$T = 2\pi\int_0^a \psi' r^2 \,\mathrm{d}r = \frac{2\pi}{\sqrt{3}}\int_0^a f^{-1}(\sqrt{3}\dot{\vartheta}r)r^2 \,\mathrm{d}r \tag{4.57}$$

and we may obtain $\dot{\vartheta}$ in terms of T for particular creep laws. For example, for the power law

$$\dot{\vartheta} = \frac{B}{a}(\sqrt{3})^{n+1}\left(3 + \frac{1}{n}\right)^n \left(\frac{T}{2\pi a^3}\right)^n$$

The torsion problem has been rigorously studied although it is not of great technical importance, except perhaps for the circular bar.

Further reading

Johnson, A.E., Henderson, J., and Khan, B., *Complex Stress Creep, Relaxation and Fracture of Metallic Alloys*, HMSO, 1962.

Johnson, A.E., Recent progress in creep mechanics, in *The Folke Odqvist Volume*, Ed. B. Broberg *et al.*, Almqvist & Wiksell, 1967, p. 289.

Mroz, Z., *Mathematical Models of Inelastic Material Behaviour*, University of Waterloo Press, 1973.

Odqvist, F.K.G., *Mathematical Theory of Creep and Creep Rupture*, Clarendon Press, 2nd Edn, 1974.

Gittus, J., *Creep, Viscoelasticity and Creep Fracture in Solids*, Applied Science Publ., 1975.

Timoshenko, S.P. and Woinowsky-Krieger, S., *Theory of Plates and Shells*, McGraw-Hill, 1959.

Chapter 5

Stress analysis for steady creep

It is apparent from the preceding chapter that stress analysis for steady creep leads to the need to solve quite complex nonlinear differential equations. The problems that we examined were exceptional since they yielded closed-form solutions; the equations could be resolved either because (i) the solution was obvious after proper combination of the equations (thick cylinder), (ii) the equations could be transformed to give an obvious solution (infinite plate), (iii) the problem was statically determinate (membrane shells) or (iv) the problem could be simplified (annular plate). In general, the equations of steady creep need to be solved approximately, either by using some numerical techniques or by a simplification to the problem. In this chapter we will describe the basic techniques of nμmerical analysis for steady creep; we examine, first, methods for the direct solution of the fundamental equations and, secondly, a more subtle approach – the use of energy methods.

5.1 Numerical methods – iteration

The differential equations of steady creep are examples of nonlinear boundary value problems on which there is a rapidly expanding literature. There are many methods available for the numerical solution of such problems; here we describe a technique which has particular advantages for steady creep, making use of the fact that it is only the constitutive relation which is nonlinear. Much of engineering mechanics is concerned with linear problems arising from the elasticity of structures. The most obvious route for nonlinear problems is to linearise the equations around an initial estimate and then solve the linear problem iteratively until the process converges. Surprisingly, this approach has been little used in steady creep – possibly owing to the attractiveness of the energy methods – even though it is the natural approach when using, for example, finite element techniques. In the early literature on creep, use was made of an iterative method known as the *method of successive approximations*. It was mostly applied to the solution of plane stress problems, such as the rotating disc. As an introduction to iterative methods, we will apply this method to the simple two-bar structure.

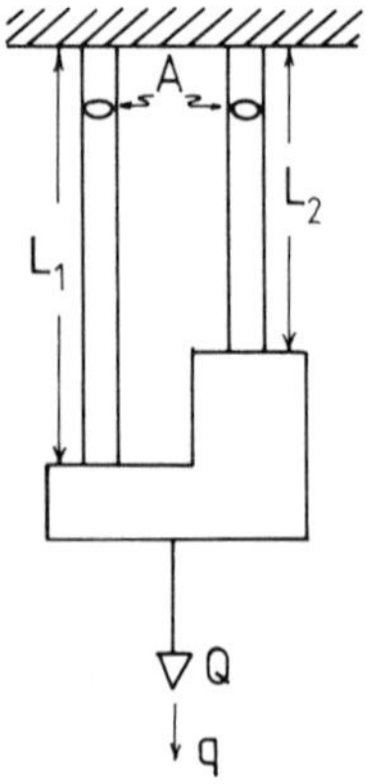

Figure 5.1 Geometry of two-bar structure

Consider two bars of length L_1 and L_2 with the same cross-sectional area rigidly attached to a horizontal bar under a tension Q (*Figure 5.1*). The basic equations are:

Equilibrium:	$\sigma_1 + \sigma_2 = Q/A$	
Strain–displacement:	$\epsilon_1 = \delta_1/L_1$	$\epsilon_2 = \delta_2/L_2$
Compatibility:	$\delta_1 = \delta_2 = q$	$\epsilon_1 L_1 = \epsilon_2 L_2$
Constitutive relation:	$\dot{\epsilon}_1 = f(\sigma_1)$	$\dot{\epsilon}_2 = f(\sigma_2)$

The iterative scheme for the method of successive approximations is as follows.

Scheme 1

(1) Guess σ_1.
(2) Use equilibrium to evaluate σ_2:

$$\sigma_2 = Q/A - \sigma_1$$

(3) Use constitutive relation to evaluate $\dot{\epsilon}_2$:

$$\dot{\epsilon}_2 = f(\sigma_2)$$

(4) Use compatibility to evaluate $\dot{\epsilon}_1$:

$$\dot{\epsilon}_1 = (L_2/L_1)\,\dot{\epsilon}_2$$

(5) Use inverted constitutive relation to obtain a new estimate of σ_1:

$$\sigma_1 = f^{-1}(\dot{\epsilon}_1)$$

(6) Return to step (2) and iterate until converged.

This process is equivalent to a solution of the nonlinear algebraic equation

$$\sigma_1 = f^{-1}\left(\frac{L_2}{L_1} f(Q/A - \sigma_1)\right) \tag{5.1}$$

in the form of a sequence of estimates $\sigma_1^{(0)}, \sigma_1^{(1)}, \ldots, \sigma_1^{(2)}, \ldots$ such that

$$\sigma_1^{(k+1)} = f^{-1}\left(\frac{L_2}{L_1} f(Q/A - \sigma_1^{(k)})\right) \tag{5.2}$$

with $\sigma_1^{(0)}$ an initial guess.

TABLE 5.1. Comparison of iterative methods.

Prandtl law: $\dot{\epsilon} = B \sinh(\alpha\sigma)$, $B = 5.39 \times 10^{-6}$, $\alpha = 0.000299$, $L_1 = 3$ in, $L_2 = 1.5$ in, $A = 0.015$ in^2, $P = 150$ lb.

Iteration 1		*Iteration 2*	*Quasilinearisation*	
σ_1	σ_2	σ_1	σ_1	σ_2
300		300	300	
351.9	700	9402	334.4	665.6
325.6	648.1	−1711	334.4	665.6
338.9	674.4	13107		
332.1	661.1	−5424	Converged	
335.6	667.9	17644	1 iteration	
333.8	664.4	−9902		
334.7	666.2	22274		
334.3	665.3			
334.5	665.7	Diverges		
334.4	665.6			
334.4	665.6			
Converged 10 iterations				

A numerical example for the Prandtl law

$$f(\sigma) = B \sinh(\alpha\sigma) \qquad f^{-1}(\dot{\epsilon}) = \frac{1}{\alpha} \ln\left\{\frac{\dot{\epsilon}}{B} + \sqrt{\left[\left(\frac{\dot{\epsilon}}{B}\right)^2 + 1\right]}\right\}$$

is given in *Table 5.1* For this law the exact solution can be obtained as

$$\dot{q} = B \sinh\left(\alpha Q/A\right) \frac{L_1 L_2}{L_1^2 + 2L_1 L_2 \cosh(\alpha Q/A) + L_2^2} \tag{5.3}$$

The iterative scheme 1 is seen to converge to the correct solution.

However, care must be taken with the method of successive approximations. We can rearrange the iterative procedure as follows.

Scheme 2

(1) Guess σ_1.

(2) Use constitutive relation to evaluate $\dot{\epsilon}_1$:

$$\dot{\epsilon}_1 = f(\sigma_1)$$

(3) Use compatibility to evaluate $\dot{\epsilon}_2$:

$$\dot{\epsilon}_2 = (L_1/L_2)\,\dot{\epsilon}_1$$

(4) Use inverted constitutive relation to evaluate σ_2:

$$\sigma_2 = f^{-1}(\dot{\epsilon}_2)$$

(5) Use equilibrium to derive a new estimate for σ_1:

$$\sigma_1 = Q/A - \sigma_2$$

(6) Return to step (2) and repeat until converged.

This process is equivalent to the solution of the nonlinear algebraic equation

$$\sigma_1 = Q/A - f^{-1}\left(\frac{L_1}{L_2} f(\sigma_1)\right) \tag{5.4}$$

in the form of a sequence of estimates $\sigma_1^{(0)}$, $\sigma_1^{(1)}$, ..., $\sigma_1^{(k)}$, ... such that

$$\sigma_1^{(k+1)} = Q/A - f^{-1}\left(\frac{L_1}{L_2} f(\sigma_1^{(k)})\right) \tag{5.5}$$

If this process is followed for the numerical example considered before, it is seen that it diverges!

Equations (5.1) and (5.4) are special cases of the general nonlinear algebraic equation

$$x = F(x) \tag{5.6}$$

in so-called *fixed point* form (if we think of F as a transformation then we require the stress $x = \sigma_1$ which is transformed into itself – it is a fixed point of the transformation). Indeed, the form of the function F need not be given exactly; for example, we may prescribe a procedure for taking one estimate and transforming it into another. The iterative scheme associated with the equation (5.6) in the form

$$x^{(k+1)} = F(x^{(k)})$$

is known as *Picard iteration.*

The lesson from the above is obvious – Picard iteration on a fixed point equation does not always converge. We would like a method which has a good chance of success; here we deal with a technique which has particular advantages for steady creep analysis. It is a generalisation of the classic *Newton method* for solving the general nonlinear equation

$$P(x) = 0 \tag{5.7}$$

of which equation (5.6) is a special case.

If $x^{(0)}$ is an initial estimate, then we construct by Newton's method a sequence of estimates $x^{(0)}, x^{(1)}, \ldots, x^{(k)}, \ldots$ by the following procedure. Approximate $P(x)$ by a linear function of x determined by the value and slope of the function P at $x^{(k)}$

$$P(x) \simeq P(x^{(k)}) + (x - x^{(k)})P'(x^{(k)})$$

where

$$P' = \mathrm{d}P/\mathrm{d}x$$

A further approximation $x^{(k+1)}$ to the solution of equation (5 7) is then obtained by solving the linear equation

$$P(x^{(k)}) + (x - x^{(k)})P'(x^{(k)}) = 0$$

i.e.

$$x^{(k+1)} = x^{(k)} - \frac{P(x^{(k)})}{P'(x^{(k)})} \tag{5.8}$$

This method usually converges quickly if a 'good' initial estimate is given. The need to start from a good estimate is of course one of the main disadvantages of iterative methods, but this can usually be overcome by a fairly obvious process known as continuation.

Although described here for a single nonlinear algebraic equation, Newton's method can be applied to systems of equations and perhaps more importantly the process by which equation (5.8) was derived can be applied to any non-linear equation, in particular to differential systems. In this form, it is sometimes known as linearisation by differentiation or *quasilinearisation* – the latter term will be used here.

We apply the quasilinearisation procedure to the simple two-bar structure as follows. Let $\sigma_1^{(k)}$, $\sigma_2^{(k)}$ be the kth in a sequence of estimates starting from an initial guess $\sigma_1^{(0)}$, $\sigma_2^{(0)}$. By quasilinearisation, replace the constitutive relation by its linearised form

$$\begin{aligned} \dot{\epsilon}_1 &= f(\sigma_1^{(k)}) + (\sigma_1 - \sigma_1^{(k)})f'(\sigma_1^{(k)}) \\ \dot{\epsilon}_2 &= f(\sigma_2^{(k)}) + (\sigma_2 - \sigma_2^{(k)})f'(\sigma_2^{(k)}) \end{aligned} \tag{5.9}$$

Since the strain–displacement, compatibility and equilibrium equations are linear, no further linearisation is required. We may now solve the resulting linear problem using well known techniques from elasticity. By inverting equation (5.9), substituting into equilibrium and using strain displacement, the following solution for $\dot{q}$ is obtained:

$$\dot{q} = \frac{f(\sigma_1^{(k)})f'(\sigma_2^{(k)}) + f(\sigma_2^{(k)})f'(\sigma_1^{(k)})}{f'(\sigma_2^{(k)})L_1 + f'(\sigma_1^{(k)})/L_2} \tag{5.10}$$

using the fact that $\sigma_1{}^{(k)}$ and $\sigma_2{}^{(k)}$ satisfy equilibrium.

From the inverted form of equation (5.9) and equation (5.10), the next estimate is

$$\begin{aligned} \sigma_1^{(k+1)} &= \sigma_1^{(k)} + \frac{\dot{q}/L_1 - f(\sigma_1^{(k)})}{f'(\sigma_1^{(k)})} \\ \sigma_2^{(k+1)} &= \sigma_2^{(k)} + \frac{\dot{q}/L_2 - f(\sigma_2{}^{(k)})}{f'(\sigma_2^{(k)})} \end{aligned} \tag{5.11}$$

This iteration is also shown in Table 5.1.

In the following sections, we will describe the quasilinearisation algorithm for two more complex problems.

5.2 Example: Steady creep of a thin tube under bending and internal pressure

The problem of a thin tube under bending and internal pressure is an important one in the analysis of high-temperature piping systems. We assume that the tube is long so that we may analyse the problem on the basis of simple bending theory, i.e. plane sections remain plane during deformation.

Consider a thin tube of circular cross-section under internal pressure p and subject to a bending moment M. The tube has thickness h and is of radius a; it is presumed that the tube is thin so that $h << a$ and the stresses and strains are constant through the thickness (*Figure 5.2*).

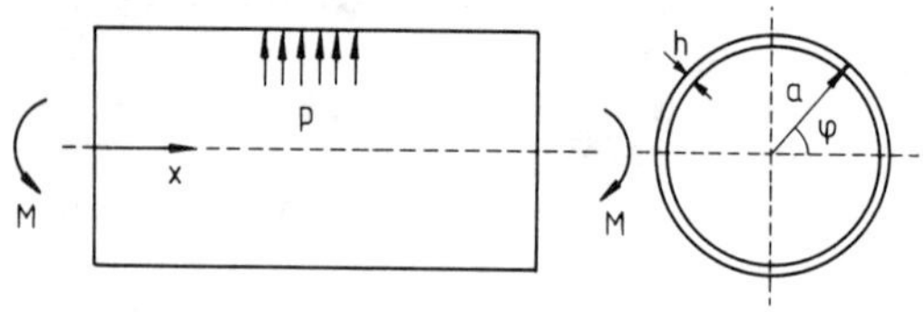

Figure 5.2 Geometry of pressurised thin tube under bending

Let φ be the angle describing the cross-section and x be the length along the tube; these may be taken as the principal directions of stress $(\sigma_\varphi, \sigma_x)$ and strain $(\epsilon_\varphi, \epsilon_x)$. Since the tube is thin, the circumferential stress is isostatic and determined from equilibrium alone

$$\sigma_\varphi = pa/h \tag{5.12}$$

The longitudinal stress is the same for each cross-section and is written in the form

$$\sigma_x = \frac{1}{2}\frac{pa}{h} + \rho(\varphi) \tag{5.13}$$

so that for pressure alone $\rho = 0$. This stress must satisfy equilibrium with the applied moment

$$M = ha^2 \int_0^{2\pi} \sigma_x \sin\varphi \, d\varphi \tag{5.14}$$

Since plane sections remain plane, the longitudinal strain rate is given by

$$\dot{\epsilon}_x = \dot{\kappa}\, a \sin\varphi \tag{5.15}$$

where $\dot{\kappa}$ is the curvature rate. This strain rate is related to the stress through the constitutive relations for plane stress, cf. equation (4.26),

$$\dot{\epsilon}_x = \frac{f(\overline{\sigma})}{\overline{\sigma}}\left(\sigma_x - \frac{1}{2}\sigma_\varphi\right)$$
$$\overline{\sigma}^2 = \sigma_x^2 - \sigma_x\sigma_\varphi + \sigma_\varphi^2 \tag{5.16}$$

Again we solve the problem iteratively using quasilinearisation: let $\sigma_x^{(0)}$, $\sigma_x^{(1)}, \ldots, \sigma_x^{(k)}, \ldots$ be a sequence of estimates of σ_x. For simplicity, write equation (5.16) as

$$\dot{\epsilon}_x = F(\sigma_x)$$

recalling that σ_φ is determined. By linearisation

$$\dot{\epsilon}_x = F(\sigma_x^{(k)}) + (\sigma_x - \sigma_x^{(k)})F'(\sigma_x^{(k)}) \tag{5.17}$$

Then from equations (5.17) and (5.15)

$$\sigma_x = \sigma_x^{(k)} + \frac{\dot{\kappa} a \sin\varphi - F(\sigma_x^{(k)})}{F'(\sigma_x^{(k)})} \tag{5.18}$$

which is substituted into equation (5.14) to give

$$\frac{M}{ha^2} = I_1^{(k)} + \dot{\kappa} a I_2^{(k)} - I_3^{(k)}$$

on defining the integrals

$$I_1^{(k)} = \int_0^{2\pi} \sigma_k^{(k)} \sin\varphi \, d\varphi \quad I_2^{(k)} = \int_0^{2\pi} \frac{\sin^2\varphi}{F'(\sigma_x^{(k)})} d\varphi \quad I_3^{(k)} = \int_0^{2\pi} \frac{F(\sigma_x^{(k)})}{F'(\sigma_x^{(k)})} \sin\varphi \, d\varphi$$

This can easily be solved, yielding the curvature rate

$$\dot{\kappa} a = \frac{M/ha^2 + I_3^{(k)} - I_1^{(k)}}{I_2^{(k)}} \tag{5.19}$$

Thus the $(k + 1)$th estimate is, from equations (5.18) and (5.19),

$$\sigma_x^{(k+1)} = \sigma_x^{(k)} + \left[\left(\frac{M/ha^2 + I_3^{(k)} - I_1^{(k)}}{I_2^{(k)}}\right) \sin\varphi - F(\sigma_x^{(k)})\right] \Big/ F'(\sigma_x^{(k)}) \tag{5.20}$$

In practice, the integrals $I_1^{(k)}$, $I_2^{(k)}$, $I_3^{(k)}$ cannot be evaluated exactly and will need to be found using some numerical quadrature (for example, Simpson's rule).

As an illustration, we consider the isothermal power law $f(\sigma) = B\sigma^n$ so that

$$F'(\sigma_x) = \bar{\sigma}^{\,n-3}[\bar{\sigma}^2 + (n-1)(\sigma_x - \tfrac{1}{2}\sigma_\varphi)^2]$$

Owing to the homogeneity of the power law, we can normalise the stress by $\sigma_0 = M/ha^2$ and the strain rate and normalised curvature $\dot{\kappa} a$ by $\dot{\epsilon}_0 = B\sigma_0^n$. The solution then depends only on the material parameter n and the loading parameter

$$\$ = pa^3/M$$

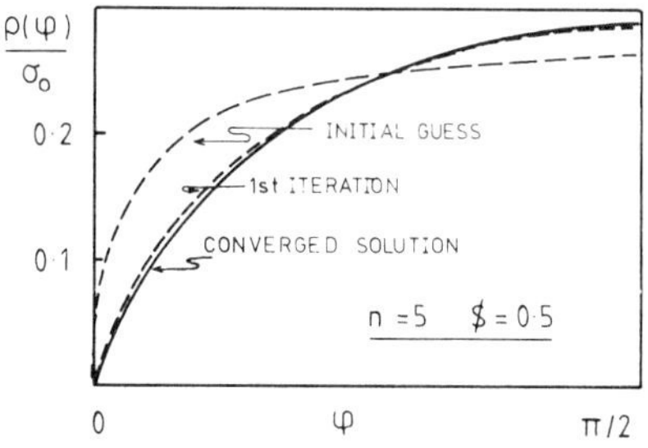

Figure 5.3 Convergence of normalised circumferential stress using quasilinearisation

In *Figure 5.3* we plot the distribution of $\rho(\varphi)/\sigma_0$ as it converges from the solution for bending alone (Section 3.1)

$$\frac{\sigma_x}{\sigma_0} = \frac{\rho(\varphi)}{\sigma_0} = \frac{1}{D_n} (\sin\varphi)^{1/n}$$

where

$$D_n = \int_0^{2\pi} (\sin \varphi)^{1+1/n} \, d\varphi$$

as an initial estimate.

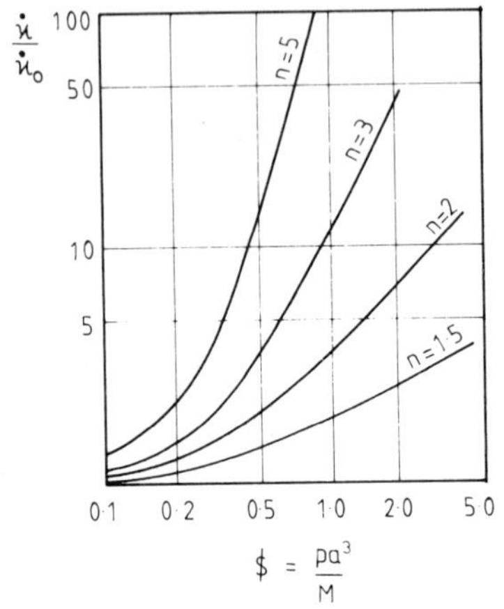

Figure 5.4 Variation of normalised curvature rate with load parameter

In *Figure 5.4* we plot the ratio of the normalised curvature rate to that of an equivalent tube in bending alone,

$$\dot{\kappa} a = B \left(\frac{M}{ha^2} \frac{1}{D_n} \right)^n$$

for various values of n against the load parameter. It is seen that the presence of internal pressure greatly increases the flexibility of the tube in creep. This result is not expected and differs from the well known elastic case where the pressure has no effect on the flexibility of the tube! Unfortunately, this severe interaction is typical of combined loading situations.

5.3 Example: Steady creep of a rotating disc

The problem of estimating the radial extension of a spinning circular disc is one of the classical problems in creep, arising from the study of turbines. Nevertheless, it presented serious difficulties to the analyst in obtaining a numerical solution since conditions of plane stress prevail.

Consider a circular disc of constant thickness with outer radius b containing a hole of radius a (*Figure 5.5*). The disc is spinning with a velocity ω; the rotational body force is then $Q = \rho\omega^2$ where ρ is the density of the disc material.

The geometry of the disc is described by the cylindrical polar coordinate system (r, φ, z); since the loading is symmetric, there is no dependence on the angle φ and under plane stress there is no dependence on the thickness coordinate z. This coordinate system may be taken as the principal directions of stress $(\sigma_r, \sigma_\varphi, \sigma_z)$ where $\sigma_z = 0$ and strain $(\epsilon_r, \epsilon_\varphi, \epsilon_z)$.

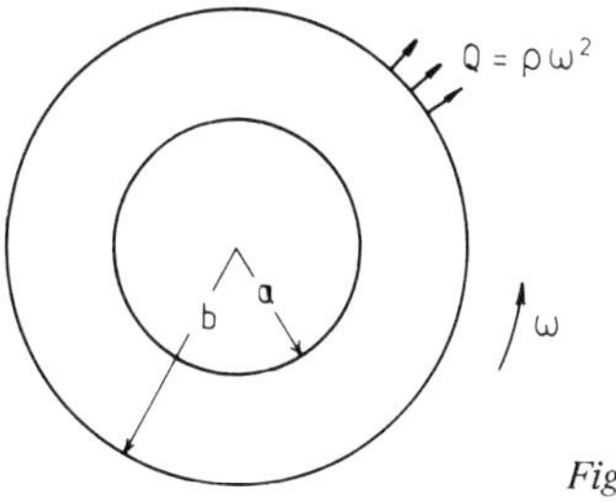

Figure 5.5 Geometry of rotating disc

If u is the radial displacement, then the strain displacement and compatibility relations are the same as in Section 4.3, i.e.

$$\epsilon_r = \mathrm{d}u/\mathrm{d}r \qquad \epsilon_\varphi = u/r \tag{5.21}$$

$$\frac{\mathrm{d}\epsilon_\varphi}{\mathrm{d}r} = \frac{1}{r}(\epsilon_r - \epsilon_\varphi) \tag{5.22}$$

The equilibrium equation now takes the form

$$\frac{\mathrm{d}\sigma_r}{\mathrm{d}r} = \frac{1}{r}(\sigma_\varphi - \sigma_r) - Qr \tag{5.23}$$

with boundary conditions

$$\sigma_r(a) = 0 \qquad \sigma_r(b) = 0 \tag{5.24}$$

assuming that both inner and outer edges are free.

The constitutive relations for plane stress are

$$\begin{aligned} \dot{\epsilon}_r &= \frac{f(\bar{\sigma})}{\bar{\sigma}}\left(\sigma_r - \frac{1}{2}\sigma_\varphi\right) = F_r(\sigma_r, \sigma_\varphi) \\ \dot{\epsilon}_\varphi &= \frac{f(\bar{\sigma})}{\bar{\sigma}}\left(\sigma_\varphi - \frac{1}{2}\sigma_r\right) = F_\varphi(\sigma_r, \sigma_\varphi) \\ \bar{\sigma}^2 &= \sigma_r^2 - \sigma_r\sigma_\varphi + \sigma_\varphi^2 \end{aligned} \tag{5.25}$$

Again, we will solve this problem using quasilinearisation. A sequence of approximate solutions $(\sigma_r^{(k)}, \sigma_\varphi^{(k)})$ $k = 0, 1, 2, \ldots$ is generated in the following manner. Linearisation of the constitutive relations (5.25) leads, on rearrangement, to

$$\begin{aligned} \dot{\epsilon}_r &= a_r + b_{rr}\sigma_r + b_{r\varphi}\sigma_\varphi \\ \dot{\epsilon}_\varphi &= a_\varphi + b_{\varphi r}\sigma_r + b_{\varphi\varphi}\sigma_\varphi \end{aligned} \tag{5.26}$$

The coefficients

$$a_r = F_r - \sigma_r\frac{\partial F_r}{\partial \sigma_r} - \sigma_\varphi\frac{\partial F_r}{\partial \sigma_\varphi} \qquad a_\varphi = F_\varphi - \sigma_r\frac{\partial F_\varphi}{\partial \sigma_r} - \sigma_\varphi\frac{\partial F_\varphi}{\partial \sigma_\varphi}$$

$$b_{rr} = \frac{\partial F_r}{\partial \sigma_r} \qquad b_{r\varphi} = \frac{\partial F_r}{\partial \sigma_\varphi} \qquad b_{\varphi r} = \frac{\partial F_\varphi}{\partial \sigma_r} \qquad b_{\varphi\varphi} = \frac{\partial F_\varphi}{\partial \sigma_\varphi}$$

are evaluated for the current estimate $(\sigma_r^{(k)}, \sigma_\varphi^{(k)})$ say. We can rearrange the equations (5.26) as

$$\sigma_\varphi = \frac{1}{b_{\varphi\varphi}}(\dot{\epsilon}_\varphi - b_{\varphi r}\sigma_r - a_\varphi)$$

$$\dot{\epsilon}_r = a_r - \frac{b_{r\varphi}}{b_{\varphi\varphi}}a_\varphi + \frac{b_{r\varphi}}{b_{\varphi\varphi}}\dot{\epsilon}_\varphi + \left(b_{rr} - \frac{b_{r\varphi}b_{\varphi r}}{b_{\varphi\varphi}}\right)\sigma_r \tag{5.27}$$

These we combine with compatibility, equation (5.22), and equilibrium, equation (5.23), to give

$$\frac{\mathrm{d}}{\mathrm{d}r}\begin{bmatrix}\sigma_r\\ \dot{\epsilon}_\varphi\end{bmatrix} = \begin{bmatrix} -\frac{1}{r} - \frac{b_{\varphi r}}{b_{\varphi\varphi}}\frac{1}{r} & \frac{1}{r}\frac{1}{b_{\varphi\varphi}} \\ \frac{1}{r}\left(b_{rr} - \frac{b_{r\varphi}b_{\varphi r}}{b_{\varphi\varphi}}\right) & -\frac{1}{r} + \frac{b_{r\varphi}}{b_{\varphi\varphi}}\frac{1}{r}\end{bmatrix}\begin{bmatrix}\sigma_r\\ \dot{\epsilon}_\phi\end{bmatrix} + \begin{bmatrix} -Qr - \frac{a_\varphi}{b_{\varphi\varphi}}\frac{1}{r} \\ \frac{1}{r}\left(a_r - \frac{b_{r\varphi}}{b_{\varphi\varphi}}a_\varphi\right)\end{bmatrix} \tag{5.28}$$

These equations are now in a standard form: a two-point linear boundary value problem in two unknowns $(\sigma_r, \dot{\epsilon}_\varphi)$, which must be solved subject to the boundary conditions at the two endpoints, equation (5.24). The solution of this problem, together with equation (5.27), then gives the next estimate $(\sigma_r^{(k+1)}, \sigma_\varphi^{(k+1)})$.

As an example, this problem has been solved for the special case of a power law $f(\sigma) = B\sigma^n$. In this case the equations can be normalised in the following manner

$$x = \frac{r}{a} \qquad Q_0 = \frac{Qa^2}{\sigma_0} \qquad \eta = \frac{b}{a}$$

and

$$S_{r,\varphi} = \frac{\sigma_{r,\varphi}}{\sigma_0} \qquad \dot{E}_{r,\varphi} = \frac{\dot{\epsilon}_{r,\varphi}}{\sigma_0} \qquad \dot{U} = \frac{\dot{u}}{a\dot{\epsilon}_0}$$

The scaling parameter σ_0 is chosen for convenience as the value of σ_ϕ at the inner edge, $r = a$, for the case $n = 1$, i.e.

$$\sigma_0 = \frac{1}{8}Qa^2\,[1 + 7(b/a)^2]$$

and $\dot{\epsilon}_0 = B\sigma_0^n$.

The sequence of linear two-point boundary value problems, equation (5.28), can be solved using a standard routine from a computer scientific subroutine package. (The particular technique used here is known as 'shooting'.)

In *Figure 5.6* is shown the convergence of S_r and S_φ starting from the solution for $n = 1$,

$$S_r = \frac{7}{16}Q_0\,(1 - x^2 + \eta^2 - \eta^2/x^2)$$

$$S_\varphi = \frac{7}{16}Q_0\,(1 - 3x^2 + \eta^2 - \eta^2/x^2) + Q_0x^2$$

as an initial guess, for a particular value of n and η.

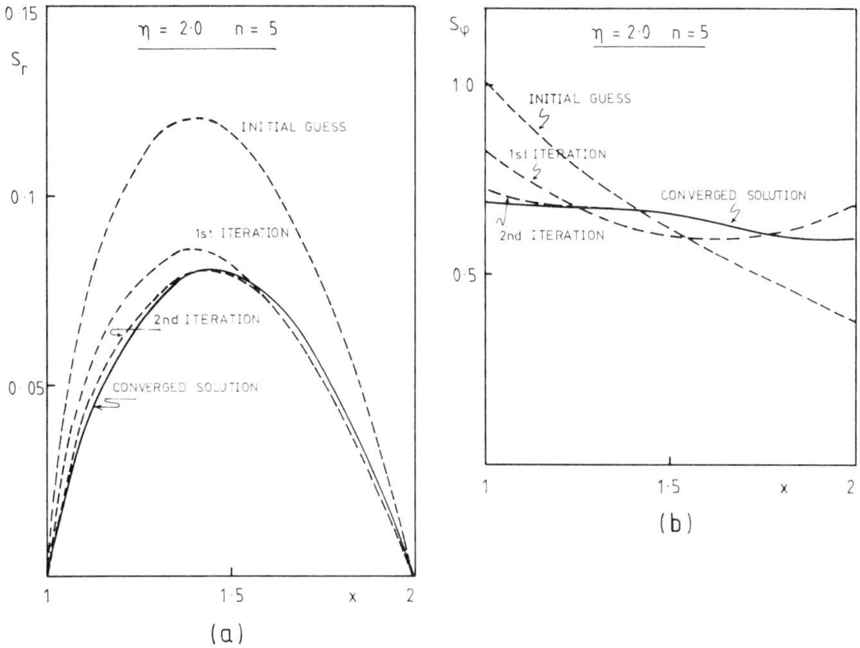

Figure 5.6 Convergence of (a) normalised radial stress, and (b) normalised circumferential stress using quasilinearisation

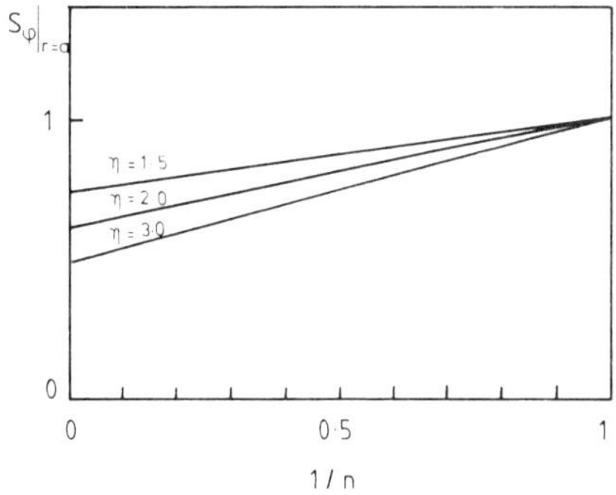

Figure 5.7 Variation of normalised circumferential stress at $r = a$ with $1/n$

In *Figure 5.7* is shown the variation in S_φ at $r = 1$ with the reciprocal of the creep index, $1/n$, for various values of η. Again it is seen that the variation is approximately linear.

Finally, it is instructive to see how we would apply the reference stress concept introduced in Section 3.1 to this problem. It is recalled that the aim is to express some characteristic deformation rate, say the displacement rate at the outside $r = b$, in the form

$$\dot{u}\big|_{r=b} = \delta \times \dot{\epsilon}_R$$

where $\dot{\epsilon}_R = B\sigma_R^n$ and σ_R is such that the scaling factor $\delta(\sigma_R, n)$ tends to a limit, and is thus independent of the exponent n, as $n \to \infty$. This is difficult in this problem since we only have numerical values for the normalised displacement rate $\dot{U}$ for several values of n and η. We can write

$$\frac{\delta}{a} = \frac{\dot{U}|_{r=b}}{\alpha_R^n}$$

where α_R is a scalar such that

$$\sigma_R = \frac{1}{8} \alpha_R Qa^2 [1 + 7(b/a)^2]$$

and it is required to determine that value of α_R which renders δ virtually independent of n. In other words, for any two large values of n, n_1 and n_2, say, it is sufficient that

$$\delta(\alpha_R, n_1) \simeq \delta(\alpha_R, n_2)$$

which is satisfied if

$$\alpha_R = \left(\frac{\dot{U}_1|_{r=b}}{\dot{U}_2|_{r=b}}\right)^{1/(n_2 - n_1)}$$

where $\dot{U}_1$ is the value of the normalised displacement rate for the exponent n_1, etc.

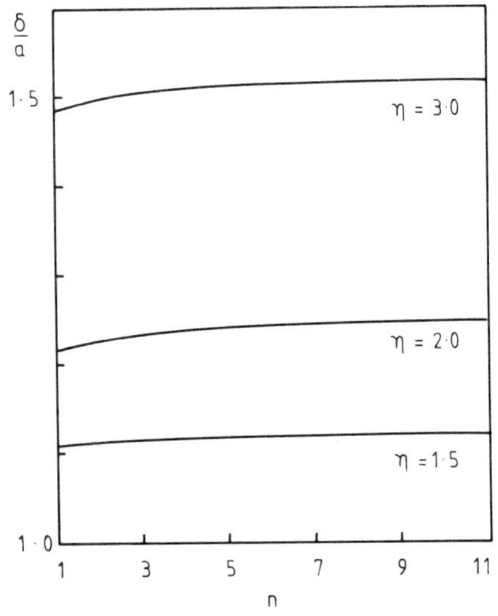

Figure 5.8 Variation of scaling factor with n

If this procedure is carried through for the present problem and we plot the scaling factor δ/a (*Figure 5.8*), then it is readily seen that, as required, the scaling factor is to all practical purposes independent of n for $n \geqslant 3$.

5.4 Energy methods

It should be apparent that the iterative methods provide an automatic means of solving steady creep problems. But they are simply standard numerical techniques for the solution of nonlinear boundary value problems applied to the differential equations of steady creep. However, there is a class of methods for resolving steady creep problems which have, for the engineer, a more intuitive mechanical interpretation – the *energy* or *variational methods*. It turns out, as may be expected, that the well known theorems of minimum potential energy and complementary energy in elasticity have counterparts in steady creep. Then approximate methods developed from these theorems for elasticity can be carried through to steady creep problems. For this reason, energy methods have become quite popular in creep mechanics. Here we will first derive the energy theorems from first principles, together with some related bounding techniques, and then discuss how they may be used to obtain approximate solutions to the steady creep problem.

5.4.1 Notation

We will derive the energy theorems in a simple manner; in order not to complicate this derivation, we will adopt a vector notation. Again using stresses and strain rates referred to their principal directions, we define the following *vector fields.*

Stress vector: $\boldsymbol{\sigma} = (\sigma_1 \quad \sigma_2 \quad \sigma_3)$

Stress deviation vector: $\mathbf{s} = (s_1 \quad s_2 \quad s_3)$

Strain-rate vector: $\dot{\boldsymbol{\epsilon}} = (\dot{\epsilon}_1 \quad \dot{\epsilon}_2 \quad \dot{\epsilon}_3)$

Associated with this vector representation we introduce the *vector product*. If **a** and **b** are vectors of dimension 3 (as above), then the vector product, a scalar, is given by

$$\mathbf{a} \cdot \mathbf{b} = a_1 b_1 + a_2 b_2 + a_3 b_3$$

and the vector derivative (or *gradient*) of a function Φ of a vector **a** by

$$\frac{\mathrm{d}\Phi}{\mathrm{d}\mathbf{a}} = \left(\frac{\partial \Phi}{\partial a_1} \quad \frac{\partial \Phi}{\partial a_2} \quad \frac{\partial \Phi}{\partial a_3}\right)$$

Note the chain rule for the derivative of a function of a function of a vector $\psi(\Phi(\mathbf{a}))$

$$\frac{\mathrm{d}\psi}{\mathrm{d}\mathbf{a}} = \frac{\mathrm{d}\psi}{\mathrm{d}\Phi} \frac{\mathrm{d}\Phi}{\mathrm{d}\mathbf{a}}$$

5.4.2 Basic assumptions

Using this notation, we may write the constitutive relations, equations (4.4), as

$$\dot{\boldsymbol{\epsilon}} = \frac{3}{2} \frac{f(\bar{\sigma})}{\bar{\sigma}} \mathbf{s} \qquad \mathbf{s} = \frac{2}{3} \frac{f^{-1}(\dot{\bar{\epsilon}})}{\dot{\bar{\epsilon}}} \dot{\boldsymbol{\epsilon}} \tag{5.29}$$

As a basic assumption, we assume that the uniaxial constitutive functions f and f^{-1} are *potential functions*; this means that there exist functions $\Omega(\bar{\sigma}) > 0$ and $W(\dot{\bar{\epsilon}}) > 0$ such that

$$f(\bar{\sigma}) = \frac{\mathrm{d}\Omega}{\mathrm{d}\bar{\sigma}} \qquad f^{-1}(\dot{\bar{\epsilon}}) = \frac{\mathrm{d}W}{\mathrm{d}\dot{\bar{\epsilon}}}$$

For example, for the power law $f = B\sigma^n$

$$\Omega = \frac{B}{n+1}\bar{\sigma}^{n+1} \qquad W = \frac{n}{n+1}\left(\frac{1}{B}\right)^{1/n}\bar{\dot{\epsilon}}^{1+1/n}$$

Then noting the relations

$$\frac{\mathrm{d}\bar{\sigma}}{\mathrm{d}\boldsymbol{\sigma}} = \frac{3}{2}\frac{1}{\bar{\sigma}}\,\mathbf{s} \qquad \frac{\mathrm{d}\bar{\dot{\epsilon}}}{\mathrm{d}\dot{\boldsymbol{\epsilon}}} = \frac{2}{3}\frac{1}{\bar{\dot{\epsilon}}}\,\dot{\boldsymbol{\epsilon}}$$

the constitutive relations, equation (5.29), can be written as

$$\dot{\boldsymbol{\epsilon}} = \frac{\mathrm{d}\Omega}{\mathrm{d}\boldsymbol{\sigma}} \qquad \mathbf{s} = \frac{\mathrm{d}W}{\mathrm{d}\dot{\boldsymbol{\epsilon}}} \tag{5.30}$$

The functionals (i.e. scalar functions of vectors)

$$W = \int_0^{\dot{\boldsymbol{\epsilon}}} \mathbf{s}\cdot\mathrm{d}\dot{\boldsymbol{\epsilon}} = \int_0^{\dot{\boldsymbol{\epsilon}}} \boldsymbol{\sigma}\cdot\mathrm{d}\dot{\boldsymbol{\epsilon}} \qquad \Omega = \int_0^{\boldsymbol{\sigma}} \dot{\boldsymbol{\epsilon}}\cdot\mathrm{d}\boldsymbol{\sigma}$$

are called the *strain energy dissipation* and the *complementary dissipation*, respectively. The energy (or work) dissipated in going from a strain rate $\dot{\boldsymbol{\epsilon}}^A$ to $\dot{\boldsymbol{\epsilon}}^B$ is

$$W(\mathrm{A},\mathrm{B}) = \int_{\dot{\epsilon}^A}^{\dot{\epsilon}^B} \boldsymbol{\sigma}\,.\,\mathrm{d}\dot{\boldsymbol{\epsilon}} = \int_{\dot{\epsilon}^A}^{\dot{\epsilon}^B} \frac{\mathrm{d}W}{\mathrm{d}\dot{\boldsymbol{\epsilon}}}\,.\,\mathrm{d}\dot{\boldsymbol{\epsilon}} = W(\mathrm{B}) - W(\mathrm{A})$$

and is said to be *path-independent*, i.e. the dissipation does not depend on the strain path taken, only on the initial and final states.

It is readily verified that

$$\Omega + W = \boldsymbol{\sigma}\,.\,\dot{\boldsymbol{\epsilon}} = \bar{\sigma}\,\bar{\dot{\epsilon}} \tag{5.31}$$

which is called the *specific rate of work*, so that alternatively

$$\Omega = \int_0^{\bar{\sigma}} \bar{\dot{\epsilon}}\,\mathrm{d}\bar{\sigma} \qquad W = \int_0^{\bar{\dot{\epsilon}}} \bar{\sigma}\,\mathrm{d}\bar{\dot{\epsilon}}$$

These may be interpreted geometrically in *Figure 5.9*; W is the area below the

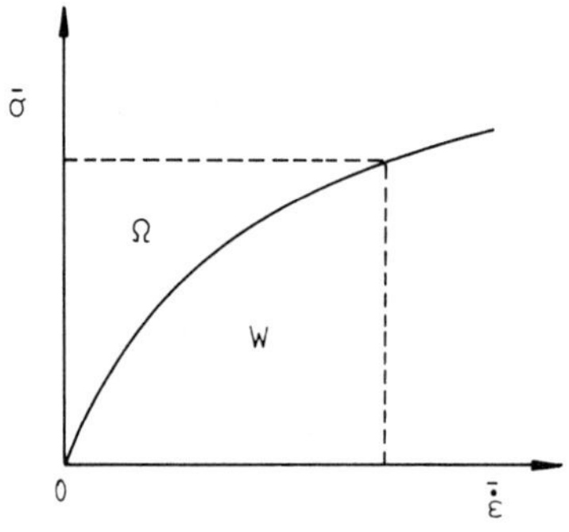

Figure 5.9 Relationship between strain-energy dissipation and complementary dissipation

$\bar{\sigma} - \bar{\dot{\epsilon}}$ curve and Ω is the area above. Also if we interpret Ω and W as surfaces in σ or $\dot{\epsilon}$ space, then $\dot{\boldsymbol{\epsilon}}$ is normal to Ω and σ is normal to W.

We derive a further property of the energy functionals Ω and W based on an observed property of steady creep behaviour. From uniaxial tests, the constitutive function f is *monotonic* – increases in stress cause increase in strain rate. Expressed mathematically, if the stress is changed from σ^A to σ^B, then the change in strain rate from $\dot{\epsilon}^A$ to $\dot{\epsilon}^B$ must satisfy the inequality

$$(\dot{\epsilon}^B - \dot{\epsilon}^A)(\sigma^B - \sigma^A) \geqslant 0$$

Generalised to vectors, this property is expressed as

$$(\dot{\epsilon}^B - \dot{\epsilon}^A)\cdot(\sigma^B - \sigma^A) \geqslant 0$$

This applies equally to small increments; if we let $\boldsymbol{\sigma}^B = \boldsymbol{\sigma}^A + \mathrm{d}\boldsymbol{\sigma}$, $\dot{\epsilon}^B = \dot{\epsilon}$, then

$$(\dot{\epsilon} - \dot{\epsilon}^A) \cdot \mathrm{d}\sigma \geqslant 0$$

Integrating from $\boldsymbol{\sigma}^A$ to σ^B (now arbitrary stresses) gives

$$\int_{\sigma^A}^{\sigma^B} (\dot{\epsilon} - \dot{\epsilon}^A) \cdot \mathrm{d}\sigma \geqslant 0$$

i.e.

$$\int_{\sigma^A}^{\sigma^B} \dot{\epsilon} \,.\, \mathrm{d}\sigma - \dot{\epsilon}^A \,.\, (\boldsymbol{\sigma}^B - \sigma^A) \geqslant 0$$

and from equation (5.30) there results the inequality

$$\Omega(\sigma^B) - \Omega(\sigma^A) - \frac{\mathrm{d}\Omega}{\mathrm{d}\boldsymbol{\sigma}^A} \cdot (\sigma^B - \boldsymbol{\sigma}^A) \geqslant 0 \qquad (5.32)$$

Similarly we can derive, nothing $\dot{\boldsymbol{\epsilon}} \cdot \sigma = \dot{\boldsymbol{\epsilon}} \cdot \mathrm{s}$ from incompressibility, that

$$W(\dot{\epsilon}^B) - W(\dot{\epsilon}^A) - \frac{\mathrm{d}W}{\mathrm{d}\dot{\epsilon}^A} \cdot (\dot{\epsilon}^B - \dot{\epsilon}^A) \geqslant 0 \qquad (5.33)$$

These inequalities express the property that Ω and W are *convex* (in the uniaxial case, they can be related to the well known definition of convexity from single-variable differential calculus in terms of the second derivative). We could equally have started from convexity; here we have demonstrated that we may begin from the more fundamental properties of existence of a potential and monotonicity.

Finally, we also note that combining equation (5.33) for $\boldsymbol{\sigma}^A$ and $\dot{\epsilon}^A$ with equation (5.32) there results

$$W(\dot{\epsilon}^B) + \Omega(\sigma^A) \geqslant \dot{\epsilon}^B \cdot \sigma^A$$

which is called *Martin's inequality*.

5.4.3 Energy theorems

Although the convexity conditions (5.32) and (5.33) are sufficient to establish the energy theorems, we need one other general result, which is independent of the material behaviour and valid for any deformable body. We state this result in the following form.

Consider a structure or solid body occupying a volume V acted on by a body force $\bar{\mathbf{b}} = (\bar{b}_1 \ \bar{b}_2 \ \bar{b}_3)$ in the principal directions and bounded by a surface S on which there is a distributed surface force $\mathbf{p} = (p_1 \ p_2 \ p_3)$; the body will deform and this results in a displacement rate field $\dot{\mathbf{u}} = (\dot{u}_1 \ \dot{u}_2 \ \dot{u}_3)$. There will be set up in the body a stress and strain rate, which are related through the prescribed constitutive relations. If σ^{p} is *any* stress field in equilibrium with the body forces in V and with the surface forces, and $\dot{\epsilon}^{\mathrm{u}}$ is *any* strain rate compatible with the displacements, then

$$\int_S \mathbf{p} \cdot \mathbf{u} \, \mathrm{d}S + \int_V \bar{\mathbf{b}} \cdot \mathbf{u} \, \mathrm{d}V = \int_V \sigma^{\mathrm{p}} \cdot \dot{\epsilon}^{\mathrm{u}} \, \mathrm{d}V \tag{5.34}$$

This is known as *Green's theorem* and can be related to the well known principle of virtual work. It is merely a 'global' statement of the strain displacement rate and equilibrium equations commonly expressed locally as differential equations.

First we specify the *structural problem* for a steadily creeping body comprised of a material which satisfies relations (5.30). We consider that in addition to the body forces the body suffers a constant load $\bar{\mathbf{p}}$ on part of the surface S_{p}

$$\mathbf{p} = \bar{\mathbf{p}} \qquad \text{on } S_{\mathrm{p}}$$

while on the remainder S_{u} the displacement rates are prescribed

$$\dot{\mathbf{u}} = \bar{\dot{\mathbf{u}}} \qquad \text{on } S_{\mathrm{u}}$$

To derive the energy theorems, we define the class of *statically admissible stresses* $\boldsymbol{\sigma}^{\mathrm{s}}$, $\mathbf{p}^{\mathrm{s}}$ in equilibrium with the body force $\bar{\mathbf{b}}$ and prescribed surface loads $\bar{\mathbf{p}}$ on S_{p} together with the class of *kinematically admissible strain rates* $\dot{\epsilon}^{\mathrm{k}}$ which are compatible with displacement rates $\dot{\mathbf{u}}^{\mathrm{k}}$ equal to the prescribed displacement rates on S_{u}.

We now establish the two basic theorems: let σ, $\dot{\epsilon}$, $\mathbf{p}$ and $\dot{\mathbf{u}}$ be the solution to the structural problem. In the convexity condition, equation (5.33), identify $\dot{\epsilon}^{\mathrm{B}}$ with any kinematically admissible strain rate $\dot{\epsilon}^{\mathrm{k}}$ and $\dot{\epsilon}^{\mathrm{A}}$ with the actual strain rate $\dot{\epsilon}$; then

$$W(\dot{\epsilon}^{\mathrm{k}}) - \boldsymbol{\sigma} \cdot \dot{\epsilon}^{\mathrm{k}} \geqslant W(\dot{\epsilon}) - \sigma \cdot \dot{\epsilon}$$

This may be integrated over the body. From Green's theorem, equation (5.34), since the actual solution is both statically and kinematically admissible

$$\int_V \sigma \cdot \dot{\epsilon}^{\mathrm{k}} \mathrm{d}V = \int_{S_{\mathrm{p}}} \bar{\mathbf{p}} \cdot \dot{\mathbf{u}}^{\mathrm{k}} \mathrm{d}S + \int_{S_{\mathrm{u}}} \mathbf{p} \cdot \bar{\dot{\mathbf{u}}} \, \mathrm{d}S + \int_V \bar{\mathbf{b}} \cdot \dot{\mathbf{u}}^{\mathrm{k}} \mathrm{d}V$$

$$\int_V \sigma \cdot \dot{\epsilon} \, \mathrm{d}V = \int_{S_{\mathrm{p}}} \bar{\mathbf{p}} \cdot \dot{\mathbf{u}} \, \mathrm{d}S + \int_{S_{\mathrm{u}}} \mathbf{p} \cdot \bar{\dot{\mathbf{u}}} \, \mathrm{d}S + \int_V \bar{\mathbf{b}} \cdot \dot{\mathbf{u}} \, \mathrm{d}V$$

Then combining the above and eliminating the integral over S_u, there results

$$\int_V W(\dot{\epsilon}^k)\,dV - \int_V \bar{b}\cdot\dot{u}^k\,dV - \int_{S_p}\bar{p}\cdot\dot{u}^k\,dS \quad \int_V W(\dot{\epsilon})dV - \int_V \bar{b}\cdot\dot{u}\,dV - \int_{S_p}\bar{p}\cdot\dot{u}\,dS \tag{5.35}$$

where the equality only holds for the actual solution. Hence we have proved the following theorem.

Theorem of minimum potential energy dissipation Amongst all kinematically admissible solutions $\dot{\epsilon}^k$, $\dot{u}^k$, the actual strain rate $\dot{\epsilon}$ and displacement rate $\mathbf{u}$ minimise the functional

$$U_p(\dot{\epsilon}^k, \dot{u}^k) = \int_V W(\dot{\epsilon}^k)\,dV - \int_V \bar{b}\cdot\dot{u}^k\,dV - \int_{S_p}\bar{p}\cdot\dot{u}^k\,dS$$

Similarly we can establish the dual result, the following theorem.

Theorem of minimum complementary energy dissipation Amongst all statically admissible solutions σ^s, $\mathbf{p}^s$, the actual stress σ and surface load $\mathbf{p}$ minimise the functional

$$U_c(\sigma^s, \mathbf{p}^s) = \int_V \Omega(\sigma^s)\,dV - \int_{S_u}\mathbf{p}\cdot\bar{\dot{u}}\,dS$$

These are seen to be analogous to the familiar theorems in elasticity.

5.4.4 Energy bounds

We finally note that we can combine the two energy inequalities to form two extended inequalities, on noting from equation (5.31) that

$$U_p(\dot{\epsilon}, \dot{u}) = -\,U_c(\sigma, \mathbf{p})$$

Hence

$$-U_c(\sigma^s, \mathbf{p}^s) \leqslant U_p(\dot{\epsilon}, \dot{u}) \leqslant U_p(\dot{\epsilon}^k, \dot{u}^k)$$

$$-U_p(\dot{\epsilon}^k, \dot{u}^k) \leqslant U_c(\sigma, \mathbf{p}) \leqslant U_c(\sigma^s, \mathbf{p}^s)$$

so that we may bound U_p and U_c in terms of known quantities. In fact, for two special cases where body forces are neglected ($\bar{b} = 0$), the bounds can be developed to yield improved results and to bound the internal energies Ω and W.

In the first case, forces are prescribed on the boundary S_p while the remainder is fixed, i.e.

$$\dot{u} = 0 \qquad \text{on } S_u$$

This is the *prescribed (surface) force problem*; for this case

$$\int_{S_p}\bar{p}\cdot\dot{u}^k\,dS - \int_V W(\dot{\epsilon}^k)\,dV \leqslant \int_V \Omega(\sigma)\,dV \leqslant \int_V \Omega(\sigma^s)\,dV \tag{5.36}$$

which bounds the complementary internal energy.

In the second case, displacements are prescribed on the boundary S_u but the remainder is free, i.e.

$$\mathbf{p} = 0 \qquad \text{on } S_p$$

which is called the *prescribed displacement problem*. (But it should be noted that steady creep does not provide a good description of this problem and elastic effects should be included.) For this case

$$\int_{S_u} \mathbf{p}^s . \bar{\mathbf{u}} \, dS - \int_V \Omega(\sigma^s) \, dV \leqslant \int_V W(\dot{\epsilon}^k) \, dV \leqslant \int_V W(\dot{\epsilon}^k) \, dV \tag{5.37}$$

which gives bounds on the internal strain energy.

In fact, for the prescribed force problem, if $\dot{\epsilon}^k$, $\dot{\mathbf{u}}^k$ is any kinematically admissible solution, then so also is $\lambda\dot{\epsilon}^k$, $\lambda\dot{\mathbf{u}}^k$ for any λ. We may then optimise the lower bound, i.e.

$$\max_\lambda \left[\lambda \int_{S_p} \bar{\mathbf{p}} . \dot{\mathbf{u}}^k \, dS - \int_V W(\lambda\dot{\epsilon}^k) \, dV \right] \leqslant \int_V \Omega(\sigma) \, dV \tag{5.38}$$

Similarly if σ^s, $\mathbf{p}^s$ is any statically admissible solution, then

$$\max_\lambda \left[\lambda \int_{S_u} \mathbf{p}^s . \bar{\mathbf{u}} \, dS - \int_V \Omega(\lambda\sigma^s) \, dV \right] \leqslant \int_V W(\dot{\epsilon}) \, dV \tag{5.39}$$

It can be shown that in both cases the resulting functionals maximise U_p and U_c. They are called *Schwinger–Levine* bounds.

5.4.5 Castigliano's theorem

Suppose that the structural problem is that of a body fixed somewhere and suffering a point load Q. The energy theorems tell us how we may solve the problem, but we are often more interested in the displacement of the point of application of the load. In elasticity, this can be done using Castigliano's theorem, and we derive its counterpart here for steady creep. The surface is fixed over S_u suffers a surface load over $S_p - S_0$ and a point load over the remainder S_0. There are no body forces. From Green's theorem, if we identify $\dot{\epsilon}^u$ with the actual solution $\dot{\epsilon} = d\Omega/d\sigma$ where σ is the actual stress in equilibrium with $\mathbf{p}$ over S then $d\sigma/dQ$ is in equilibrium with the unit load $\mathbf{1}$ over S_0 and zero load over $S_p - S_0$, which we identify with σ^p; then

$$\int_{S_0} \mathbf{1} . \dot{\mathbf{u}} \, dS = \int_V \frac{d\Omega}{d\sigma} \cdot \frac{d\sigma}{dQ} \, dV = \frac{d}{dQ} \int_V \Omega \, dV$$

The right-hand side is the displacement rate in the direction of Q, so

$$\dot{q} = \frac{d}{dQ} \int_V \Omega \, dV$$

This is Castigliano's theorem for steady creep. We can derive the displacement rate from the actual, or an approximate, solution. However, it is possible to derive bounds on the displacement rate for the special case of the power law, which is done in the next section.

5.4.6 Displacement bounds for the special case of a power law

The power law has the special property that

$$W(\dot{\epsilon}) = \frac{n}{n+1}\,\dot{\epsilon}\cdot\sigma \qquad \Omega(\sigma) = \frac{1}{n+1}\,\dot{\epsilon}\cdot\sigma$$

due to homogeneity.* This allows us to bound the *global compliance*

$$\int_S \mathrm{p}\cdot\dot{\mathrm{u}}\;\mathrm{d}S = \int_V \sigma\cdot\dot{\epsilon}\;\mathrm{d}V$$

from Green's theorem (5.34) for no body force. This has practical implications since the bounds on Ω and W can be regarded directly as a means of bounding the global compliance. The bounds can be obtained in an obvious way, e.g. for the prescribed force problem

$$\left(\frac{n}{n+1}\,\frac{\int_{S_\mathrm{p}}\bar{\mathrm{p}}\cdot\dot{\mathrm{u}}^\mathrm{k}\;\mathrm{d}S}{\int_V W(\dot{\epsilon}^\mathrm{k})\;\mathrm{d}V}\right)^n \int_{S_\mathrm{p}}\bar{\mathrm{p}}\cdot\dot{\mathrm{u}}^\mathrm{k}\;\mathrm{d}S \leqslant \int_S \bar{\mathrm{p}}\cdot\dot{\mathrm{u}}\;\;\mathrm{d}S \leqslant (n+1)\int_V \Omega(\sigma^\mathrm{s})\;\mathrm{d}V$$

using the Schwinger–Levine bound after optimisation. We have therefore bounded the compliance.

If the prescribed loading is simply a point load Q resulting in a displacement $\dot{q}$ at that point in the direction of the load, then

$$\int_S \bar{\mathrm{p}}\cdot\dot{\mathrm{u}}\;\;\mathrm{d}S = Q\dot{q}$$

and we may thus directly obtain an *upper bound* on the displacement rate

$$\dot{q} \leqslant \frac{B}{Q}\int_V (\bar{\sigma}^\mathrm{s})^{n+1}\;\mathrm{d}V \tag{5.40}$$

We can also obtain a lower bound, although it is not as obvious. We interpret the single-load problem as a prescribed displacement rate problem with $\dot{q}$ given and Q to be determined; so from equation (5.37) using Green's theorem

$$Q \leqslant \frac{1}{\dot{q}}\left(\frac{1}{B}\right)^{1/n}\int_V (\bar{\dot{\epsilon}}^\mathrm{k})^{1+1/n}\;\mathrm{d}V \tag{5.41}$$

where $\dot{\epsilon}^\mathrm{k}$ is compatible with $\dot{q}$. This is an upper bound on the load required to produce this displacement rate. Now, in general, there is a one-to-one relation between Q and $\dot{q}$ (*Figure 5.10*). If we construct an upper bound curve, equation (5.40), then it will lie below the actual curve. Similarly, if we construct a curve using equation (5.41), it will lie above the actual curve. We then have a *lower*

* For a power law $\dot{\epsilon} = B\sigma^n$, if $\sigma_2 = \lambda\sigma_1$ then $\dot{\epsilon}_2 = \lambda^n\dot{\epsilon}_1$.

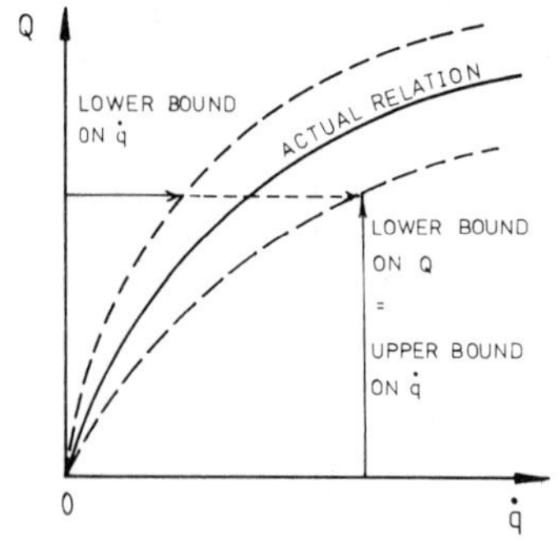

Figure 5.10 Hill's estimates

bound curve for $\dot{q}$ from equation (5.41). This is best seen by an example, which we consider later.

These upper and lower $Q/\dot{q}$ curves are called *Hill's estimates*. They apply to 'point' load problems. However, there is a method of bounding the displacement at a point in a body under an arbitrary load. We start from Martin's inequality for the power law

$$\frac{n}{n+1}\,\sigma^{B}\,.\,\dot{\epsilon}^{B} + \frac{1}{n+1}\,\sigma^{A}\,.\,\dot{\epsilon}^{A} \geqslant \sigma^{A}\,.\,\dot{\epsilon}^{B} \tag{5.42}$$

Identify σ^B, $\dot{\epsilon}^B$ with the actual solution σ, $\dot{\epsilon}$ and $\sigma^A = \sigma^*$ a stress statically admissible with surface forces $\mathbf{p}^*$ as yet arbitrary, with $\dot{\epsilon}^* = \mathrm{d}\Omega/\mathrm{d}\sigma^*$ from the constitutive relation. Integrating equation (5.42) over the volume and using Green's theorem there results

$$\frac{1}{n+1}\int_V \sigma^*\,.\,\epsilon^*\,\mathrm{d}V \geqslant \int_{S_\mathrm{p}} \left(\mathbf{p}^* - \frac{n}{n+1}\,\bar{\mathbf{p}}\right)\,.\,\dot{\mathbf{u}}\;\mathrm{d}S \tag{5.43}$$

for the prescribed force problem.

This result may now be used to bound the displacement at some point, x say, of the body. Since $\mathbf{p}^*$ is arbitrary, it may be chosen such that over S_p

$$\mathbf{p}^* = \frac{n}{n+1}\,\bar{\mathbf{p}}$$

except at x where

$$\mathbf{p}^*\Big|_x = Q^*\mathbf{n} + \frac{n}{n+1}\,\bar{\mathbf{p}}\Big|_x$$

Then if $\dot{q}$ is the displacement rate at x in the direction of Q^*, from equation (5.43)

$$\frac{1}{Q^*}\,\frac{1}{n+1}\int_V \sigma^*\,.\,\dot{\epsilon}^* \geqslant \dot{q} \tag{5.44}$$

Since σ^*, $\dot{\epsilon}^*$ depend on Q^*, we may optimise the left-hand side with respect to Q^*. The result is known as *Martin's estimate*. We can also derive a lower bound, but it requires more calculation than the upper bound and often gives poor results.

5.4.7 Numerical solution of steady creep problems using the energy theorems

Let us deal with the theorem of minimum complementary energy dissipation. It tells us that the solution to a steady creep problem minimises the functional U_c of statically admissible solutions. Intuitively, we can obtain a good approximation to the problem if we can construct a *minimising sequence* $\sigma_1^s, \sigma_2^s, \ldots, \sigma_m^s, \ldots$ such that

$$U_c(\sigma_m^s) \to U_c(\sigma) \qquad \text{as } m \to \infty$$

Therefore, terms of the minimising sequence can be taken as approximate solutions to the steady creep problem.

The classical way of constructing a minimising sequence is to use the *Rayleigh–Ritz method*. In this, suppose that $\sigma_1^*, \sigma_2^*, \ldots, \sigma_j^*, \ldots$ are a set of known statically admissible stresses – these may in fact be quite simple functions. The only conditions we impose are that these stresses are 'linearly independent', i.e. no member of σ_j^* can be obtained by a linear combination of the others, and that it is 'complete' in the sense that any stress can sensibly be approximated by a linear combination of these, i.e.

$$\sigma_m^s = \sum_{j=1}^{m} c_j \, \sigma^*j \tag{5.45}$$

where the scalar coefficients c_j are arbitrary. Then it can be shown that the sequence formed by equation (5.45) is minimising; the coefficients are determined from the minimum condition.

Thus we can form a sequence of approximate solutions to the steady creep problem, given by equation (5.45) where the coefficients are determined from the equations

$$\frac{\partial U_c}{\partial c_j} = 0 \qquad j = 1, 2, \ldots, m \tag{5.46}$$

Obviously the values of the coefficients c_j will vary as the number of terms m is increased.

Determination of the coefficients requires the solution of a set of nonlinear simultaneous equations (5.46); in elasticity, the equations are linear. In practice, most computing facilities have available on a scientific subroutine package a direct minimisation procedure which requires only the functional to be specified – not its derivatives. If this approach is used, then the steady creep problem poses no more computational difficulties than the elastic solution except that the numerical routine may take longer to converge. Also, in practice, the Rayleigh–Ritz method need not be interpreted so rigorously. The approximating statically admissible stress in the energy theorem can be determined from engineering judgement – this will of course depend on the problem. Obviously the preceding development carries through to the theorems of minimum total potential energy dissipation. We can form a minimising sequence $\dot{\mathbf{u}}_1^k, \dot{\mathbf{u}}_2^k, \ldots, \dot{\mathbf{u}}_m^k, \ldots$ by the Rayleigh–Ritz method

$$\dot{\mathbf{u}}_m^k = \sum_{j=1}^{m} c_j \, \dot{\mathbf{u}}_j^*$$

where the $\dot{\mathbf{u}}_j^*$ are kinematically admissible and linearly independent.

5.5 Example: Steady creep of a cantilever beam

We begin a practical study of the energy methods with the simple example of a cantilever beam loaded uniformly with p per unit length (*Figure 5.11*), for which an exact solution is available. Let x denote distance along the beam with the clamped end at $x = 0$ and the free end at $x = l$. The problem is analysed using

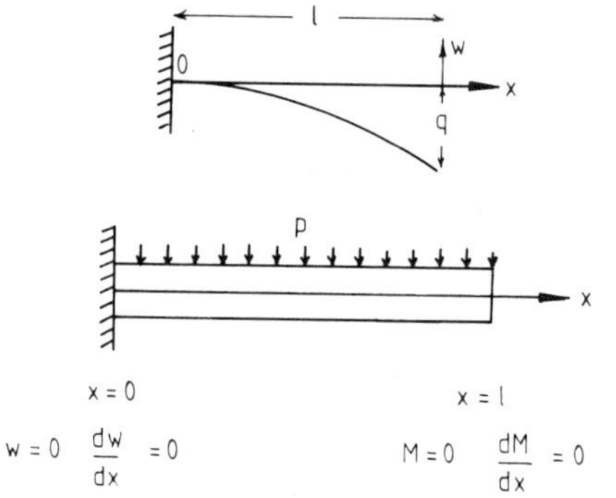

Figure 5.11 Geometry of cantilever beam

simple beam bending theory: let M denote the bending moment such that $M > 0$ for a beam bent concave from below, $\dot{\kappa}$ the curvature rate for steady creep and w the deflection rate, positive upwards (*Figure 5.11*).

The equilibrium equations are

$$\frac{d^2 M}{dx^2} = p$$

with boundary conditions at the free end

$$M = 0 \qquad \frac{dM}{dx} = 0 \qquad x = l$$

The strain–displacement rate equations are

$$\dot{\kappa} = -\frac{d^2 \dot{w}}{dx^2}$$

with boundary conditions at the clamped end

$$\dot{w} = 0 \qquad \frac{d\dot{w}}{dx} = 0 \qquad x = 0$$

The constitutive relations are taken from the moment–curvature rate equation for a beam in bending (Section 3.1) assuming a power law

$$\dot{\kappa} = K_n M^n$$

where for simplicity n is an odd integer and

$$K_n = B/I_n^n$$

in the previous notation. (We may regard this as a generalised constitutive relation – we are not considering the actual stress and strain, rather than the generalised quantities M and $\dot{\kappa}$.)

Immediately from equilibrium

$$M = \frac{1}{2} p(l - x)^2 \tag{5.47}$$

and using the constitutive relation together with strain–displacement we have the equation

$$\frac{d^2 \dot{w}}{dx^2} = -K_n (p/2)^n (l - x)^{2n} \tag{5.48}$$

which can be integrated with the boundary conditions to yield

$$\dot{w} = -K_n (p/2)^n \left(\frac{l^{2n+1}}{2n+1} x + \frac{(l-x)^{2n+2}}{(2n+1)(2n+2)} - \frac{l^{2n+2}}{(2n+1)(2n+2)} \right) \tag{5.49}$$

Therefore, the magnitude of the deflection at the tip ($x = l$) is

$$\dot{q} = K_n (p/2)^n \frac{l^{2n+2}}{2n+2} \tag{5.50}$$

We now apply the energy theorems. Since the problem is statically determinate, there is nothing instructive to be gained from minimum complementary dissipation. The theorem of minimum total potential energy dissipation here states that amongst all kinematically admissible curvature rates and displacement rates $\dot{\kappa}^k$, $\dot{w}^k$, the actual $\dot{\kappa}$, $\dot{w}$ minimise

$$U_p(\dot{\kappa}^k, \dot{w}^k) = \int_0^l \left[\left(\frac{1}{K_n} \right)^{1/n} \frac{n}{n+1} (\dot{\kappa}^k)^{1+1/n} - p\dot{w}^k \right] dx$$

Since $\dot{\kappa}^k$, $\dot{w}^k$ must be kinematically admissible, they must satisfy

$$\dot{\kappa}^k = -\frac{d^2 \dot{w}^k}{dx^2} \qquad x = 0 \qquad \frac{d\dot{w}^k}{dx} = 0 \qquad \dot{w}^k = 0$$

We first examine what happens if we assume that

$$\dot{w}^k = a\omega(x)$$

where $\omega(x)$ is the spatial distribution of the elastic solution

$$w = -\frac{1}{EI} (p/2) \left(\frac{l^3}{3} x + \frac{(l-x)^4}{12} - \frac{l^4}{12} \right)$$

i.e.

$$\omega(x) = -\left(\frac{l^3}{3} x + \frac{(l-x)^4}{12} - \frac{l^4}{12} \right)$$

and a is a constant to be determined from the minimum condition. Substitution into U_p followed by minimisation yields

$$a = K_n (p/2)^n \left(\frac{3n+2}{5n}\right)^n l^{2n-2}$$

and hence the approximate magnitude of the tip deflection rate is

$$\dot{q} = K_n \, (p/2)^n \left(\frac{3n+2}{5n}\right)^n \frac{l^{2n+2}}{4}$$

This may be compared with the exact solution, eqution (5.50); their ratio is shown in *Table 5.2* for various n values. Obviously the comparison is poor, particularly for the higher values of n. This situation occurs because if the moment is evaluated from the constitutive relation then equilibrium is increasingly violated

TABLE 5.2. Cantilever beam – use of elastic distribution.

n	$\dot{q}_{\text{app}}/\dot{q}_{\text{exact}}$
1	1.00
3	0.79
5	0.44
7	0.21

for increasing values of n. We have made a poor choice of trial function. To improve this, we could use the Rayleigh–Ritz method – for example, expanding $\dot{w}^{\text{k}}$ in a power series in (x/l) (to ensure convergence)

$$\dot{w}^{\text{k}} = \sum_{j=2}^{\infty} a_j \, (x/l)^j$$

which satisfies the requirements of the method. This could be substituted into U_p and minimised for the unknown coefficients a_j , i.e.

$$\frac{\partial U_p}{\partial a_j} = 0 \qquad j = 2, 3, \ldots$$

In fact, provided we take at least $2n + 2$ terms, recalling that n is presumed to be an odd integer, we can regain the exact solution (since the latter can be expanded as a polynomial in (x/l) of degree $2n + 2$). Hence, by 'turning the handle' on the Rayleigh–Ritz method we can, in this example, get the exact answer!

Finally, we will use this example to derive Martin's estimate for the tip displacement. Rather than interpret the general bound, equation (5.44), we derive Martin's inequality for this example as

$$\int_0^l \dot{\kappa}_{\text{B}} M_{\text{A}} \, dx \leqslant \frac{K_n}{n+1} \int_0^l M_{\text{A}}^{n+1} \, dx + \frac{n}{n+1} \left(\frac{1}{K_n}\right)^{1/n} \int_0^l \dot{\kappa}_{\text{B}}^{1+1/n} \, dx$$

which can be written as, defining $\dot{\kappa}_{\text{B}} = K_n M_{\text{B}}^n$,

$$\int_0^l \dot{\kappa}_B M_A \, dx \leqslant \frac{K_n}{n+1} \int_0^l M_A^{n+1} \, dx + \frac{n}{n+1} K_n \int_0^l M_B^{n+1} \, dx \qquad (5.51)$$

As before choose $M_B = M$, the exact solution; then

$$K_n \int_0^l M^{n+1} \, dx = -\int_0^l \frac{d^2 \dot{w}}{dx^2} M \, dx = -\int_0^l \dot{w} \frac{d^2 M}{dx^2} \, dx = -p\int_0^l \dot{w} \, dx$$

on integrating by parts (equivalent to Green's theorem) and using the boundary conditions.

Similarly

$$\int_0^l \dot{\kappa} M_A \, dx = -\left(\frac{d\dot{w}}{dx} M_A\right)_l + \left(\dot{w} \frac{dM_A}{dx}\right)_l - \int_0^l \dot{w} \frac{d^2 M_A}{dx^2} \, dx$$

Note that we have put no conditions on M_A.

On combining these, the inequality (5.51) becomes

$$-\left(\frac{d\dot{w}}{dx} M_A\right)_l + \left(\dot{w} \frac{dM_A}{dx}\right)_l + \int_0^l \left(\frac{np}{n+1} - \frac{d^2 M_A}{dx^2}\right) \dot{w} \, dx \leqslant \frac{K_n}{n+1} \int_0^l M_A^{n+1} \, dx$$

This gives a bound on $\dot{w}$ at $x = l$ if we choose (*Figure 5.12*)

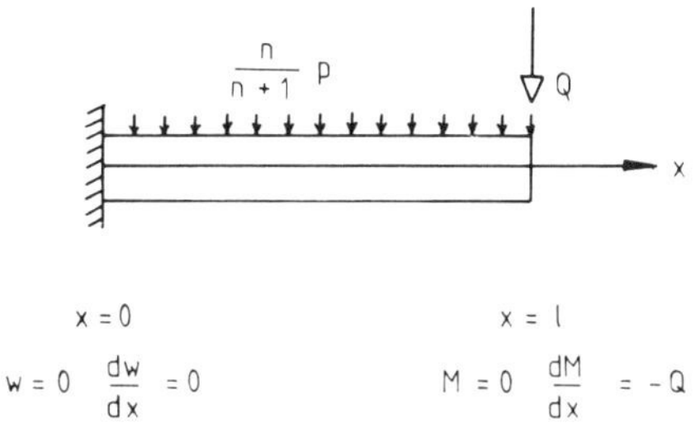

Figure 5.12 Loading system for Martin's estimate

$$\frac{d^2 M_A}{dx^2} = \frac{n}{n+1} p$$

$$x = l \qquad M_A = 0 \qquad \frac{dM_A}{dx} = -Q$$

for arbitrary Q, so that

$$M_A = \frac{n}{n+1} \frac{1}{2} (l - x)^2 p + Q(l - x)$$

and finally

$$\dot{q} = -\dot{w}\Big|_{x=l} \leqslant \frac{K_n}{n+1} \frac{1}{Q} \int_0^l (l-x)^{n+1} \left(Q + \frac{n}{n+1} p \frac{(l-x)}{2}\right)^{n+1} dx \qquad (5.52)$$

Any chosen Q may be substituted into the expression and it is preferable to optimise the right-hand side to obtain the best possible bound. This cannot be done in closed form but is readily done numerically in an approximate manner. *Table 5.3* shows the approximate minimising value of $2Q/pl$ and the ratio of the actual solution to Martin's estimate. Clearly, in this case, it is very good.

TABLE 5.3. Cantilever beam – Martin's upper bound.

n	$\dot{q}_{\text{app}}/\dot{q}_{\text{exact}}$	$2Q/pl$
1	1.016	0.39
3	1.008	0.25
5	0.996	0.15

5.6 Example: Steady creep of an annular plate

We return here to the problem of the bending of thin plates out of their planes, introduced in Section 4.6. The generalised constitutive relations, equations (4.41) and (4.42), can be written in potential form

$$\dot{\kappa}_1 = \frac{\partial \Omega}{\partial M_1} \qquad \dot{\kappa}_2 = \frac{\partial \Omega}{\partial M_2}$$

$$M_1 = \frac{\partial W}{\partial \dot{\kappa}_1} \qquad M_2 = \frac{\partial W}{\partial \dot{\kappa}_2} \tag{5.53}$$

where the *generalised potentials* are defined as

$$\Omega = \frac{B}{n+1} \frac{1}{D_n^n} \bar{M}^{n+1}$$

$$W = \frac{n}{n+1} \frac{1}{B^{1/n}} \frac{3}{4} D_n \bar{\dot{\kappa}}^{1+1/n}$$

The particular problem we will consider is that of the annular plate for which we obtained an approximate solution in Section 4.6 (*Figure 5.13*). Here we will derive a more accurate solution using energy methods.

The theorem of minimum total potential dissipation becomes: amongst all kinematically admissible curvature rates $\bar{\dot{\kappa}}^{\text{k}}$ compatible with the lateral deflection $\dot{w}^{\text{k}}$ which satisfies the support conditions at the outer edge, the actual solution minimises

$$U_{\text{p}} = \int_a^b W(\bar{\dot{\kappa}}^{\text{k}}) r \, \mathrm{d}r - \int_a^b p \dot{w}^{\text{k}} \, r \, \mathrm{d}r$$

The theorem of minimum complementary dissipation becomes: amongst all statically admissible moments $\bar{M}^{\text{s}}$ in equilibrium with the pressure p and satisfying the conditions on the inner and outer edges, the actual solution minimises

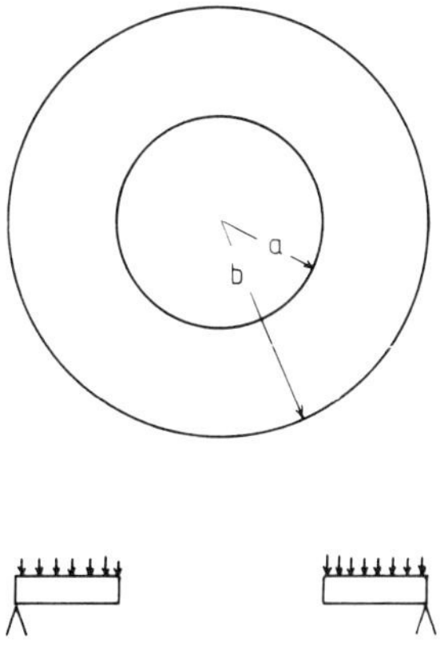

Figure 5.13 Geometry of annular plate

$$U_c = \int_a^b \Omega(\bar{M}^s)\, r\, dr$$

We construct an approximate solution using the Rayleigh–Ritz method. The trial solutions are found by expanding $\dot{w}$ and M_r in power series in $(1 - r/b)$

$$\dot{w} = \dot{w}_0 \sum_{k=1}^{N} a_k (1 - r/b)^k$$
$$M_r = M_0 \sum_{k=1}^{N} b_k (1 - r/b)^k \tag{5.54}$$

where we choose

$$\dot{w}_0 = \frac{Bb^2}{D_n^n}(pb^2)^n \qquad M_0 = pb^2$$

(for reasons which will become clear later).

These series satisfy the required boundary conditions, $M_r = 0$, $\dot{w} = 0$ at the outer edge $r = b$. From the condition at the free edge, $r = a$, $M_r = 0$, we must have

$$b_1 = -\sum_{k=2}^{N} b_k (1 - a/b)^k$$

and consequently

$$\frac{M_r}{M_0} = \sum_{j=1}^{N} b_{j+1} (1 - r/b)[(1 - r/b)^j - (1 - a/b)^j] \tag{5.55}$$

The circumferential moment M_φ is obtained from equilibrium as

$$M_\varphi = \frac{d}{dr}(rM_r) + \frac{1}{2} pr^2 + c$$

where c is a constant of integration. From the condition at the inner edge

$$r = a \qquad Q_r = \frac{1}{r}\left(M_\varphi - \frac{d}{dr}(rM_r)\right) = 0$$

it follows that we must have

$$c = -\frac{1}{2}pa^2$$

If we substitute the series approximations (5.54) and (5.55) into the energy theorems, defining

$$\xi = r/b \qquad \eta = a/b < 1$$

there results on rearrangement

$$\frac{U_p}{U_0} = \frac{n}{n+1}\left(\frac{4}{3}\right)^{1/n}\int_\eta^1 f_p(\xi\,; a_k)^{(1+1/n)/2}\,\xi\,d\xi - \int_\eta^1 \sum_k a_k(1-\xi)^k \xi\,d\xi$$

$$\frac{U_c}{U_0} = \frac{1}{n+1}\int_\eta^1 f_c(\xi; b_k)^{(n+1)/2}\,\xi\,d\xi$$

where $U_0 = \dot{w}_0 M_0$ and

$$f_p = \left(\sum_{k=2}^N a_k k(k-1)(1-\xi)^{k-2}\right)^2 - \frac{1}{\xi}\left(\sum_{k=2}^N a_k k(k-1)(1-\xi)^{k-2}\right)$$
$$\left(\sum_{k=1}^N a_k k(1-\xi)^{k-1}\right) + \frac{1}{\xi}\left(\sum_{k=1}^N a_k k(1-\xi)^{k-1}\right)^2$$

$$f_c = \left(\sum_{k=1}^N b_k(1-\xi)^k\right)^2 + \left(\sum_{k=1}^N b_k(1-\xi)^k\right)$$
$$\left(\sum_{k=1}^N b_k(1-\xi-k)(1-\xi)^{k-1} + \frac{1}{2}(\xi^2-\eta^2)\right)$$
$$+\left(\sum_{k=1}^N b_k(1-\xi-k)(1-\xi)^{k-1} + \frac{1}{2}(\xi^2-\eta^2)\right)^2$$

It is seen that the minimising values of a_k, b_k should be functions only of η and n (thus justifying our original choice of $\dot{w}_0$ and M_0).

The Rayleigh–Ritz procedure is as follows: with a given number of terms in the series expansions for $\dot{w}$ and M_r, the functionals U_p and U_c are minimised with respect to the a_k or b_k. This can be done using any standard scientific subroutine package for minimisation without derivatives. The number of terms in each series is increased until convergence is obtained. We will then have a close approximation to the true solution.

In *Figures 5.14* and *5.15* the convergence of $\dot{w}/\dot{w}_0$ and M_r/M_0 for an increasing number of terms is shown. It can be seen that the Rayleigh–Ritz

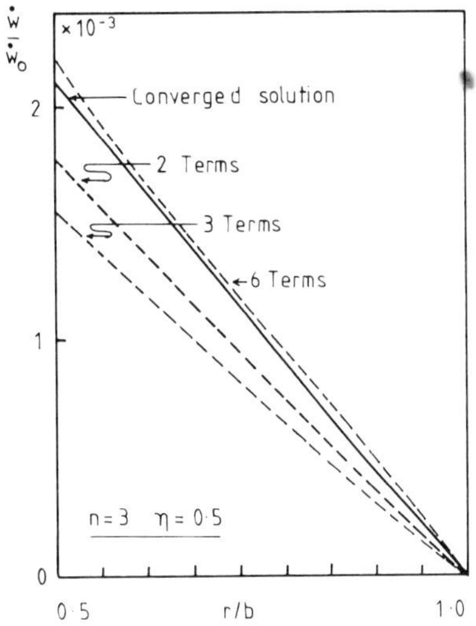

Figure 5.14 Convergence of normalised displacement using Rayleigh–Ritz method

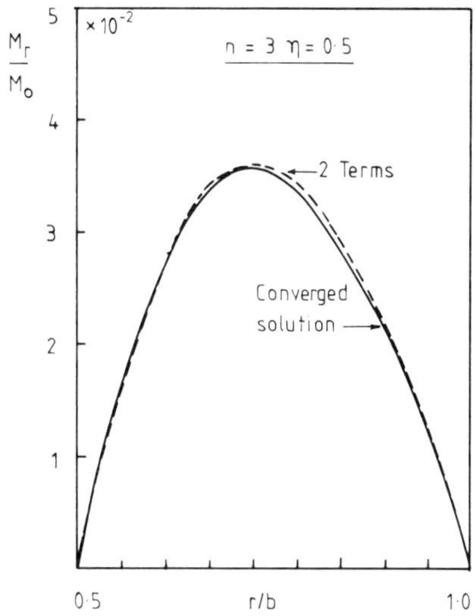

Figure 5.15 Convergence of normalised radial bending moment using Rayleigh–Ritz method

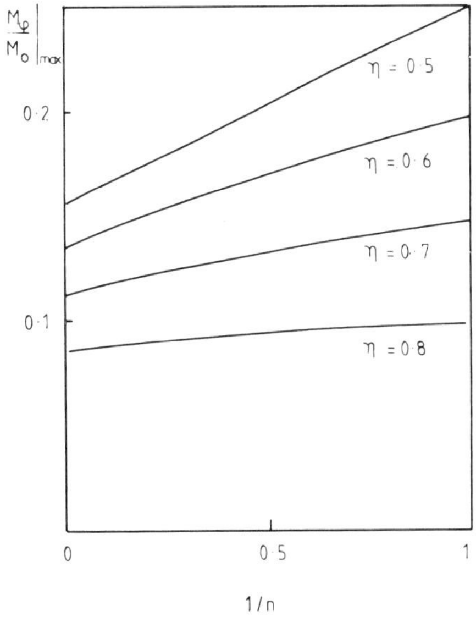

Figure 5.16 Variation of normalised circumferential bending moment with $1/n$

procedure applied to the theorem of minimum complementary dissipation converges faster than when applied to that of minimum total potential dissipation, but this is probably due to the nature of the problem and the prescribed boundary conditions.

In *Figure 5.16* the variation of M_φ/M_0 at the inner edge with the reciprocal of the creep exponent $1/n$ is shown for various values of η. Again, this is seen to be very nearly linear.

We can now compare this more accurate solution with the simple approximation obtained in Section 4.6. The derived maximum displacement (at $r = a$), equation (4.50), can be written in the form

$$\frac{\dot{w}_{max}}{\dot{w}_0} = \frac{3}{4}(1-\eta)M^n\left(\frac{1-1/n}{1-\eta^{1-1/n}}\right)^n$$

where M is as previously defined. In *Figure 5.17* this approximation is compared to the solution obtained using the complementary energy method for various values of n. It can be seen that the simple solution is perfectly adequate.

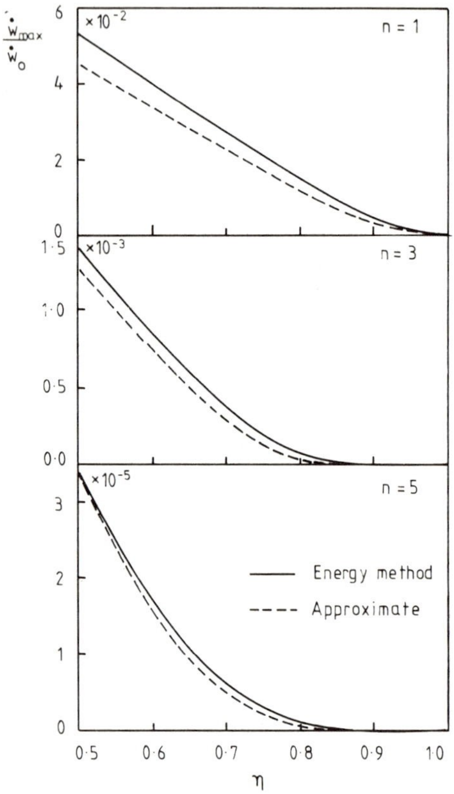

Figure 5.17 Comparison between energy method and simple approximation

5.7 Approximate generalised models for power-law creep

In Sections 4.5 and 4.6 we examined the consequences of two special cases of the theory of thin shells – membrane shells and bending of plates. For these, we were able to develop suitable generalised constitutive relations between the mid-surface strains or curvatures and stress resultants or moments. This simplification greatly eased the complexity of the stress analysis. Let us consider how far we can proceed in constructing a generalised constitutive relation for the general bending theory of shells.

Let the mid-surface be described by two principal directions and let the stresses and strains in these directions be

$$\sigma = (\sigma_1 \quad \sigma_2) \qquad \epsilon = (\epsilon_1 \quad \epsilon_2)$$

in vector notation. Similarly, let us define the mid-surface stress resultant and moment vectors

$$\mathbf{N} = (N_1\ N_2) = \int_{-h/2}^{h/2} \sigma\ \mathrm{d}z$$

$$\mathbf{M} = (M_1\ M_2) = \int_{-h/2}^{h/2} \sigma z\ \mathrm{d}z$$

as before. Assuming that plane sections remain plane during bending of the shell, the mid-surface strain and curvature are given by

$$\epsilon = \in + z\kappa$$

where

$$\in = (\in_1\ \in_2) \qquad \kappa = (\kappa_1 \quad \kappa_2)$$

Since plane stress conditions apply, we can write the inverted constitutive relation for steady creep as

$$\sigma = \mathrm{d}W/\mathrm{d}\dot{\boldsymbol{\epsilon}}$$

But

$$\frac{\mathrm{d}W}{\mathrm{d}\dot{\epsilon}} = \frac{\mathrm{d}W}{\mathrm{d}\dot{\in}} = \frac{1}{z}\ \frac{\mathrm{d}W}{\mathrm{d}\dot{\kappa}}$$

Therefore

$$\mathbf{N} = \int_{-h/2}^{h/2} \frac{\mathrm{d}W}{\mathrm{d}\dot{\epsilon}}\ \mathrm{d}z = \int_{-h/2}^{h/2} \frac{\mathrm{d}W}{\mathrm{d}\dot{\in}}\ \mathrm{d}z = \frac{\mathrm{d}W_h}{\mathrm{d}\dot{\in}}$$

$$\mathbf{M} = \int_{-h/2}^{h/2} \frac{\mathrm{d}W}{\mathrm{d}\dot{\epsilon}}\ z\ \mathrm{d}z = \int_{-h/2}^{h/2} \frac{\mathrm{d}W}{\mathrm{d}\dot{\kappa}}\ \mathrm{d}z = \frac{\mathrm{d}W_h}{\mathrm{d}\dot{\kappa}}$$

where we have defined the functional

$$W_h = \int_{-h/2}^{h/2} W(\dot{\epsilon})\ \mathrm{d}z$$

as a generalised steady creep potential for a thin shell.

Also

$$\int_{-h/2}^{h/2} (W + \Omega)\ \mathrm{d}z = \int_{-h/2}^{h/2} \sigma\, .\, \dot{\epsilon}\ \mathrm{d}z = \mathbf{N}\, .\, \dot{\in} + \mathbf{M}\, .\, \dot{\kappa}$$

and defining the functional

$$\Omega_h = \mathbf{N} \cdot \dot{\epsilon} + \mathbf{M} \cdot \dot{\kappa} - W_h$$

we have the *generalised constitutive relation*

$$\dot{\epsilon} = \frac{\partial \Omega_h}{\partial \mathbf{N}} \qquad \dot{\boldsymbol{\kappa}} = \frac{\partial \Omega_h}{\partial \mathbf{M}} \tag{5.56}$$

It should be obvious that if we were to use these relations, for example, in the energy theorems, some work is needed since the integral through the thickness is required at any step in the calculation to evaluate W_h. The integrals can be evaluated in closed form for the special cases $\dot{\kappa} = 0$ for membrane shells, and $\dot{\epsilon} = 0$ for the bending of plates.

Thus it is clear that, in general for thin shells, the advantages of the generalised formulation with reference to the mid-surface have been lost, since effectively the stress and strain rates should be calculated at each point through the thickness. We can only regain these advantages if some approximate formulation is used. The approach adopted here is to produce approximations to the specific complementary dissipation Ω_h . Any accuracy lost in this is more than regained by the simplification of the calculations. (Moreover, it is to be recalled that the classical local formulation is only an approximate global model of the real microstructural behaviour.)

Mechanical models of material behaviour which relate not to local stresses and strains but more to generalised quantities (which depend on geometry, e.g. thinness) are called *generalised models*. The theory of shells is only one such model, being a reduction of a three-dimensional continuum to a two-dimensional model based on simplifications arising from thinness. The other most well known is that of beams, being a one-dimensional model based on the simplification that its cross-sectional dimensions are small compared to its length. We have already met this, e.g. Section 3.1. Such models are useful in the stress analysis of frameworks or piping systems.

Generalised models are accepted in elasticity and used without comment. The linearity of elasticity theory allows 'exact' generalised constitutive relations to be derived. But, as we have seen, the nonlinearity of steady creep only allows exact relations to be derived for special cases. More often we will need approximate generalised constitutive relations. We will consider these here in more detail for steady creep assuming a power law throughout. Some guidance as to the proper choice of such approximations can be had from a useful property of the power law.

5.7.1 The theorem of nesting surfaces

We return to the general notation used in Section 5.4. Let $\mathbf{Q} = (Q_1, Q_2, \ldots)$ be a system of generalised forces corresponding to generalised displacement rates $\dot{\mathbf{q}} = (\dot{q}_1, \dot{q}_2, \ldots)$ acting on a structure and resulting in a stress $\boldsymbol{\sigma}$ and strain rate $\dot{\boldsymbol{\epsilon}}$. The material is assumed to exhibit steady isothermal creep such that there exists a convex potential Ω

$$\dot{\boldsymbol{\epsilon}} = \mathrm{d}\Omega/\mathrm{d}\sigma$$

where for steady creep

$$\Omega = \frac{B}{n+1}\,\bar{\sigma}^{\,n+1}$$

The structure is assumed fixed on some part of its surface.

These 'generalised' forces and displacement rates can represent actual external point loads and corresponding deflections or, more importantly, quantities such as mid-surface stress resultants and moments and strain rates and curvature for thin bodies. In the former case, we are concerned with integration of the stresses and strains over the whole body; in the latter, we are concerned only with integration in the principal direction through the thickness. In the case of beams, the integration would be over the area of the cross-section. We represent each integration here as being over a 'reduced' volume v so that Green's theorem is

$$\mathbf{Q}\,.\,\dot{\mathbf{q}} = \int_v \sigma\,.\,\dot{\epsilon}\;\mathrm{d}v$$

We derive our result from the theorem of minimum complementary energy over the reduced volume

$$\int_v \bar{\sigma}^{n+1}\;\mathrm{d}v \leqslant \int_v (\bar{\sigma}^{\mathrm{s}})^{n+1}\;\mathrm{d}v$$

where $\bar{\sigma}^{\mathrm{s}}$ is any statically admissible effective stress in equilibrium with the generalised forces $\mathbf{Q}$. Notice that we are not necessarily using the whole volume of the body, just an integration in one, two or three of the spatial dimensions of the body.

We first establish the following result.

Theorem Consider two identical bodies A and B of the same dimensions and under the same loading but composed of different materials satisfying the power laws

$$\dot{\epsilon} = B\sigma^{n_{\mathrm{A}}} \qquad \dot{\epsilon} = B\sigma^{n_{\mathrm{B}}}$$

then if $n_{\mathrm{A}} \geqslant n_{\mathrm{B}}$

$$\left(\frac{1}{v}\int_v \bar{\sigma}_{\mathrm{A}}^{\,n_{\mathrm{A}}+1}\;\mathrm{d}v\right)^{1/(n_{\mathrm{A}}+1)} \geqslant \left(\frac{1}{v}\int_v \bar{\sigma}_{\mathrm{B}}^{\,n_{\mathrm{B}}+1}\;\mathrm{d}v\right)^{1/(n_{\mathrm{B}}+1)}$$

where $\bar{\sigma}_{\mathrm{A}}$, $\bar{\sigma}_{\mathrm{B}}$ are the effective stresses associated with the actual stresses.

The proof of this is straightforward. Let us define σ_{A}, σ_{B} by

$$\frac{1}{v}\int_v \bar{\sigma}_{\mathrm{A}}^{\,n_{\mathrm{A}}+1}\;\mathrm{d}v = \sigma_{\mathrm{A}}^{\,n_{\mathrm{A}}+1} \qquad \frac{1}{v}\int_v \bar{\sigma}_{\mathrm{B}}^{\,n_{\mathrm{B}}+1}\;\mathrm{d}v = \sigma_{\mathrm{B}}^{\,n_{\mathrm{B}}+1}$$

and consider their ratio

$$\beta = \sigma_{\mathrm{B}}/\sigma_{\mathrm{A}}$$

From complementary energy, for body B

$$\int_v \bar{\sigma}_B^{n_B+1}\,dv \leqslant \int_v (\bar{\sigma}_B^s)^{n_B+1}\,dv$$

for any statically admissible stress $\bar{\sigma}_B^s$. Since both bodies are under the same loading, we may take $\bar{\sigma}_B^s = \bar{\sigma}_A$. Then

$$\int_v \bar{\sigma}_B^{n_B+1}\,dv \leqslant \int_v \bar{\sigma}_A^{n_B+1}\,dv$$

Dividing both sides by $v\,\sigma_B^{n+1}$ and noting that $\sigma_B = \beta\sigma_A$, we obtain

$$\frac{1}{v}\int_v \left(\frac{\bar{\sigma}_B}{\sigma_B}\right)^{n_B+1} dv \leqslant \frac{1}{\beta^{n_B+1}}\;\frac{1}{v}\int_v \left(\frac{\bar{\sigma}_A}{\sigma_A}\right)^{n_B+1} dv$$

By definition, the left-hand side is unity; hence

$$\beta^{n_B+1} \leqslant \frac{1}{v}\int_v \left[\left(\frac{\bar{\sigma}_A}{\sigma_A}\right)^{n_A+1}\right]^N dv$$

where

$$N = \frac{n_B + 1}{n_A + 1}$$

Let us now consider n_A to be fixed and write the right-hand side as

$$I(N) = \frac{1}{v}\int_v s^N\,dv$$

Obviously $I(0) = I(1) = 1$ and using Holder's inequality in the form

$$I(N) = \frac{1}{v}\int_v s^N\,dv \leqslant \left(\frac{1}{v}\int_v s\,dv\right)^N = 1$$

i.e.

$$\beta^{n_B+1} \leqslant 1 \qquad \beta \leqslant 1$$

and on rearrangement the result is proved.

We have demonstrated that the functional

$$\mathscr{F}_n(\sigma) = \left(\frac{1}{v}\int_v \bar{\sigma}^{\,n+1}\,dv\right)^{1/(n+1)}$$

is strictly monotonic increasing with increasing n and is therefore bounded below by that for $n = 1$ and above by that for $n \to \infty$; $\bar{\sigma}$ is the effective stress associated with the actual stress in a body under certain generalised loads.

The use of the result to construct generalised models comes from the following observation.

From Castigliano's theorem, we have the form of a generalised constitutive relation

$$\dot{\mathbf{q}} = \mathrm{d}\Omega/\mathrm{d}\mathbf{Q} \tag{5.57}$$

where

$$\Omega(\mathbf{Q}) = \frac{B}{n+1} \int_v \bar{\sigma}^{\,n+1} \, \mathrm{d}v$$

which we write in the form

$$\Omega(\mathbf{Q}) = \frac{Bv}{n+1} \bar{Q}_n^{\,n+1} \tag{5.58}$$

where by definition the *effective generalised stress* is

$$\bar{Q}_n = \left(\frac{1}{v} \int_v \bar{\sigma}^{\,n+1} \, \mathrm{d}v \right)^{1/(n+1)}$$

The theorem of nesting surfaces tells us that $\bar{Q}_n(\mathbf{Q})$ is monotonic increasing with n . Geometrically, if we consider hypersurfaces $\bar{Q}_n$ = constant, then they must 'nest' inside each other for increasing n and are enveloped above by $\bar{Q}_1$, analogous to linear elasticity, and below by $\bar{Q}_\infty$. The latter is a yield surface in $\mathbf{Q}$ constructed on the assumption that the condition of plasticity is $\bar{Q}_\infty$ = constant – the plastic analogy.

Thus if we know the bounding surfaces and special results for single loads, we may approximate the form of $\bar{Q}_n$. This will depend on the problem at hand and we examine three examples.

5.7.2 Example: A statically determinate structure

Consider the two-bar structure shown in *Figure 5.18* consisting of two bars of equal length L and cross-sectional area A pin-jointed to each other and to a rigid foundation. The structure supports loads Q_1 and Q_2 at the central pin.

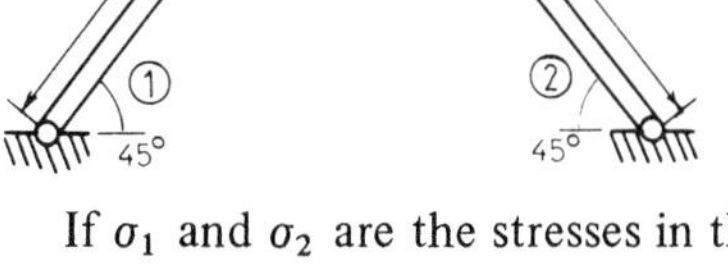

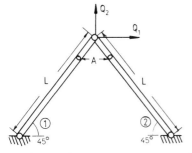

Figure 5.18 Geometry of statically determinate two-bar structure

If σ_1 and σ_2 are the stresses in the bars, then from equilibrium

$$\sigma_1 = \frac{Q_1 + Q_2}{\sqrt{2}\,A} \qquad \sigma_2 = \frac{Q_1 - Q_2}{\sqrt{2}\,A}$$

thus $v = 2LA$

$$\bar{Q}_n^2 = \left(\frac{1}{v} \int_v \sigma^{n+1} \, \mathrm{d}v \right)^{2/(n+1)} = \left[\frac{1}{2} \left(\frac{Q_1 + Q_2}{\sqrt{2}A} \right)^{n+1} + \frac{1}{2} \left(\frac{Q_1 - Q_2}{\sqrt{2}A} \right)^{n+1} \right]^{2/(n+1)}$$

The vertical and horizontal deflection rates are given by

$$\dot{q}_1 = \partial\Omega/\partial Q_1 \qquad \dot{q}_2 = \partial\Omega/\partial Q_2$$

$$\Omega = \frac{B}{n+1}(2LA)\,\bar{Q}_n^{\,n+1}$$

If we examine Q_n further, we see that

$$n = 1 \qquad \bar{Q}_n^2 = \left(\frac{Q_1}{\sqrt{2}A}\right)^2 + \left(\frac{Q_2}{\sqrt{2}A}\right)^2$$

$$n \to \infty \qquad \bar{Q}_n^2 \to \left(\frac{Q_1}{\sqrt{2}A} + \frac{Q_2}{\sqrt{2}A}\right)^2 \qquad Q_1 > 0, Q_2 > 0, \text{etc.}$$

using the result that

$$a, b > 0 \qquad \lim_{m\to\infty}(a^m + b^m)^{1/m} = \begin{cases} a & a \geqslant b \\ b & b \geqslant a \end{cases}$$

and clearly $\bar{Q}_n^2$ is an increasing function of n bounded by $\bar{Q}_1^2$ and $\bar{Q}_\infty^2$. The 'nesting' surfaces $\bar{Q}_n^2 = 1$ are shown in *Figure 5.19*. For this problem, they touch for $(Q_1/\sqrt{2}A), (Q_2/\sqrt{2}A) = \pm 1$. Surfaces for other n values are also shown.

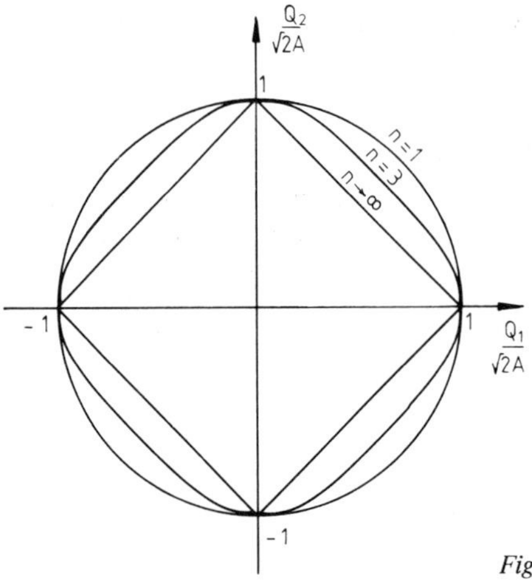

Figure 5.19 Nesting surfaces

5.7.3 Example: A beam under tension and bending

As a more complex example, we consider the problem of a beam of symmetric cross-section of area A and height $2y_0$, under the action of a bending moment M and an axial load N (*Figure 5.20*).

Let x be the longitudinal direction and y be the direction along the axis of symmetry. Then from equilibrium

$$M = \int_A \sigma y \,\mathrm{d}A$$

$$N = \int_A \sigma \,\mathrm{d}A$$

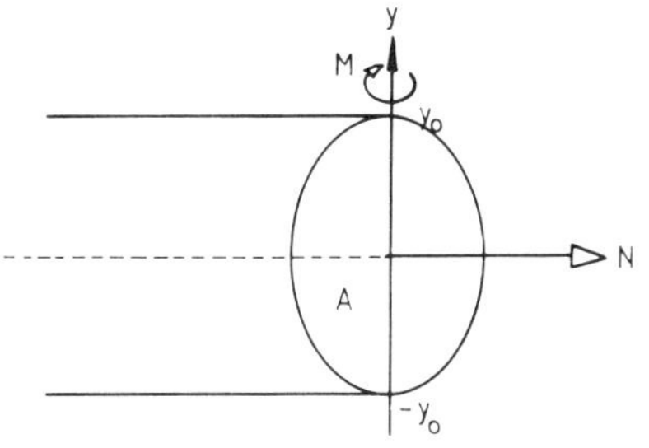

Figure 5.20 Geometry of beam under tension and bending

and assuming plane sections remain plane

$$\epsilon = \kappa y + \delta$$

where κ is the curvature and δ the extension.

This problem can be solved given the constitutive relation for power-law creep

$$\dot{\epsilon} = B\sigma^n$$

but not in closed form. We could use quasilinearisation to obtain a numerical solution for particular cases.

Here we construct an approximate solution using nesting surfaces; define

$$\dot{\kappa} = \partial\Omega/\partial M \qquad \dot{\delta} = \partial\Omega/\partial N$$

where

$$\Omega = \frac{B}{n+1} A \bar{Q}_n^{n+1}$$

We will approximate $\bar{Q}_n$.

It is readily verified that for $n = 1$ we have

$$\bar{Q}_1^2 = \mu_1^2 \left(\frac{M}{y_0 A}\right)^2 + \left(\frac{N}{A}\right)^2$$

where

$$\mu_1^2 = y_0 A/I_1 \qquad I_1 = \int_A y^2 \, \mathrm{d}A$$

and that for $n \to \infty$ analogous to the *approximate* limit surface for this problem

$$\bar{Q}_\infty^2 \simeq \mu_\infty^2 \left(\frac{M}{y_0 A}\right)^2 + \left(\frac{N}{A}\right)^2$$

where

$$\mu_\infty^2 = y_0 \left(A/I_\infty\right)^2 \qquad I_\infty = \int_A y \, \mathrm{d}A$$

For bending alone, $N = 0$, we have exactly

$$\bar{Q}_n^2 = \mu_n^2 \left(\frac{M}{y_0 A}\right)^2$$

$$\mu_n^2 = y_0 \left(A/I_n\right)^{n/(n+1)} \qquad I_n = \int_A y^{1+1/n} \, \mathrm{d}A$$

Assuming a rectangular cross-section of height $2d$, breadth b, the above parameters are

$$y_0 = d \qquad A = 2bd$$

$$I_n = bd^{2+1/n}\,\frac{2n}{2n+1} \qquad \mu_n = (2 + 1/n)^{n/(n+1)}$$

and we can draw the nesting surfaces (*Figure 5.21*) $\bar{Q}_n^2 = 1$, which are ellipses touching at $M = 0$.

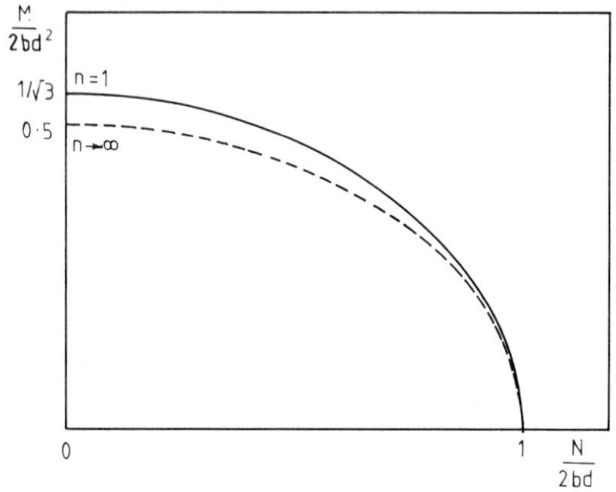

Figure 5.21 Nesting surfaces

In order to approximate $\bar{Q}_n^2$ for any n, we assume that the elliptical form is retained and passes through the known point $N = 0$. Then for the rectangular cross-section

$$\bar{Q}_n^2 \simeq \mu_n^2 \left(\frac{M}{2\,bd^2}\right)^2 + \left(\frac{N}{2bd}\right)^2$$

and for the general cross-section

$$\bar{Q}_n^2 \simeq \mu_n^2 \left(\frac{M}{y_0 A}\right)^2 + \left(\frac{N}{A}\right)^2 \tag{5.59}$$

This approximate solution can be compared to an accurate solution obtained using quasilinearisation (in a manner similar to Section 5.2). The ratio of curvature rate $\dot{\kappa}$ to that of an equivalent beam under bending alone (Section 3.1)

$$\dot{\kappa}_0 = B(M/I_n)^n$$

is plotted for various values of n against the load parameter $\Pi = Nd/M$ in *Figure 5.22.*

The approximate solution

$$\dot{\kappa}/\dot{\kappa}_0 = (1 + \Pi^2/\mu_n^2)^{(n-1)/2}$$

clearly underestimates the exact solution with an increasing error for increasing values of the exponent n. A better approximation could be obtained if an improved representation of the plastic limit surface was used.

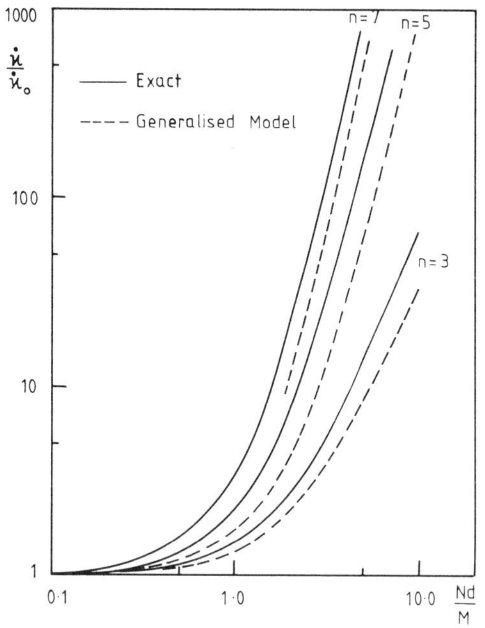

Figure 5.22 Comparison of exact and generalised model: variation of normalised curvature rate with load parameter

5.7.4 Approximate energy functionals for thin shells

We return to the problem described at the beginning of this section – the development of an approximate energy functional for thin shells. From equation (5.56) we wish to approximate Ω_h in the form

$$\Omega_h = \frac{B}{n+1} h \bar{Q}_n^{n+1} \tag{5.60}$$

noting that in the preceding development

$$v = h \qquad \int_v \mathrm{d}v = \int_{h/2}^{h/2} \mathrm{d}z$$

$$\mathbf{Q} = (\mathbf{N} \quad \mathbf{M}) \qquad \dot{\mathbf{q}} = (\dot{\mathrm{E}} \quad \dot{\kappa})$$

For the special cases of membrane shells, equation (4.37), we have

$$\bar{Q}_n^2 = \frac{1}{h}(N_1^2 - N_1 N_2 + N_2^2)$$

On the other hand, from the bending theory of plates, equation (4.42), we have

$$\bar{Q}_n^2 = \frac{4}{h^4} \mu_n^2 (M_1^2 - M_1 M_2 + M_2^2)$$

$$\mu_n = (2 + 1/n)^{n/(n+1)}$$

Also for $n = 1$ from linear elasticity

$$\bar{Q}_1^2 = \frac{1}{h^2}(N_1^2 - N_1 N_2 + N_2^2) + \frac{12}{h^4}(M_1^2 - M_1 M_2 + M_2^2)$$

For $n \to \infty$, analogous to the yield surface for a thin shell, no exact form is possible, but we can use Ilyushin's approximation

$$\bar{Q}^2_\infty \simeq \frac{1}{h^2}(N_1^2 - N_1 N_2 + N_2^2) + \frac{16}{h^4}(M_1^2 - M_1 M_2 + M_2^2)$$

The surfaces $\bar{Q}_1^2$ and $\bar{Q}_\infty^2$ touch for $\mathbf{M} = 0$ and are seen to be ellipses. In addition, the case $\mathbf{N} = 0$ is known exactly. Then, as in the case of a beam under tension and bending, this elliptical form is retained as an approximation to the real surface, and we have

$$\bar{Q}_n^2 \simeq \frac{1}{h^2}(N_1^2 - N_1 N_2 + N_2^2) + \frac{4}{h^4}\mu_n^2 (M_1^2 - M_1 M_2 + M_2^2) \tag{5.61}$$

where

$$\mu_n = (2 + 1/n)^{n/(n+1)}$$

The resulting approximate generalised model, equations (5.60) and (5.61), can now be used in the stress analysis of thin-shell problems. It can be inverted to give

$$W_h = \frac{n}{n+1} \frac{1}{B^{1/n}} h \, \bar{\dot{q}}_n^{\,1+1/n}$$

$$\bar{\dot{q}}_n^2 \simeq \frac{4}{3}\left((\dot{\epsilon}_1^2 + \dot{\epsilon}_1\dot{\epsilon}_2 + \dot{\epsilon}_2^2) + \frac{h^4}{4}\frac{1}{\mu_n^2}(\dot{\kappa}_1^2 + \dot{\kappa}_1\dot{\kappa}_2 + \dot{\kappa}_2^2)\right) \tag{5.62}$$

In general, it is difficult to assess the accuracy of this model here. It turns out that it gets increasingly inaccurate to about 20% in the predicted strain rates with increasing n. To overcome this, a more accurate plastic yield surface can be used as a starting point.

Finally, it should be noted that the approximate potential Ω_h is convex, and consequently the energy theorems and Hill's and Martin's estimates remain valid.

5.8 Example: Steady creep of a curved pipe under in-plane bending

The smooth curved pipe is one of the most important components in any pipework system. It is known in elasticity to be more 'flexible' than an equivalent length of straight pipe of the same cross-section when both are subject to the same bending loads. This flexibility has been exploited in almost all designs for pipework systems.

The present analysis of a curved pipe under in-plane bending can be treated as a special case of a thin shell of revolution loaded symmetrically, with a slight modification. The bend has radius a, cross-sectional radius b and thickness h, small compared to b (*Figure 5.23*). It is subject to a pure in-plane bending

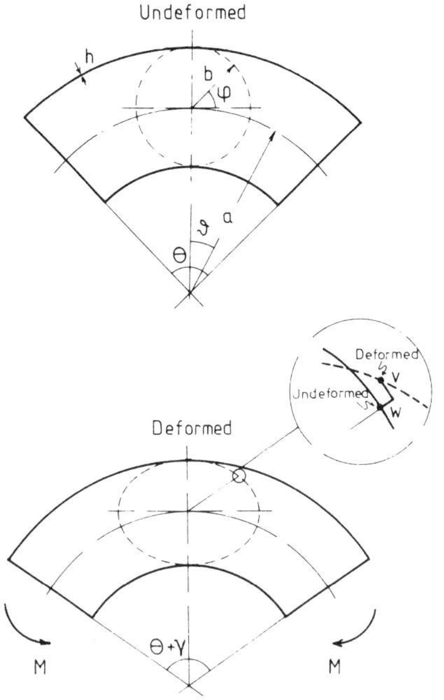

Figure 5.23 Geometry of curved pipe

moment M. The bend can be described geometrically as that part of a torus formed by rotating a circle about some axis and is therefore part of a shell of revolution. The mid-surface of this torus is described by the coordinate system (φ, ϑ) as in Section 4.5, where $-\theta/2 \leqslant \vartheta \leqslant \theta/2$. The following assumptions are made to simplify the problem.

(a) Each section of the bend behaves in the same manner; then all stress and strain quantities are functions of φ alone.

(b) The circumferential stress is constant through the thickness:

$$M_\vartheta = 0$$

(c) The circumferential strain is constant through the thickness:

$$\kappa_\vartheta = 0$$

(d) The mid-surface meridional strain is zero:

$$\epsilon_\varphi = 0$$

(e) The bend has a long radius, i.e. $a >> b$.

If w and v are the mid-surface radial and tangential displacements (*Figure 5.23*), then the strain–displacement relations are

$$\epsilon_\varphi = \frac{1}{b}\left(\frac{dv}{d\varphi} + w\right)$$

$$\epsilon_\vartheta = \frac{1}{a}(kb \sin\varphi + w \sin\varphi + v \cos\varphi) \tag{5.63}$$

$$\kappa_\varphi = -\frac{1}{b^2}\left(w + \frac{d^2 w}{d\varphi^2}\right)$$

$$\kappa_\vartheta = 0$$

The equilibrium equations are, for a shell of revolution,

$$\frac{d}{d\varphi}(rN_\varphi) - N_\vartheta b \cos\varphi - rQ_\varphi = 0$$

$$N_\vartheta b \sin\varphi + N_\varphi r + \frac{d}{d\varphi}(rQ_\varphi) = 0 \qquad (5.64)$$

$$\frac{d}{d\varphi}(rM_\varphi) - Q_\varphi rb - M_\vartheta b \cos\varphi = 0$$

where

$$r = a + b \sin\varphi$$

The constant k is the 'fractional end rotation' of the bend resulting from the moment – the ratio of the induced end rotation γ to the angle of the bend

$$k = \gamma/\theta$$

It can be used to give a measure of the flexibility of the bend if it is compared to that of a straight pipe.

For the given applied moment, we can construct Hill's estimate for the end rotation

$$\dot{\gamma} \leqslant \frac{Bh}{M}\int_0^{2\pi} (\bar{Q}_n^{\,s})^{n+1}\,(a\theta)\,b\,d\varphi \qquad (5.65)$$

where $\bar{Q}_n^{\,s}$ is a statically admissible generalised effective stress, equation (5.61). We use the same approximation as that common in elasticity, expanding in a Fourier series

$$\frac{N_\vartheta}{h} = \frac{M}{\pi h b^2}\sum_{j=1}^{\infty} c_{2j-1} \sin[(2j-1)\varphi]$$

which satisfies the obvious symmetry requirements. Further, at the end of the bend, N_ϑ must be in equilibrium with the applied moment; hence

$$c_1 = 1$$

From equilibrium, equations (5.64),

$$M_\varphi = (b^2/a)\int \sin\varphi \int N_\vartheta \,d\varphi\,d\varphi$$

$$N_\varphi = (b/a)\cos\varphi \int N_\vartheta\,d\varphi$$

$$Q_\varphi = -N_\varphi \sin\varphi/\cos\varphi$$

where the constants of integration can be ignored.

Alternatively, we can consider the end rotation to be given; then Hill's estimate for the required applied in-plane moment to produce this is

$$M \leqslant \frac{1}{\dot{\gamma}}\,\frac{1}{B^{1/n}}\,h\int_0^{2\pi}(\bar{\dot{q}}_n^{\mathrm{k}})^{1+1/n}\,(a\theta)\,b\,\mathrm{d}\varphi \tag{5.66}$$

where $\bar{\dot{q}}_n^{\mathrm{k}}$ is a kinematically admissible generalised strain rate. Again, using the elastic form we have for the radial displacement rate

$$\dot{w} = \frac{b\dot{\gamma}}{\theta}\sum_{j=1}^{\infty} D_j \cos(2j\varphi)$$

which satisfies the symmetry conditions. Since $\in_\varphi = 0$ from equation (5.63)

$$\dot{v} = -b\int \dot{w}\;\mathrm{d}\varphi$$

and the constant of integration is zero from symmetry.

The bounds from equations (5.65) and (5.66) can be rearranged to give the extended inequalities

$$\left(\frac{3}{4}\right)^{(n+1)/2}\frac{a\theta}{b}\left(\frac{M}{hb^2}\right)^n\frac{1}{I_k^n} \leqslant \frac{\dot{\gamma}}{B} \leqslant \frac{a\theta}{b}\left(\frac{M}{\pi hb^2}\right)^n\frac{I_s}{\pi}$$

$$I_k = 4\int_0^{\pi/2}\left\{\left[\sin\varphi + \Sigma D_i\left(\sin\varphi\cos(2j\varphi) - \frac{1}{2j}\cos\varphi\sin(2j\varphi)\right)\right]^2 + \left(\frac{n}{2n+1}\right)^{2n/(n+1)}\frac{\lambda^2}{4}\,[\Sigma D_j(4j^2-1)\cos(2j\varphi)]^2\right\}^{(1+1/n)/2}\mathrm{d}\varphi \tag{5.67}$$

$$I_s = 4\int_0^{\pi/2}\left\{(\sin\varphi + \Sigma\, c_{2j-1}\sin[(2j-1)\varphi])^2 + (2+1/n)^{2n/(n+1)}\,\frac{1}{4\lambda^2}\left[\cos(2\varphi) + \Sigma c_{2j-1}\left(\frac{\cos(2j\varphi)}{(2j-1)j} - \frac{\cos[(2j-2)\varphi]}{(2j-1)(j-1)}\right)\right]^2\right\}^{(n+1)/2}\mathrm{d}\varphi$$

where $\lambda = ah/b^2$ is the 'pipe bend parameter'.

We can get arbitrarily close to the exact solution using a Rayleigh–Ritz procedure, increasing the number of terms in the series and minimising or maximising the bounds. Again, this was achieved through a standard computer scientific subroutine package for various values of n and λ.

We can define a creep *flexibility factor* as

$$K_n = \frac{\text{end rotation rate of the pipe bend}}{\text{end rotation rate of a straight pipe}} = \frac{\dot{\gamma}}{\dot{\gamma}_0}$$

where, from Section 3.1,

$$\frac{\dot{\gamma}_0}{B} = \frac{a\theta}{b}\left(\frac{M}{D_n h_b{}^2}\right)^n \qquad D_n = \int_0^{2\pi} (\sin\varphi)^{1+1/n}\, \mathrm{d}\varphi$$

Convergence of the bounds on the flexibility factor in a typical case is shown in *Figure 5.24*: fully converged results are shown in *Figure 5.25*. It can be seen immediately that the curved pipe is considerably more flexible than an equivalent straight pipe.

Comparison is also shown with a more exact analysis, not reported here, using the local stress–strain relations and integrating through the thickness. It can be seen that the generalised thin-shell model is very good in this case. Incidentally, it should be noted that we have in fact developed a generalised beam-bending model for the curved pipe, since

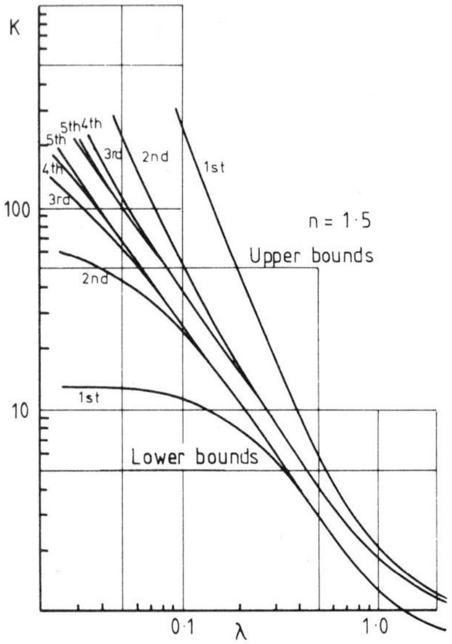

Figure 5.24 Convergence of Hill's estimates for flexibility factor

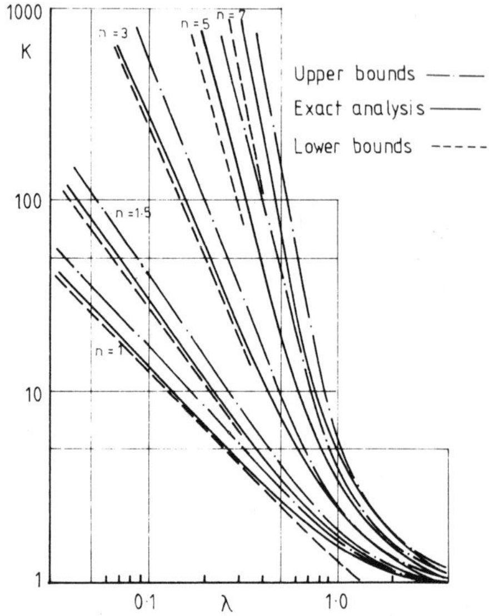

Figure 5.25 Flexibility factor for a curved pipe

$$\frac{\dot{\gamma}}{\theta} = \frac{a}{b} K_n(\lambda)\left(\frac{M}{D_n h b^2}\right)^n$$

This can be used in the stress analysis of piping systems for creep using simple beam theory.

Further reading

Odqvist, F.K.G., *Mathematical Theory of Creep and Creep Rupture,* Clarendon Press, 2nd Edn, 1974.

Kachanov, L.M., *The Theory of Creep,* 1960 (Transl. Material Lending Library for Science and Technology, 1967).

Bailey, R.W., The utilization of creep test data in engineering design, *Proc. I. Mech. E.,* **131**, 1935.

Hill, R., New horizons in the mechanics of solids, *J. Mech. Phys. Solids,* **5**, 66–74, 1956.
Calladine, C.R., A rapid method for estimating the greatest stress in a structure subject to creep, *Proc. I. Mech. E. Conf. on Thermal Loading and Creep*, I. Mech. E., 1964.
Calladine, C.R. and Drucker, D.C., A bound method for creep analysis of structures: direct use of solutions in elasticity and plasticity, *J. Mech. Eng. Sci.,* **4**, 1–11, 1962.
Martin, J.B., A note on the determination of an upper bound on displacement rates for steady creep problems, *Trans. ASME, J. Appl. Mech.,* **33**, 216–17, 1966.
Rozenblium, V.I., Approximate equations of creep of thin shells, *J. Appl. Math. Mech.,* **17**, 217–225, 1963.
Mackenzie, A.C., Generalised stress/strain relations for creep of thin shells: a comparison of 'exact' and approximate relations, *Appl. Sci. Res.,* **20**, 252–65, 1973.
Murakami, S. *et al.*, Application of extended Newton method to the creep analysis of shells of revolution, *Ing. Arch.,* **41**, 235–57, 1972.
Spence, J., Creep analysis of smooth curved pipes under in-plane bending, *J. Mech. Eng. Sci.,* **15**, 252–65, 1973.
Bellman, R. and Kalaba, R.E., *Quasilinearisation and Nonlinear Boundary Value Problems,* Elsevier, 1965.
Roberts, S.M. and Shipman, J.S., *Two-point Boundary Value Problems: Shooting Methods,* Elsevier, 1972.
Mikhlin, S.G. and Smolitsky, K.I., *Approximate Methods for the Solution of Differential and Integral Equations*, Elsevier, 1967.
Ortega, J.M. and Rheinbolt, W.C., *Iterative Solution of Nonlinear Equations in Several Variables,* Academic Press, 1970.

Chapter 6

Reference stress methods in steady creep

It should be apparent, even for the elementary examples discussed here, that the solution of the steady creep problem presents no particular difficulty to the analyst. If a suitable linear analysis method is available, then quasilinearisation, or an energy formulation, can be applied routinely to solve the equivalent steady creep problem. The solutions that were presented in the previous chapter assumed an idealised material behaviour, principally following a power law of creep. In practice, creep data for real materials are not seen to conform to such an idealised constitutive realtionship, essentially because of the unavoidable scatter which is always present. Thus, in a practical situation, the material parameters can only be estimated, and are therefore subject to error. We must then face up to the problem discussed in Section 3.1 in an analysis of the beam in bending. There, it was demonstrated that the nonlinearity of the constitutive equation can result in a relatively small error in the material parameters giving rise to a large error in the estimated deformation rates. Since the constitutive relation is merely an idealisation, this means that there is an unacceptably wide range of possible answers! The designer, and particularly the analyst constructing an elaborate solution, should *never* lose sight of this fundamental difficulty. It assumes an even greater importance if the analysis is expensive.

In Section 3.1 this difficulty was partially avoided by directly relating the structural behaviour to the uniaxial test data – the basis of the so-called *reference stress method*. The reference stress σ_R corresponding to a power law of steady creep has the property that the proportionality constant, or *scaling factor* δ, relating some characteristic deformation rate, say, $\dot{q}$, to a uniaxial test

$$\dot{q} = \delta(\sigma_R, n) \times \dot{\epsilon}_R$$

where $\dot{\epsilon}_R = B\sigma_R^n$, is virtually independent of the material exponent n. **It** was discovered for the beam that the reference stress could be chosen such that, for a large range of n, the scaling factor could be approximated by its limiting value (*Figure 6.1*)

$$\delta(\sigma_R, n) \simeq \lim_{n\to\infty} \delta(\sigma_R, n) \tag{6.2}$$

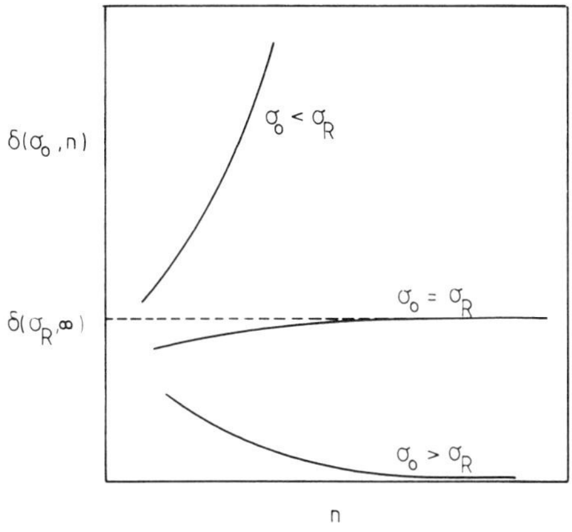

Figure 6.1 Definition of a reference stress

The uniaxial reference strain rate $\dot{\epsilon}_R$ could then be estimated from the uniaxial test data, *thus avoiding any estimation of the material parameters.*

It is the purpose of this chapter to examine the reference stress method in more detail. In general, it will be found that the simple approximation of equation (6.2) will not be sufficient to deal with all cases and that a different approach will be required. To begin with, let us establish that the above approach will work for other structural problems.

First consider the two-bar structure shown in *Figure 5.1*, consisting of two parallel bars of length L_1, L_2 at the same temperature and of the same cross-sectional area A subject to a vertical load Q. This was analysed in Section 5.1. The vertical deflection rate can be written as

$$\dot{q} = \delta(\sigma_0, n) \times \dot{\epsilon}_0 \tag{6.3}$$

for some σ_0, $\dot{\epsilon}_0 = B\sigma_0^n$, where the scaling factor is

$$\delta = \frac{L_1 L_2}{(L_1^{1/n} + L_2^{1/n})^n} \left(\frac{Q}{\sigma_0 A}\right)^n$$

It can be verified that, provided

$$\sigma_0 = \sigma_R = Q/2A$$

then in the limit

$$\delta(\sigma_R, n) \to \sqrt{(L_1 L_2)}$$

(This is perhaps best seen by simply working out the scaling factor for various values of n and σ_0.) The variation with n is plotted in *Figure 6.2*. Obviously to a good approximation

$$\delta(\sigma_R, n) \simeq \delta(\sigma_R, \infty)$$

as before, and the reference stress method works.

Secondly, consider the internally pressurised thick cylinder of inner radius a and outer radius b examined in Section 4.3. The internal expansion rate can be written as

$$\dot{u}_a = \delta(\sigma_0, n) \times \dot{\epsilon}_0 \tag{6.4}$$

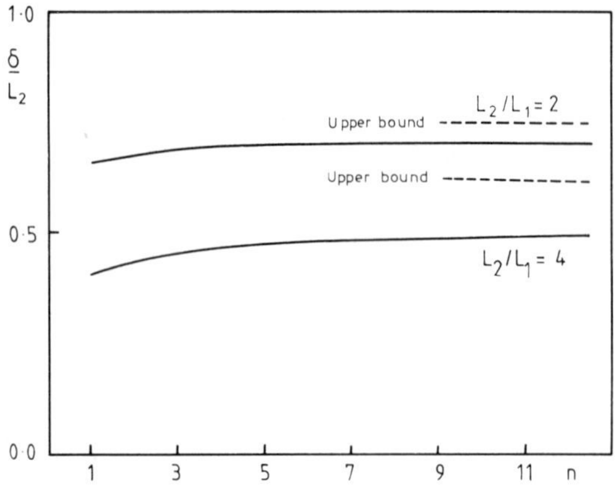

Figure 6.2 Scaling factor for two-bar structure

for arbitrary σ_0, where the scaling factor is

$$\delta = \frac{(\sqrt{3})^{n+1}}{2}\, a\left(\frac{b}{a}\right)^2\left(\frac{p/\sigma_0}{n[(b/a)^{2/n}-1]}\right)^n$$

It can be verified that, provided we choose

$$\sigma_0 = \sigma_R = (\sqrt{3}/2)\,p/\ln(b/a)$$

we have in the limit

$$\delta(\sigma_R, n) \to (\sqrt{3}/2)\,b$$

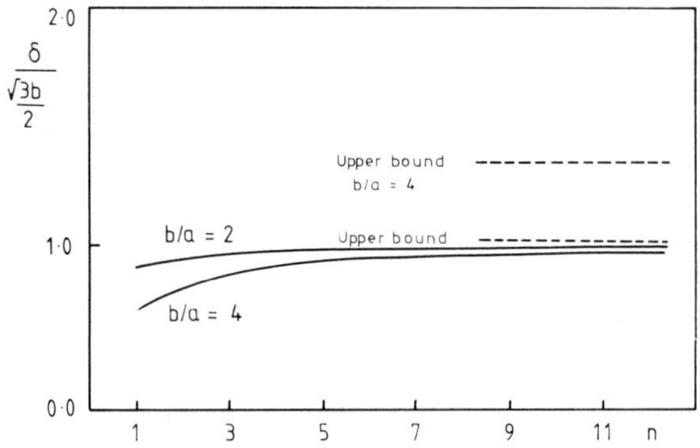

Figure 6.3 Scaling factor for thick cylinder

The variation with n is shown in *Figure 6.3.* Therefore, we can again write approximately that

$$\delta(\sigma_R, n) \simeq \delta(\sigma_R, \infty)$$

It would seem that the reference stress method does have general applicability. We will find that there are some problems where it does *not* work, but first we will try to discover why it should.

6.1 Existence of a reference stress for the power law

Throughout this introduction, we assume that there is available some means of solving the steady creep problem. Our purpose in the reference stress method is to minimise the uncertainty as to the material behaviour.

In the preceding discussion, we have made three presumptions:

(1) The constitutive relation is a power law.
(2) There exists a reference stress such that the scaling factor has a finite, non-zero limit.
(3) If this limit exists, the error in using it as an approximation to the scaling factor is small (and this will of course depend on the likely value of the exponent n).

Later we will describe another interpretation of the reference stress method which removes these restrictions. With the present interpretation, the second point is the most critical. In fact, we will demonstrate that there does exist a reference stress such that the scaling factor has a finite limit, but there remains the possibility that this limit is zero.

Consider a structure undergoing isothermal creep subject to a single point (or generalised) load Q, resulting in a generalised deflection rate $\dot{q}$ as discussed in Section 5.7.1. Then we have

$$\dot{q} = \frac{\mathrm{d}}{\mathrm{d}Q} \frac{Bv}{n+1} \bar{Q}_n^{n+1}$$

where $\bar{Q}_n$ is the effective generalised stress, equation (5.58). We can write

$$\dot{q} = \delta(\sigma_0, n) \times \dot{\epsilon}_0 \tag{6.5}$$

so that the scaling factor is

$$\delta = v \frac{\mathrm{d}\bar{Q}_n}{\mathrm{d}Q} \left(\frac{\bar{Q}_n}{\sigma_0}\right)^n$$

We now establish the following fundamental result.

Theorem If we choose

$$\sigma_R = \lim_{n \to \infty} \bar{Q}_n = \bar{Q}_\infty$$

then

$$\lim_{n \to \infty} \delta(\sigma_R, n) = v \frac{\mathrm{d}\sigma_R}{\mathrm{d}Q} \exp\left(\frac{1}{\sigma_R} \left.\frac{\mathrm{d}\bar{Q}_n}{\mathrm{d}\mu}\right|_{\mu=0}\right)$$

where $\mu = 1/n$. The exponential term is either finite or zero.

The proof is rather mathematical: From the theorem of nesting surfaces (Section 5.7.1), we know that $\bar{Q}_n$ is monotonic increasing with n and bounded, i.e.

$$\bar{Q}_1 \leqslant \bar{Q}_n \leqslant \bar{Q}_\infty$$

Since the limit $\bar{Q}_\infty$ exists,

$$\lim_{n \to \infty} \frac{d\bar{Q}_n}{dQ} = \frac{d\bar{Q}_\infty}{dQ} = \frac{d\sigma_R}{dQ}$$

and therefore

$$\lim_{n \to \infty} \delta = \nu \frac{d\sigma_R}{dQ} \lim_{n \to \infty} \left(\frac{\bar{Q}_n}{\bar{Q}_\infty}\right)^n$$

Now from the theorem of nesting surfaces, the term

$$(\bar{Q}_n/\bar{Q}_\infty)^n \leqslant 1$$

and it is also positive (since $\bar{Q}_n$ and $\bar{Q}_\infty$ have the same sign). A fundamental theorem of limits (called the Bolzano–Weierstrasse theorem) now tells us that since this term is bounded, a limit must exist. It can be evaluated in the following manner: write

$$(\bar{Q}_n/\bar{Q}_\infty)^n = \exp\,[n \ln (\bar{Q}_n/\bar{Q}_\infty)]$$

and consider

$$\lim_{\mu \to 0} \ln (\bar{Q}_n/\bar{Q}_\infty)/\mu$$

where $\mu = 1/n$. Now as $\mu \to 0$, $n \to \infty$ and both numerator and denominator have zero limit. But from L'Hospital's rule, which can be stated as:

"If $\lim_{x \to 0} f(x) = \lim_{x \to 0} g(x) = 0$ while $\lim_{x \to 0} g'(x) \neq 0$

then

$$\lim_{x \to 0} \frac{f(x)}{g(x)} = \lim_{x \to 0} \frac{f'(x)}{g'(x)}\text{"}$$

we have

$$\lim_{u \to 0} \frac{\ln(\bar{Q}_n/\bar{Q}_\infty)}{\mu} = \frac{1}{\sigma_R} \left.\frac{d\bar{Q}_n}{d\mu}\right|_{\mu = 0}$$

Therefore

$$\lim_{n \to \infty} \delta = \nu \frac{d\sigma_R}{dQ} \exp\left(\frac{1}{\sigma_R} \left.\frac{d\bar{Q}_n}{d\mu}\right|_{\mu = 0}\right)$$

Further, since $\bar{Q}_n$ is monotonic increasing, by the theorem of nesting surfaces, it follows

$$\left.\frac{d\bar{Q}_n}{d\mu}\right|_{\mu = 0} < 0$$

and therefore either the limit is finite or zero. The result is therefore proved.

For a single load, it is easily shown that

$$\sigma_R = \bar{Q}_\infty = (Q/Q_L)\,\sigma_y \tag{6.6}$$

where Q_L is the plastic limit load for the structure with a fictitious yield σ_y. Then

$$\delta(\sigma_R,\infty) = \frac{\nu\sigma_R}{Q}\exp\left(\frac{1}{\sigma_R}\frac{d\bar{Q}_n}{d\mu}\bigg|_{\mu=0}\right) \leqslant \frac{\nu\sigma_R}{Q}$$

since the term in the exponential is negative. We therefore have a simple upper bound for the scaling factor which requires only the knowledge of the reference stress, i.e. the limit load for the structure. Also from the theorem of nesting surfaces, it follows that for all n

$$\delta(\sigma_R, n) \leqslant \nu\sigma_R/Q$$

For the three problems we have been considering the upper bounds can be evaluated as

Beam in bending:

$$\nu = 2bd \qquad \delta(\sigma_R, n) \leqslant 2l/d$$

Two-bar structure:

$$\nu = A(L_1 + L_2) \qquad \delta(\sigma_R, n) \leqslant \frac{1}{2}(L_1 + L_2)$$

Thick cylinder:

$$\nu = \pi(b^2 - a^2) \qquad \delta(\sigma_R, n) \leqslant \frac{b^2 - a^2}{2a}\,\frac{\sqrt{3}}{2\ln(b/a)}$$

If these are compared with the exact limits, e.g. *Figures 6.2 and 6.3*, it is seen that these simple bounds are often poor, although acceptable.

The above result does *not* guarantee that the limit of the scaling factor is non-zero, although we have not encountered an example so far where this situation does occur. Unfortunately, such do exist, as now discovered.

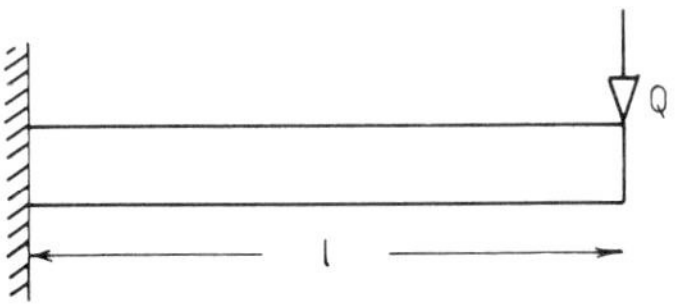

Figure 6.4 Geometry of point-loaded cantilever beam

Consider a cantilever beam of length l and rectangular cross-section $2d \times b$ with an end load Q(*Figure 6.4*). Let $\dot{w}(l)$ be the end deflection rate resulting from this load; the problem can be resolved in a manner similar to Section 5.5, and we may write

$$\dot{w}(l) = \delta(\sigma_0, n) \times \dot{\epsilon}_0 \tag{6.7}$$

where

$$\delta(\sigma_0, n) = \frac{l^2/d}{n+2}\left(\frac{Ql}{\sigma_0 bd^2}\right)^n\left(1 + \frac{1}{2n}\right)^n$$

It can be verified, either analytically or by simple numerical evaluation, that if $\sigma_0 < Ql/bd^2$, $\delta \to \infty$ as $n \to \infty$, but if $\sigma_0 \geqslant Ql/bd^2$, $\delta \to 0$. That is, no finite, non-zero limit exists! The method therefore leads to a trivial result in this case.

In summary, the success of this interpretation of the reference stress method will depend on the problem in hand. Indeed, even if a non-zero limit is found, success still depends on whether it can be used with little error as an approximation to the actual scaling factor. We will show that this is not true for non-isothermal creep, but before doing so we consider the application of the present result to combined loading.

6.2 Reference stresses for combined loading with a power law

Although the present form of the reference stress method is not always successful, when it is, it leads to particularly powerful results useful in design. If used in conjunction with the approximate generalised models described in Section 5.7.1, it becomes even more attractive. Let us examine the reference stress method for combined loading.

Consider an isothermal structure under a number of generalised loads Q_k, $k = 1, 2, \ldots$ with corresponding deflection rates $\dot{q}_k$, $k = 1, 2, \ldots$. Then from Section 5.7.1.

$$\dot{q}_k = \delta_k(\sigma_0, n) \times \dot{\epsilon}_0$$

where the scaling factor is

$$\delta_k = \nu \frac{\partial \bar{Q}_n}{\partial Q_k} \left(\frac{\bar{Q}_n}{\sigma_0} \right)^n$$

As in the previous section, if we choose

$$\sigma_R = \lim_{n \to \infty} \bar{Q}_n = \bar{Q}_\infty$$

then

$$\lim_{n \to \infty} \delta_k(\sigma_R, n) = \nu \frac{\partial \sigma_R}{\partial Q_k} \exp\left(\frac{1}{\sigma_R} \left. \frac{d\bar{Q}_n}{d\mu} \right|_{\mu = 0} \right) \tag{6.8}$$

where $\mu = 1/n$ and the exponential term is either finite or zero. The results for a single load can therefore be extended to combined loading. Similarly we have the bound on the limit

$$\delta_k(\sigma_R, \infty) \leqslant \nu \frac{\partial \sigma_R}{\partial Q_k}$$

although this is not necessarily true for all n.

We will examine two problems here. First consider the statically determinate structure discussed in Section 5.7.2, consisting of two pin-jointed bars of equal length L and cross-sectional area A subject to horizontal and vertical forces Q_1 and Q_2 (*Figure 5.18*). The scaling factors for the corresponding deflection rates $\dot{q}_1$ and $\dot{q}_2$ are

$$\delta_k(\sigma_0, n) = \nu \frac{\partial \bar{Q}_n}{\partial Q_k} \left(\frac{\bar{Q}_n}{\sigma_0}\right)^n \qquad k = 1, 2$$

where $\nu = 2LA$. For the case $Q_1, Q_2 > 0$ we take as reference stress $\lim_{n \to \infty} \bar{Q}_n$ (see Section 5.7.2)

$$\sigma_R = \frac{1}{\sqrt{2}A} (Q_1 + Q_2)$$

and examine

$$\delta_1(\sigma_R, n) = 2LA \frac{\partial \bar{Q}_n}{\partial Q_1} \left(\frac{\bar{Q}_n}{\sigma_R}\right)^n$$

where

$$\bar{Q}_n = \left[\frac{1}{2}\left(\frac{Q_1 + Q_2}{\sqrt{2}A}\right)^{n+1} + \frac{1}{2}\left(\frac{Q_1 - Q_2}{\sqrt{2}A}\right)^{n+1}\right]^{1/(n+1)}$$

In *Figure 6.5* we plot the quantities $(1/\sqrt{2}L)\nu\,(\partial \bar{Q}_n/\partial Q_1)$ and $(\bar{Q}_n/\sigma_R)^n$ against n for various values of the load ratio Q_1/Q_2. The former is ultimately increasing to unity while the latter is decreasing to 0.5. The upper bound on the latter is

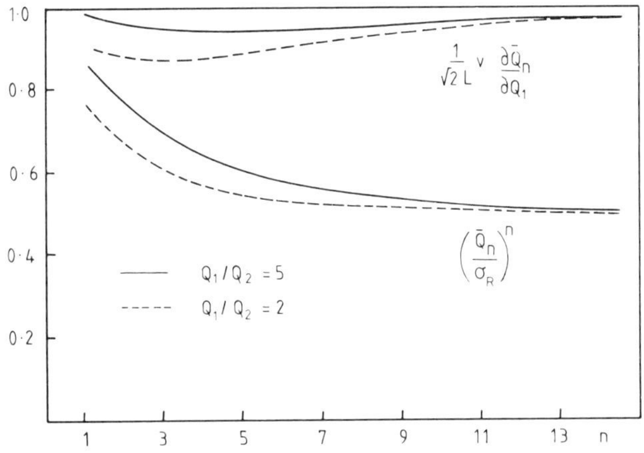

Figure 6.5 Variation of groupings in scaling factor for statically determinate structure

equal to unity and is thus extremely poor. Finally, in *Figure 6.6* we plot the scaling factor $\delta_1/\sqrt{2}L$ against n for various Q_1/Q_2. The reference stress approximation

$$\delta_1(\sigma_R, n) \simeq \delta_1(\sigma_R, \infty) = L/\sqrt{2}$$

is clearly acceptable with the appropriate choice of reference stress.

Secondly, consider the rectangular cross-section beam $2b \times d$ subject to a bending moment M and an axial tension N discussed in Section 5.7.3 (*Figure 5.20*). The scaling factors for the curvature rate $\dot{\kappa}$ and extension rate $\dot{\delta}$ are

$$\delta_\kappa(\sigma_0, n) = \nu \frac{\partial \bar{Q}_n}{\partial M} \left(\frac{\bar{Q}_n}{\sigma_0}\right)^n$$

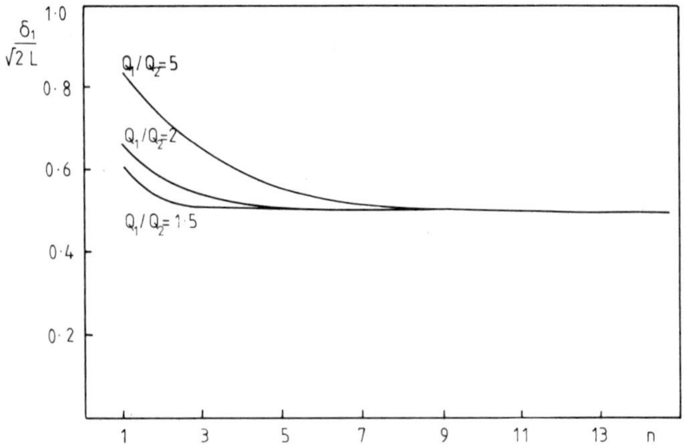

Figure 6.6 Scaling factor for statically determinate structure

$$\delta_{\dot{\delta}}(\sigma_o, n) = \nu \frac{\partial \bar{Q}_n}{\partial N} \left(\frac{\bar{Q}_n}{\sigma_o}\right)^n$$

where $\nu = 2bd$. The reference stress is chosen as the approximation (equation (5.59), $n \to \infty$)

$$\sigma_R = \sqrt{\left[\frac{1}{4}\left(\frac{N}{bd}\right)^2 + \left(\frac{M}{bd^2}\right)^2 \mu^2_\infty\right]}$$

and we examine

$$\delta_{\dot{\kappa}}(\sigma_R, n) = 2\,bd\,\frac{\partial \bar{Q}_n}{\partial M}\left(\frac{\bar{Q}_n}{\sigma_R}\right)^n$$

where

$$\bar{Q}_n = \sqrt{\left[\frac{1}{4}\left(\frac{N}{bd}\right)^2 + \left(\frac{M}{bd^2}\right)^2 \mu^2_n\right]}$$

$$\mu_n = (2 + 1/n)^{n/(n+1)}$$

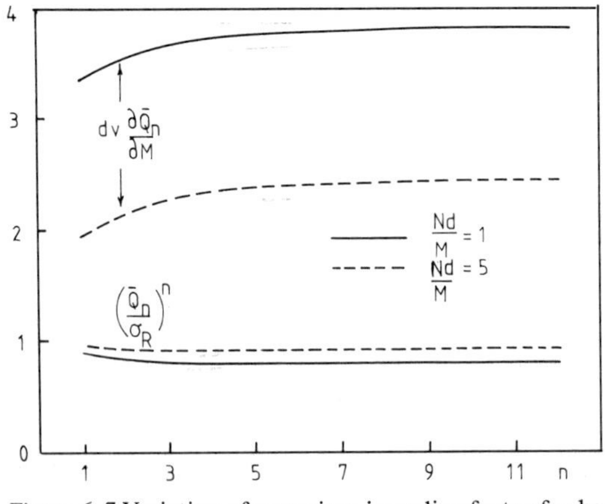

Figure 6.7 Variation of groupings in scaling factor for beam under tension and bending

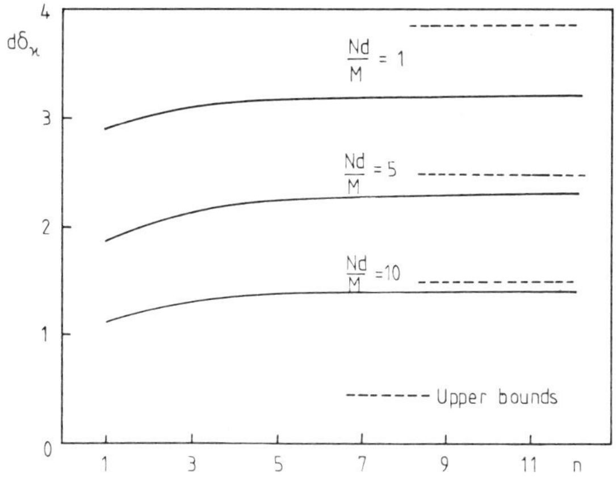

Figure 6.8 Scaling factor for beam under tension and bending

In *Figure 6.7* we plot the quantities $\nu\partial\bar{Q}_n/\partial M$ and $(\bar{Q}_n/\sigma_R)^n$ against the exponent n for various values of the load parameter Nd/M; both are increasing functions of n (unlike the previous example). The scaling factor δ_κ is shown in *Figure 6.8* and once again the approximation

$$\delta_\kappa(\sigma_R, n) \simeq \delta_\kappa(\sigma_R, \infty)$$

is adequate for values of $n \geqslant 3$. Also shown is the upper bound, which is not as poor for this problem; in fact, it improves for increasing values of the load parameter.

6.3 Non-isothermal power-law creep

Our discussion of the reference stress method has so far been concerned with isothermal creep. We know from Section 3.2 that the method is not successful for non-uniformly heated structures. In this section we examine this possibility further.

First consider the two-bar structure, but assume that the two bars are at different temperatures T_1 and T_2. The vertical deflection rate $\dot{q}$ induced by the load Q is now given simply by

$$\dot{q} = \frac{Ce^{-\gamma/T}(Q/A)^n\, L_1 L_2}{(L_1^{1/n} + \{\exp[\gamma(1/T_1 - 1/T_2)]L_2\}^{1/n})^n}$$

assuming, without loss in generality, that $T_2 > T_1$ and that the uniaxial law is

$$\dot{\epsilon} = Ce^{-\gamma/T}\sigma^n$$

We attempt to express this as the result of a reference stress test held at the maximum temperature so that

$$\dot{q} = \delta(\sigma_0, n)\,\dot{\epsilon}_0$$

for arbitrary σ_0 where now $\dot{\epsilon}_0 = Ce^{-\gamma/T}\sigma_0^n$. The scaling factor is then

$$\delta(\sigma_0, n) = \frac{(Q/\sigma_0 A)^n L_1 L_2}{(L_1^{1/n} + \{\exp[\gamma(1/T_2 - 1/T_1)] L_2\}^{1/n})^n} \tag{6.9}$$

If we choose $\sigma_R = Q/2A$ as before, then in the limit

$$\lim_{n \to \infty} \delta(\sigma_R, n) = \sqrt{\left(\frac{L_1 L_2}{\exp[\gamma(1/T_2 - 1/T_1)]}\right)}$$

As a specific example, we will assume that

$$0 \leqslant \gamma \left(\frac{1}{T_1} - \frac{1}{T_2}\right) \leqslant 10$$

as a typical range for our purposes. In *Figure 6.9* we plot the variation in δ with the exponent n for $L_2/L_1 = 2$ and various values of $\gamma(1/T_1 - 1/T_2)$. Clearly, when there is a significant temperature difference, the limit scaling factor is definitely *not* an acceptable approximation for the actual scaling factor over a range of n.

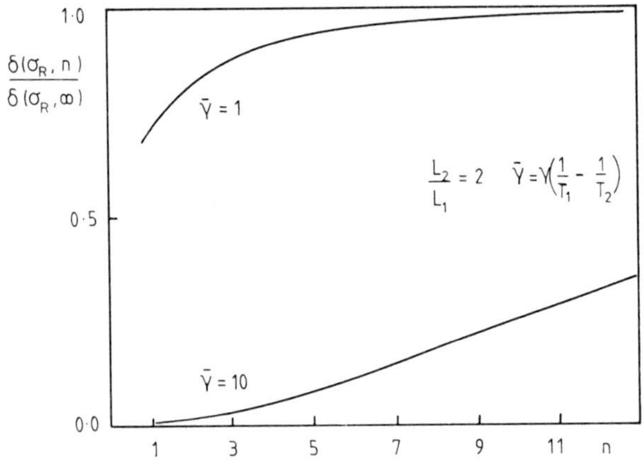

Figure 6.9 Scaling factor for non-uniformly heated two-bar structure

Secondly, consider the internally pressurised cylinder with a radial temperature distribution examined in Section 4.3. The tube has an inner temperature T_a and an outer temperature T_b. Then the inner radial extension is

$$\dot{u}_a = C\exp(-\gamma/T_b)\exp[-\gamma(T_b - T_a)/T_b^2]\frac{(\sqrt{3})^{n+1}}{2}a\left(\frac{b}{a}\right)^2\left(\frac{p}{n_T[(b/a)^{2/n_T} - 1]}\right)^n$$

where

$$n_T = \frac{n}{1 + (\gamma/T_b^2)(T_b - T_a)/2\ln(b/a)}$$

If we attempt to express this as the result of a reference stress test held at the maximum temperature T_b, then

$$\dot{u}_a = \delta(\sigma_0, n) \times \dot{\epsilon}_0$$

where

$$\dot{\epsilon}_0 = C e^{-\gamma/T_b} \sigma_0^n$$

The scaling factor is

$$\delta(\sigma_0, n) = a \exp[-\gamma(T_b - T_a)/T_b^2] \frac{(\sqrt{3})^{n+1}}{2} \left(\frac{b}{a}\right)^2 \left(\frac{p/\sigma_0}{n_T[(b/a)^{2/n_T} - 1]}\right)^n \quad (6.10)$$

As in the isothermal case, if we choose

$$\sigma_R = \frac{\sqrt{3}}{2} \frac{p}{\ln(b/a)}$$

then the scaling factor has a limit

$$\delta(\sigma_R, \infty) = \frac{\sqrt{3}}{2} b \exp[-\gamma(T_b - T_a)/T_b^2] \left(\frac{b}{a}\right)^{-(\gamma/T_b^2)(T_b - T_a)/2 \ln(b/a)}$$

As a typical range we again take

$$0 \leqslant \gamma\left(\frac{1}{T_a} - \frac{1}{T_b}\right) \simeq \frac{\gamma}{T_b^2}(T_b - T_a) \leqslant 10$$

In *Figure 6.10* we plot the variation in δ with the exponent n for the case $b/a = 2$ and various values of $(\gamma/T_b^2)(T_b - T_a)$. Again there is a great variation in the scaling factor with n.

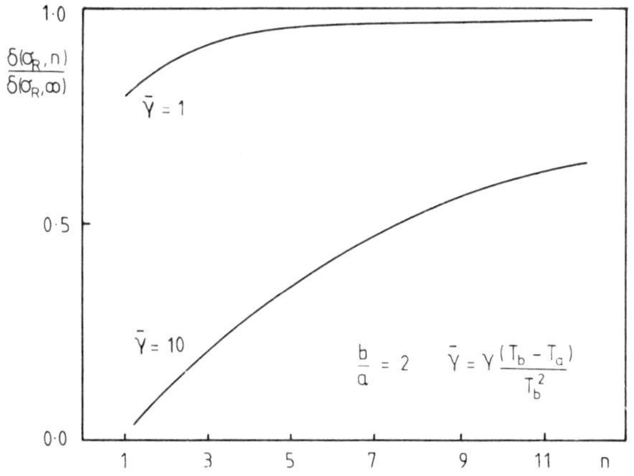

Figure 6.10 Scaling factor for non-uniformly heated thick cylinder

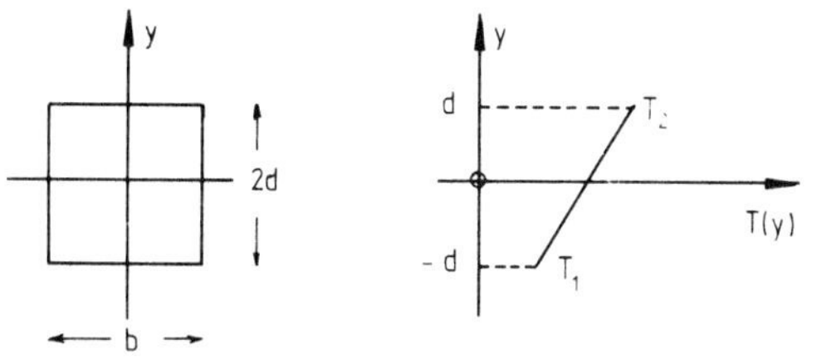

Figure 6.11 Temperature distribution in rectangular cross-section beam

Finally, consider the rectangular cross-section beam subject to a temperature distribution $T(y)$, $-d \leqslant y \leqslant d$ (*Figure 6.11*). It can easily be shown that the curvature rate is given by

$$\dot{\kappa} d = C e^{-\gamma/T_M} \frac{1}{\gamma_n^n} \left(\frac{M}{bd^2}\right)^n$$

where

$$\gamma_n = \int_{-1}^{1} \exp\left[\frac{\gamma}{n}\left(\frac{1}{T} - \frac{1}{T_M}\right)\right] z^{1+1/n}\, dz$$

and T_M is the maximum temperature. For a linear temperature distribution

$$T = T_1 + \frac{(T_2 - T_1)}{2d}(y + d)$$

and using the approximation

$$\gamma\left(\frac{1}{T_2} - \frac{1}{T_1}\right) \simeq \frac{\gamma}{T_2^2}(T_2 - T_1) = \bar{\gamma}$$

the integral γ_n is given by

$$\gamma_n = \exp\left(\frac{\bar{\gamma}}{2n}\right) \int_{-1}^{1} z^{1+1/n} \exp\left(-\frac{\bar{\gamma}}{2n} z\right) dz$$

which should, of course, be evaluated numerically. Expressed as the result of a reference stress test

$$\dot{\kappa} = \delta(\sigma_0, n) \times \dot{\epsilon}_0$$

the scaling factor is

$$\delta(\sigma_0, n) = \frac{1}{d} \exp(-\bar{\gamma}/2) \frac{1}{J_n^n} \left(\frac{M}{\sigma_0 bd^2}\right)^n \qquad (6.11)$$

where

$$J_n = \int_{-1}^{1} z^{1+1/n} \exp\left(-\frac{\bar{\gamma}}{2n} z\right) dz$$

It can be shown that if we choose

$$\sigma_R = M/bd^2$$

as in the isothermal case, then

$$\delta(\sigma_R, \infty) = (1/d) \exp(-\bar{\gamma}/2)\sqrt{e}$$

In *Figure 6.12* we plot the variation of the scaling factor with n for the range

$$0 \leqslant \bar{\gamma} \leqslant 10$$

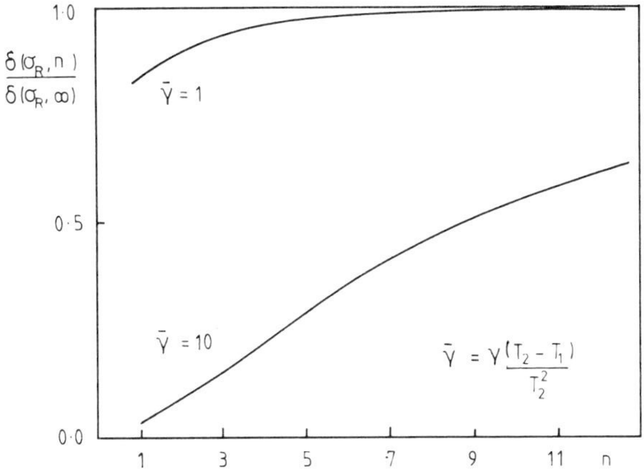

Figure 6.12 Scaling factor for non-uniformly heated beam

as in the previous examples. Once more, the limit scaling factor is not a good approximation for a range of the exponent n.

It should be apparent by now that the present form of the reference stress method is inadequate for non-isothermal creep.

6.4 Reference temperatures

Some of the difficulty associated with non-isothermal problems can be removed to an extent using the concept of a 'reference temperature'. The aim here is to find that temperature which equates the characteristic displacement rate for the isothermal and non-isothermal problems. For example, denoting this temperature by T_0, we may define

Two-bar structure:

$$\frac{\gamma}{n}\left(\frac{1}{T_0}-\frac{1}{T_2}\right)=\ln\left(\frac{L_1^{1/n}+e^{\beta}L_2^{1/n}}{L_1^{1/n}+L_2^{1/n}}\right)$$

Thick cylinder:

$$\frac{\gamma}{n}\left(\frac{1}{T_0}-\frac{1}{T_b}\right)=\beta+\ln\left(\frac{n_T}{n}\,\frac{(b/a)^{2/n_T}-1}{(b/a)^{2/n}-1}\right)$$

Beam in bending

$$\frac{\gamma}{n}\left(\frac{1}{T_0}-\frac{1}{T_2}\right)=\ln\left(\frac{\gamma_n}{I_n}\right)$$

where

$$\beta=\frac{\gamma}{n}\left(\frac{1}{T_1}-\frac{1}{T_2}\right)$$

or

$$\beta = \frac{\gamma}{n}\frac{(T_b - T_a)}{T_b^2}$$

as appropriate. This grouping has been introduced already (Section 3.6). In *Figures 6.13–6.15* are plotted the variations of the grouping

$$\frac{\gamma}{n}\left(\frac{1}{T_o} - \frac{1}{T_{max}}\right)$$

with n treating β as a constant. Clearly this tends to a limit as $n \to \infty$ and is reasonably constant over the whole range. In *Figures 6.16–6.18* this is plotted against β; for $\beta \geqslant 2$ it is virtually a straight line.

These results could be used in the following manner. The isothermal characteristic displacement can be represented by a reference stress test. The temperature at which this test is held – the *reference temperature* T_R – or the

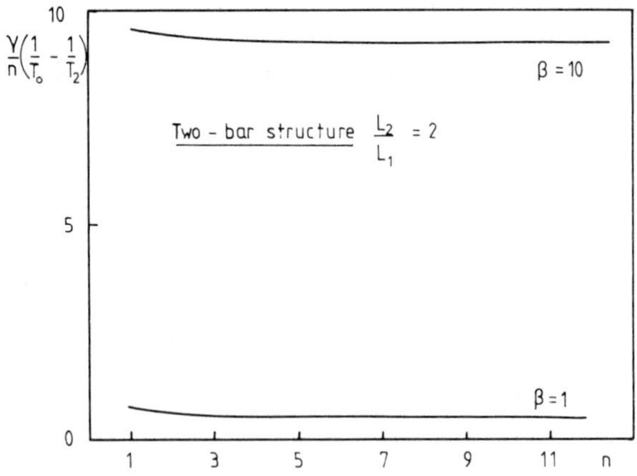

Figure 6.13 Reference temperature for two-bar structure: variation with n

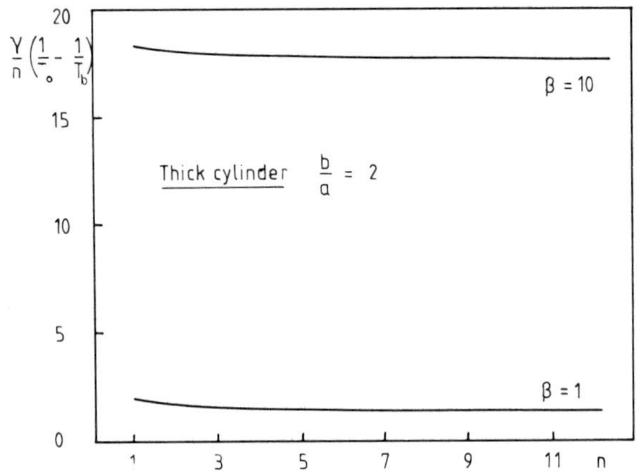

Figure 6.14 Reference temperature for thick cylinder: variation with n

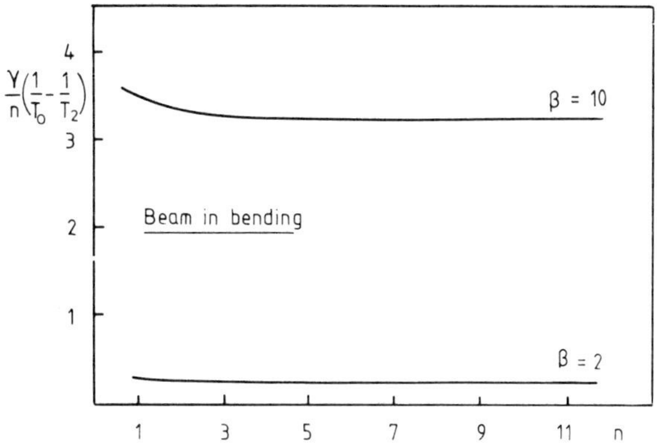

Figure 6.15 Reference temperature for beam: variation with n

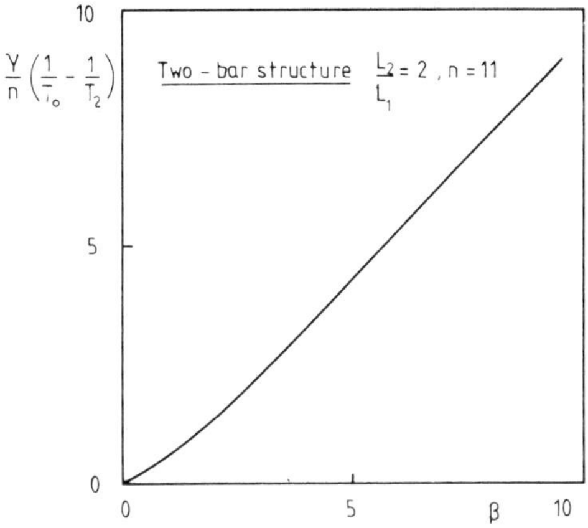

Figure 6.16 Reference temperature for two-bar structure; variation with β

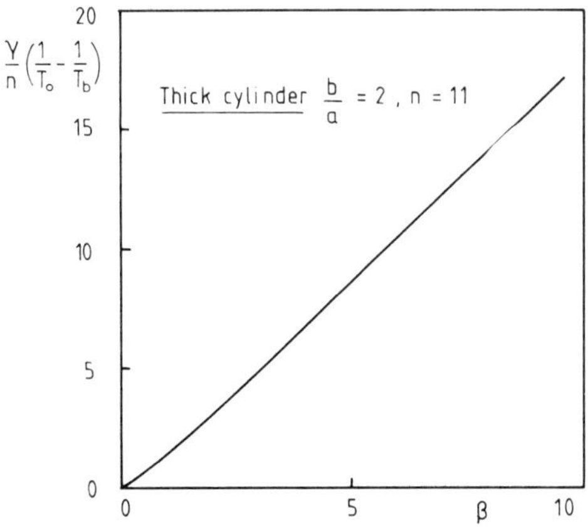

Figure 6.17 Reference temperature for thick cylinder: variation with β

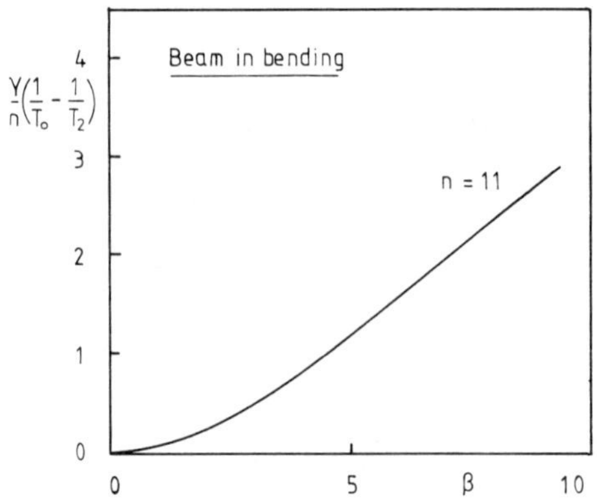

Figure 6.18 Reference temperature for beam: variation with β

grouping $(\gamma/n)(1/T_R - 1/T_{max})$ can then be given to a good approximation by the exponent n multiplied by a factor which is virtually independent of n but dependent on the grouping $(\gamma/n)(1/T_R - 1/T_{max})$. For small variations in the exponent n, these are sensibly linear.

6.5 A local reference stress method

The inherent disadvantages with the previous interpretation of the reference stress method have been illustrated in the preceding sections. This is *not* to say that no attempt should be made to use it in practice – when it works, it can lead to very powerful results. Although we have seen that success of the method depends on the problem in hand, perhaps the greatest disadvantage lies in its inability to be extended to deal with material constitutive relations other than the power law. In the following, we will develop an alternative approach which retains the basic essence of the reference stress method described above – to express some characteristic response of a structure as proportional to a uniaxial creep test in such a way that the constant of proportionality (the scaling factor) is largely independent of variations in the material parameters – but which is more widely applicable.

Let us re-examine the discussion given in Section 3.1. So far we have been trying to render the scaling factor completely independent of the exponent n by suitable choice of a reference stress. This, as we have seen, may not always be successful. In practice, we will have some idea of the likely (or, more rigorously, 'mean') value of the exponent. All we really need to do is to try to make the scaling factor insensitive to variations about this expected value. In order to differentiate between this interpretation and the previous one, we will call this a *local* reference stress method. We may then call the previous one a *global* reference stress method. In the numerical example considered in Section 3.1, we would like to make the grouping $\dot{\kappa}d/\dot{\epsilon}_0$ virtually constant over the range $n = 4.7 \pm 0.2$. One way of doing this – taking the limit as $n \to \infty$ – was the basis of the global approach. The local approach can be deduced in the following manner.

Denote by $\bar{n}$ the expected value of the material parameter n, and let us expand the scaling factor in a Maclaurin series about $\bar{n}$ i.e.

$$\delta(\sigma_0, n) = \delta(\sigma_0, \bar{n}) + (n - \bar{n}) \left.\frac{d\delta}{dn}\right|_{\bar{n}} + o((n - \bar{n})^2)$$

If we consider only *small* variations about $\bar{n}$, we may ignore terms of the order $(n - \bar{n})^2$. Now if we choose $\sigma_0 = \sigma_R$ so that

$$\left.\frac{d\delta}{dn}\right|_{\bar{n}} = 0 \tag{6.12}$$

then

$$\delta(\sigma_R, n) \simeq \delta(\sigma_R, \bar{n})$$

to second order.

We may consider equation (6.12) as a *defining* equation for the local reference stress. Let us apply it to our three examples:

For the beam in bending, if $\sigma_0 = \alpha M/bd^2$ as before, then

$$\delta(\sigma_0, n) = \frac{1}{d}\left(1 + \frac{1}{2n}\right)^n \frac{1}{\alpha^n}$$

$$\frac{d\delta}{dn} = \delta\left[-\ln\alpha + \ln\left(1 + \frac{1}{2n}\right) - \frac{1/2n}{1 + 1/2n}\right]$$

which vanishes at $n = \bar{n}$ provided

$$\alpha = \bar{\alpha} = \left(1 + \frac{1}{2\bar{n}}\right)\exp\left(-\frac{1}{2\bar{n} + 1}\right)$$

The scaling factor is then

$$\delta(\sigma_R, n) = \frac{1}{d}\left(\frac{2n + 1}{2n}\right)^n \frac{1}{\bar{\alpha}^n}$$

and if we plot variations of this with n for a given $\bar{n}$, we see that to a good approximation as intended (*Figure 6.19*)

$$\delta(\sigma_R, n) \simeq \delta(\sigma_R, \bar{n}) = \frac{1}{d}\left(\frac{2\bar{n} + 1}{2n}\right)^{\bar{n}} \frac{1}{\bar{\alpha}^{\bar{n}}}$$

in the vicinity of the mean $\bar{n}$.

Similarly, for the two-bar structure, the defining condition (6.12) is satisfied if

$$\sigma_R = \bar{\alpha}\,\frac{Q}{A} \qquad \bar{\alpha} = \exp\left\{\frac{(L_2/L_1)^{1/\bar{n}}\ln(L_2/L_1)^{1/\bar{n}}}{1 + (L_2/L_1)^{1/\bar{n}}} - \ln\left[1 + \left(\frac{L_2}{L_1}\right)^{1/\bar{n}}\right]\right\}$$

whence (*Figure 6.20*)

$$\delta(\sigma_R, n) = \frac{L_1 L_2}{(L_1^{1/n} + L_2^{1/n})^n}\,\frac{1}{\bar{\alpha}^n} \simeq \delta(\sigma_R, \bar{n})$$

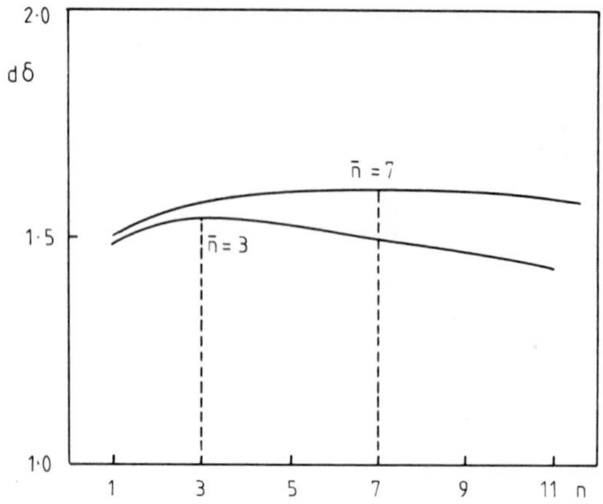

Figure 6.19 Variation of scaling factor using local reference stress method for beam in bending

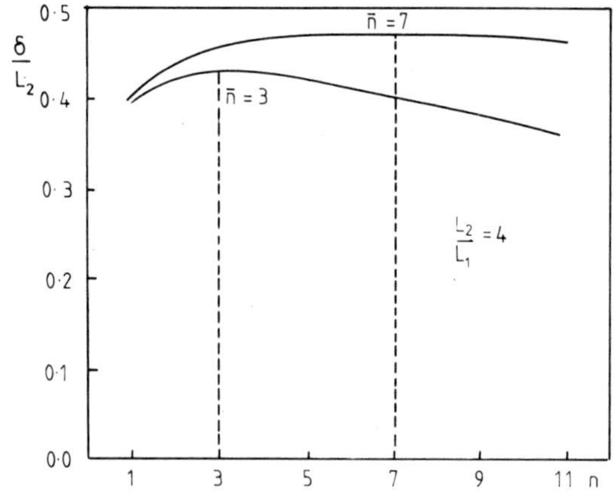

Figure 6.20 Variation of scaling factor using local reference stress method for two-bar structure

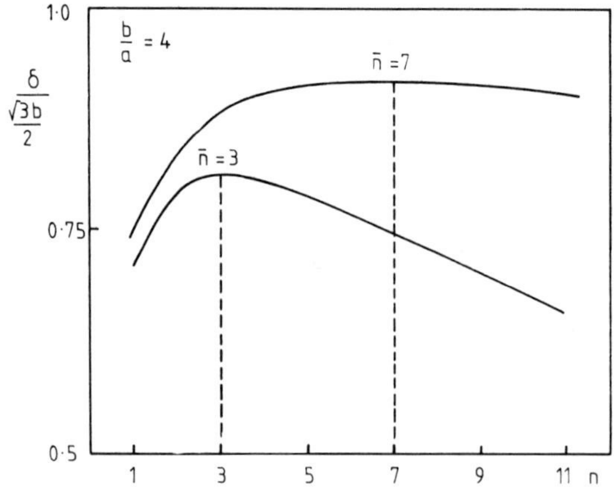

Figure 6.21 Variation of scaling factor using local reference stress method for thick cylinder

And for the thick cylinder

$$\sigma_R = \bar{\alpha}\,\frac{\sqrt{3}\,p}{2\ln(b/a)} \qquad \bar{\alpha} = \frac{(b/a)^{2/\bar{n}}\ln(b/a)^{2/\bar{n}}}{(b/a)^{2/\bar{n}}-1}\,\frac{1}{e}\,\exp\left(\frac{\ln(b/a)^{2/\bar{n}}}{(b/a)^{2/\bar{n}}-1}\right)$$

whence (*Figure 6.21*)

$$\delta(\sigma_R, n) = \frac{\sqrt{3}}{2}\,b\left(\frac{b}{a}\right)\left(\frac{2\ln(b/a)/\bar{\alpha}}{n[(b/a)^{2/n}-1]}\right)^n \simeq \delta(\sigma_R, \bar{n})$$

It can also be verified that these reference stresses are not too different from those derived using the global method. The advantage that the local approach has over the global can be seen if we examine those problems where the latter failed.

Consider the cantilever beam under a point load Q (*Figure 6.4*). The scaling factor (6.7) has been obtained as

$$\delta(\sigma_o, n) = \frac{1}{d}\,\frac{l^2}{n+2}\left(\frac{Ql}{\sigma_o bd^2}\right)^n\left(\frac{2n+1}{2n}\right)^n$$

It can be shown that the condition of equation (6.12) is satisfied if

$$\sigma_R = \bar{\alpha}\,\frac{Ql}{bd^2} \qquad \bar{\alpha} = \left(1+\frac{1}{2\bar{n}}\right)\exp\left(\frac{-1/2\bar{n}}{1+1/2\bar{n}} - \frac{1}{\bar{n}+2}\right)$$

and indeed (*Figure 6.22*) in the vicinity of $\bar{n}$

$$\delta(\sigma_R, n) = \frac{1}{d}\,\frac{l^2}{n+2}\left(\frac{2n+1}{2n}\right)^n\frac{1}{\bar{\alpha}^n} \simeq \delta(\sigma_R, \bar{n})$$

Therefore using the local approach the reference stress method is successful in this problem!

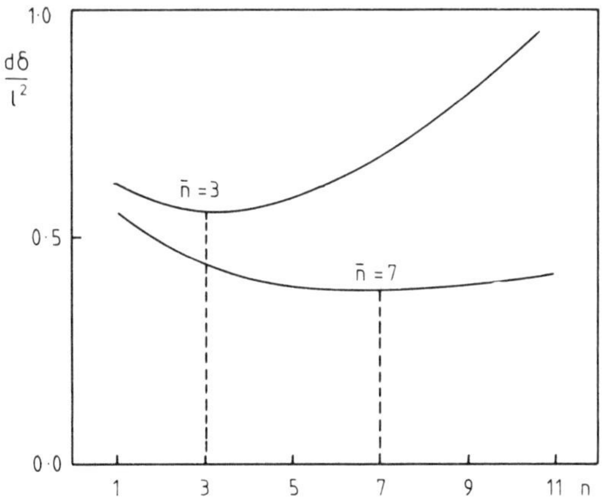

Figure 6.22 Variation of scaling factor using local reference stress method for cantilever beam

Turning to non-isothermal creep problems, we again find that the local approach is successful. For example, in the case of the two-bar structure, the defining condition (6.12) is satisfied if

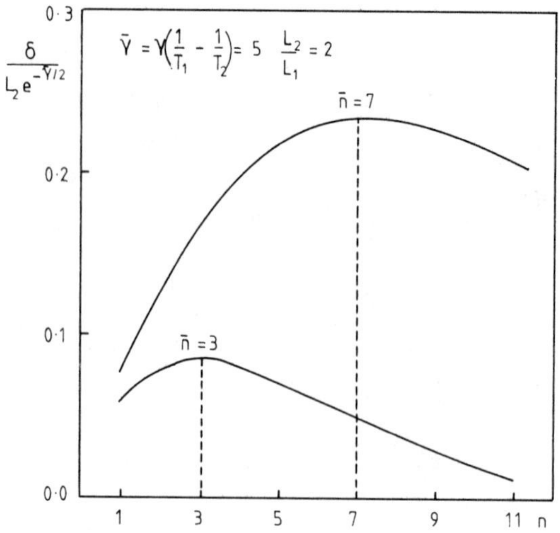

Figure 6.23 Variation of scaling factor using local reference stress method for non-uniformly heated two-bar structure

$$\sigma_R = \bar{\alpha}\,\frac{Q}{A} \qquad \bar{\alpha} = \exp\left(\frac{\eta^{1/\bar{n}}\ln \eta^{1/\bar{n}}}{1+\eta^{1/\bar{n}}} - \ln(1+\eta^{1/\bar{n}})\right)$$

$$\eta = \exp[\gamma(1/T_1 - 1/T_2]\ L_2/L_1$$

whence (*Figure 6.23*)

$$\delta(\sigma_R, n) = \frac{L_1 L_2}{(L_1^{1/n} + \{\exp[\gamma(1/T_1 - 1/T_2)][L_2\}^{1/n})^n} \simeq \delta(\sigma_R, \bar{n})$$

in the vicinity of $\bar{n}$.

The present approach can also be used when the constitutive relation is other than a power law – but is such that the scaling factor depends only on a *single* material parameter, e.g.

$$\dot{\epsilon}_0 = Bf(\sigma_0, p) \tag{6.13}$$

where f is a known functional form and B, p are material parameters. If we consider a characteristic displacement rate

$$\dot{u} = g(p)$$

then using the reference stress approach

$$\dot{u} = \delta(\sigma_0, p) \times \dot{\epsilon}_0 \tag{6.14}$$

where the scaling factor is defined by

$$\delta(\sigma_0, p) = g(p)/f(\sigma_0, p)$$

If $\bar{p}$ is the expected value of the material parameter p, then the defining condition for the local reference stress is

$$\left.\frac{d\delta}{dp}\right|_{\bar{p}} = 0 \tag{6.15}$$

that is $\sigma_0 = \sigma_R$ satisfies the nonlinear equation

$$g'(\bar{p})f(\sigma_R, \bar{p}) - g(\bar{p})f'(\sigma_R, \bar{p}) = 0 \tag{6.16}$$

where

$$(\quad)' = d(\quad)/dp$$

Then in the vicinity of $\bar{p}$

$$\delta(\sigma_R, p) \simeq \delta(\sigma_R, \bar{p})$$

As an example, consider the two-bar structure composed of a material which is described by Prandtl's law (Section 5.1)

$$\dot{\epsilon} = B \sinh(\gamma\sigma)$$

Then the deflection rate is

$$\dot{q} = B \sinh(\gamma Q/A) \frac{L_1 L_2}{[L_1^2 + 2L_1 L_2 \cosh(\gamma Q/A) + L_2^2]^{1/2}}$$

and the scaling factor is

$$\delta(\sigma_0, n) = \frac{\sinh(\gamma Q/A)}{\sinh(\gamma\sigma_0)} \frac{L_1 L_2}{[L_1^2 + 2L_1 L_2 \cosh(\gamma Q/A) + L_2^2]^{1/2}}$$

Performing the differentiation in the defining condition (6.15), we obtain the following equation for σ_R:

$$\sigma_R \coth(\bar{\gamma}\sigma_R) = \frac{Q}{A}\left[\coth \frac{\bar{\gamma}Q}{A} - \frac{L_1 L_2 \sinh(\bar{\gamma}Q/A)}{[L_1^2 + 2L_1 L_2 \cosh(\bar{\gamma}Q/A) + L_2^2]}\right]$$

This has no closed-form solution and should be solved numerically. Finally,

$$\delta(\sigma_R, \gamma) \simeq \delta(\sigma_R, \bar{\gamma})$$

in the vicinity of $\bar{\gamma}$ (*Figure 6.24*).

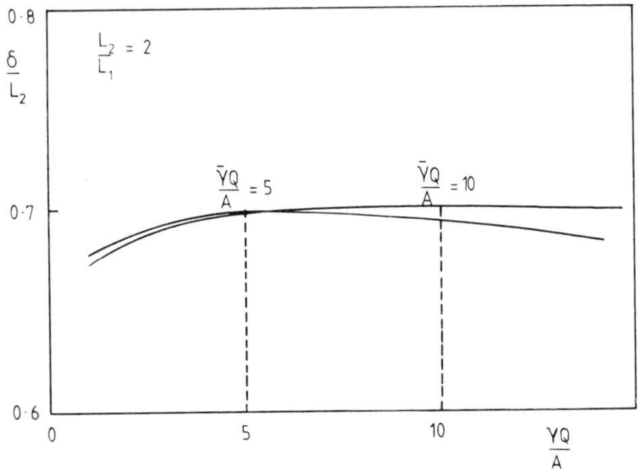

Figure 6.24 Variation of scaling factor for Prandtl's law

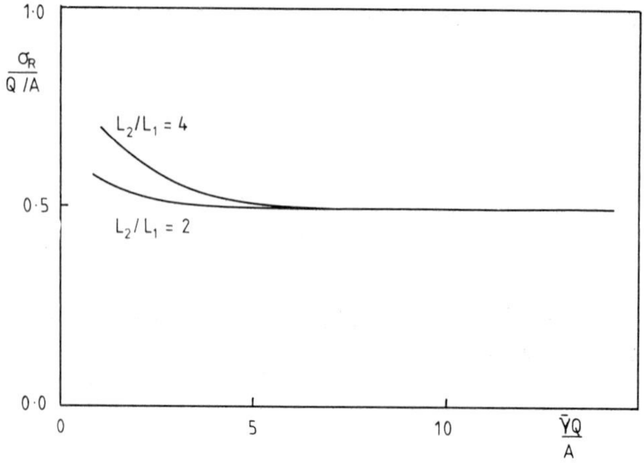

Figure 6.25 Variation of reference stress with $\bar{\gamma}Q/A$ for Prandtl's law

It should be noted here that we have not been able to write immediately $\sigma_R = \alpha Q/A$ for some scalar α; but if we plot σ_R against $\bar{\gamma}\, Q/A$ (*Figure 6.25*) it is seen that the relationship is approximately constant. Indeed $\sigma_R \simeq Q/2A$ as for the power law!

6.6 Approximate reference stress methods

There is one obvious disadvantage here – it is necessary to evaluate the derivative in the defining condition, equation (6.15), in a form suitable to solve the equation (6.16). Clearly, in general, this is not always possible. Can we do this approximately? The following approach is suggested.

If p_L and p_U are two values of the material parameter defining the range of expected values

$$p_L \leqslant \bar{p} \leqslant p_U$$

then we may approximate the derivative in equation (6.15) by

$$\left.\frac{d\delta}{dp}\right|_{\bar{p}} \simeq \frac{\delta_U - \delta_L}{p_U - p_L}$$

The defining condition is then replaced by

$$\delta_U = \delta_L \tag{6.17}$$

so that the reference stress approximation is

$$\delta(\sigma_R, p) \simeq \delta(\sigma_R, p_U) = \delta(\sigma_R, p_L) \tag{6.18}$$

This approximate local reference stress method can obviously be used in conjunction with numerical stress analysis where we evaluate the scaling factor for particular values of the material parameters (and are consequently called *point estimates*).

For example, consider the beam in bending with a power law; the condition, equation (6.17), leads to $\sigma_R = \alpha_R M/bd^2$

$$\left(1 + \frac{1}{2n_L}\right)^{n_L} \frac{1}{\alpha_R^{n_L}} = \left(1 + \frac{1}{2n_U}\right)^{n_U} \frac{1}{\alpha_R^{n_U}}$$

i.e.

$$\alpha_R = \left[\left(\frac{2n_U + 1}{2n_U}\right)\left(\frac{2n_L}{2n_L + 1}\right)\right]^{1/(n_U - n_L)}$$

A comparison between α_R calculated above and the previous 'exact' values given in Section 6.5, $\bar{\alpha}$, for various values of $\bar{n}$ and the range $\bar{n} \pm \Delta n$ is shown in *Figure 6.26*. The approximate technique clearly works very well. The reader should recall that we have used this technique before in Section 5.3 to develop

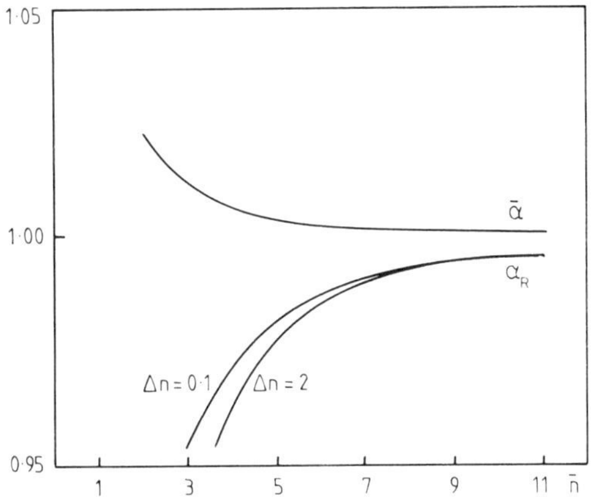

Figure 6.26 Comparison of exact and approximate reference parameters

a reference stress for the rotating disc. This method can also be applied to other constitutive relations with a single parameter. How we deal with multiparameter relations is dealt with in the final sections.

6.7 A general local reference stress method

The local reference stress approach so far applies only to constitutive relations which give a scaling factor dependent on a single material parameter. The question is: what can be done about multiparameter relations? To deal with these, we need to adopt a wider interpretation of the aims of the method – but one that is still consistent with our original motivation to overcome difficulties which arise due to the uncertainty about the actual values of the material parameters. We must return to fundamentals.

The prime cause of the uncertainty is the inherent scatter in the basic material data, coupled with the fact that the functional form of the constitutive relation is merely phenomenological. Owing to scatter, the material parameters for a particular functional form are *random variables*. In particular, any *estimates* of these parameters for a set of scattered creep data are also random variables.

Consider a material creep law for steady creep given by

$$\dot{\epsilon} = f(\sigma_0, p_1, p_2, \ldots)$$

where $p_1, p_2, \ldots$ are the material parameters. As usual, we want to evaluate some characteristic displacement rate

$$\dot{u} = g(p_1, p_2, \ldots)$$

using a reference stress approach by defining the scaling factor

$$\dot{u} = \delta \times \dot{\epsilon}_0 \tag{6.19}$$

where for arbitrary σ_0

$$\delta(\sigma_0, p_1, p_2 \ldots) = g(p_1, p_2, \ldots)/f(\sigma_0, p_1, p_2, \ldots)$$

The scaling factor is a function of the random variables $p_1, p_2, \ldots$ and is therefore itself a random variable. In the local reference stress method discussed above, we attempted to choose the reference stress $\sigma_o = \sigma_R$ such that variations in the scaling factor arising from variations in the (single) material parameter were small. Possible variations in the scaling factor were examined by expanding in a Maclaurin series. How do we measure variations in the scaling factor for a multi-parameter law? The introduction of random variables simplifies this, and in fact leads to a self-consistent interpretation of the method. The variation of the scaling factor can be measured using the *statistical variance* of the scaling factor, denoted by $V[\delta]$. (Here we recommend the reader to any good text on elementary probability and statistics.)

The reference stress is chosen such that

$$V[\delta(\sigma_R)] = \text{minimum} \tag{6.20}$$

It is usually possible to choose this so that the variance vanishes, i.e.

$$V[\delta(\sigma_R)] = 0 \tag{6.21}$$

This being the case, from Chebychev's inequality

$$\text{Prob}\ [\delta = E[\delta]\ ] = 1$$

or, informally

$$\delta = E[\delta]$$

being the statistical *expected value* of the scaling factor.

So with this choice of reference stress

$$\dot{u} = E[\delta] \times \dot{\epsilon}_R \tag{6.22}$$

and since $E[\delta]$ is determined – the reference stress test contains *all* the effects of the uncertainties in the material data.

The local approach for a single parameter is contained in this interpretation as a special case. Let p be the material parameter and let $\bar{p}$ and ν_p^2 be its statistical mean and variance. If the latter is small, then

$$E[\delta] \simeq \delta(\sigma_0, \bar{p}) + \left.\frac{d^2\delta}{dp^2}\right|_{\bar{p}} \frac{\nu_p^2}{2}$$

$$V[\delta] \simeq \left.\frac{d\delta}{dp}\right|_{\bar{p}} \nu_p^2$$

and the condition (6.21) reduces to the previous condition (6.15). The approximate scaling factor used is, however, slightly different.

6.8 Further approximations

In order to evaluate the expected value and variance of the scaling factor, the *statistical distributions* of the material parameters should be available. This is usually not the case. At best, variance and expected values of the material parameters can be estimated. Thus, given a means of calculating point estimates of the characteristic displacement for any set of material parameters, can we estimate the required variance and expected values of the scaling factor? We present here two such approximations.

For a scaling factor dependent on a single parameter p whose mean is $\bar{p}$ and variance ν_p^2, then to first order

$$E[\delta] \simeq \frac{1}{2}(\delta_1 + \delta_2)$$
$$V[\delta] \simeq \frac{1}{4}(\delta_1 - \delta_2)^2 \tag{6.23}$$

where

$$\delta_1 = \delta(\bar{p} + \nu_p) \qquad \delta_2 = \delta(\bar{p} - \nu_p)$$

It is seen that the condition (6.21) reduces to that of equation (6.18).

For a scaling factor dependent on two parameters p_1 and p_2 with means $\bar{p}_1$ and $\bar{p}_2$, variances $\nu_1{}^2$ and $\nu_2{}^2$ and correlation coefficient ρ, then also to first order

$$E[\delta] \simeq \frac{1}{4}(\delta_1 + \delta_2 + \delta_3 + \delta_4) + \frac{1}{4}\rho(\delta_1 - \delta_2 + \delta_3 - \delta_4) \tag{6.24}$$

$$V[\delta] \simeq c_1\delta_1^2 + c_2\delta_2^2 + c_3\delta_3^2 + c_4\delta_4^2 - (c_1\delta_1 + c_2\delta_2 + c_3\delta_3 + c_4\delta_4)^2$$

where

$$\delta_1 = \delta(\bar{p}_1 - \nu_1, \bar{p}_2 - \nu_2) \qquad \delta_2 = \delta(\bar{p}_1 + \nu_1, \bar{p}_2 - \nu_2)$$

$$\delta_3 = \delta(\bar{p}_1 + \nu_1, \bar{p}_2 + \nu_2) \qquad \delta_4 = \delta(\bar{p}_1 - \nu_1, \bar{p}_2 + \nu_2)$$

$$c_1 = c_3 = (1 + \rho)/4 \qquad c_2 = c_4 = (1 - \rho)/4$$

Similar expressions can be established for any number of material parameters.

The defining conditions for a reference stress, equation (6.20), can be approximately satisfied by minimising either equations (6.23) or (6.24) as appropriate. The corresponding expected values can then be computed.

TABLE 6.1. Parameter estimation.
For 304SS (Heat 924296), at 593° C, σ (psi), time (h).

		Estimated parameters	
Norton	$\dot{\epsilon} = B_N \sigma^n$	B_N	2.8291×10^{-26}
		$\bar{n}$	4.6875
		ν_n	0.1569
Prandtl	$\dot{\epsilon} = B_P \sinh(\alpha\sigma)$	B_P	2.4162×10^{-8}
		$\bar{\alpha}$	2.955×10^{-4}
		ν_α	0.9×10^{-5}
Garofalo	$\dot{\epsilon} = B_G[\sinh(\gamma\sigma)]^m$	B_G	2.4027×10^{-7}
		γ	8.4077×10^{-5}
		m	3.0295
		ν_γ	37×10^{-5}
		ν_m	0.85
		ρ	−0.98

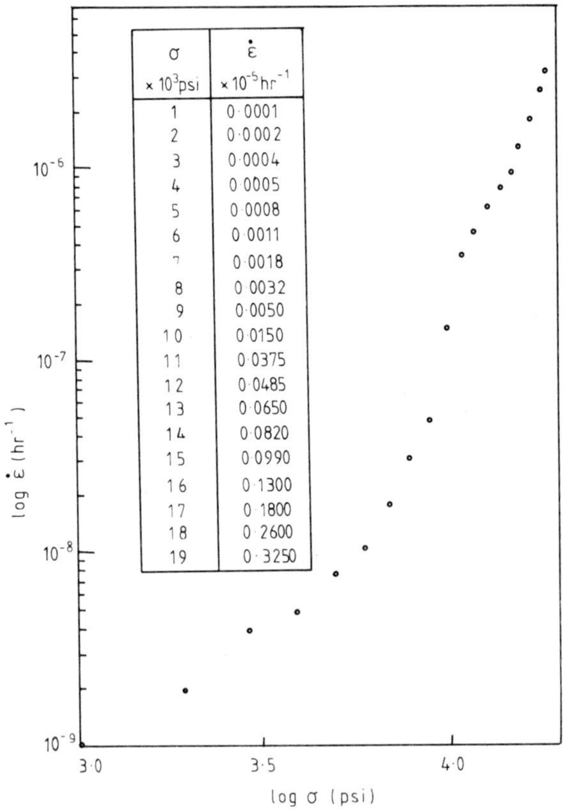

Figure 6.27 Steady-state creep data for 304SS (see Corum, J.M., Material property data for elastic–plastic–creep analyses of benchmark problems, in *Pressure Vessels and Piping: Verification and Qualification of Inelastic Analysis Computer Programs*, ASME Special Publ., 1975)

By way of an example, we consider a set of material data for 304SS to which we have estimated the material parameters (*Table 6.1*) for the Norton, Prandtl and Garofalo steady creep laws, given in *Figure 6.27* using a standard statistical package. The reference stresses and scaling factors for our three examples – two-bar structure, beam in bending and thick cylinder – have been evaluated for specific numerical cases and are shown in *Tables 6.2–6.5*. These estimations are clearly very close.

TABLE 6.2. Reference stress and scaling factors for two-bar structure.

L_1 = 3 in, L_2 = 1.5 in, A = 0.015 in², Q = 250 lbf.

	σ_R (psi)	δ (in)
Norton	8356	2.07
Prandtl	8364	2.10
Garofalo	8321	2.11

TABLE 6.3. Reference stress and scaling factors for two-bar structure.

L_1 = 12 in, L_2 = 48 in, A = 2 in², Q = 50000 lbf.

	σ_R (psi)	δ (in)
Norton	12636	21.68
Prandtl	12517	23.85
Garofalo	12246	25.20

TABLE 6.4. Reference stress and scaling factors for beam.

b = 1 in, d = 1 in, M = 10 000 lb in.

	σ_R (psi)	δ (in^{-1})
Norton	10 047	1.57
Prandtl	10 067	1.61
Garofalo	9 962	1.63

TABLE 6.5. Reference stress and scaling factors for thick cylinder.

a = 6 in, b = 9 in, p = 10 000 psi.

	σ_R (psi)	δ (in)
Norton	21 386	7.70
Prandtl	21 358	7.80
Garofalo	21 400	7.70

6.9 Summary

The reference stress method is attractive and has some important consequences for design. Although originally concerned with the reduction of the effect of uncertainties in estimated parameters (and this is how it has been treated here), it has alternatively acquired the status of a simplified method. It is useful to summarise the pertinent aspects of the method in this context.

It can readily be verified that the global, limit-type reference stress for a power law (Section 6.1) is a good approximation to the local reference stress (Section 6.5 *et seq.*) in all the examples we have considered. As we have seen in the last section, this statement is largely true *no matter what the assumed material model*. The importance of this observation is that the limit-type reference stress can then be evaluated from an equivalent *plastic* analysis, *without recourse to any creep analysis* (equation (6.6) and Section 6.2). Of course, the reference stress alone is insufficient to predict some characteristic deflection rate – the result of the reference stress test must be multiplied by some scaling factor. The appropriate limiting scaling factor is developed in Section 6.2; *alternatively, the scaling factor can be well approximated by assuming a power law and then using its value for an estimated value of n.* This, of course, requires a creep analysis, but we still achieve our original aim of reducing the sensitivity of the component behaviour to uncertainty in the material model. However, *the method has interesting consequences for design if the scaling factor can be approximated without recourse to a creep analysis.* Two possibilities have been suggested. The limiting scaling factor has an upper bound which can be evaluated from knowledge of the reference stress alone (Section 6.2). Alternatively, the scaling factor can be approximated by its value for $n = 1$ (corresponding to incompressible elasticity); in some cases this gives a lower bound, but this is not guaranteed for combined loading. Examination of the examples provided here suggests that these estimates can often be poor. It is then up to the designer to choose whether or not these simplifications can be used. For the time being, little guidance can be given, but caution is advised.

Further reading

Anderson, R.G., *et al.*, Deformation of uniformly loaded beams obeying complex creep laws, *J. Mech. Eng. Sci.*, **5**, 238–44, 1963.

Mackenzie, A.C., On the use of a single uniaxial test to estimate deformation rates in some structures undergoing creep, *Int. J. Mech. Sci*, **10**, 441–53, 1968.

Sim, R.G., Reference stress concepts in the analysis of structures during creep, *Int. J. Mech. Sci*, **12**, 561–73, 1970.

Johnsson. A., An alternative definition of reference stress for creep, *Proc. I. Mech. E. Conf on Creep and Fatigue in Elevated Temperature Applications*, I. Mech. E., 1974, Paper C205/73.

Johnsson, A., Reference stress for structures obeying the Prandtl and Dorn creep laws, *J. Mech. Eng. Sci*, **16**, 298–305, 1974.

Boyle, J.T., Approximations in the reference stress method for creep design, *Int. J. Mech. Sci*, **22**, 73–82, 1980.

Meyer, P.L., *Introductory Probability and Statistical Applications*, Addison–Wesley, 1966.

Rosenblueth, E., Point estimates for probability moments, *Proc. Natl Acad. Sci.*, **72**, 3812–14, 1975.

Chapter 7

Stress analysis for transient creep

We have spent some time describing methods for steady-state creep analysis, since these are the most frequently used calculation techniques for creep. They form the basis for a number of approximate methods which we will develop in the next chapter. Nevertheless, it may be necessary to know with more accuracy the effects of including stress redistribution in our calculations. This is not so critical for steadily loaded structures, but assumes more importance for variable loading, or if it is thought that a structure is likely to fail.

In spite of the above remarks, one crucial point still needs to be made. Apart from the simplest of examples, wherein a closed-form solution is available, for steady loading if some form of numerical solution is necessary it is usually *easier* to perform the transient analysis evaluating the stress redistribution from the initial elastic to the steady state. Thus extra information concerning the performance of the structure is gained at no extra cost. The reason for this can be found in the nature of the mathematical problems which arise. As we have seen, steady creep gives rise to nonlinear boundary value problems which must be solved iteratively from an initial guessed solution. This may be time-consuming both in preparation of the problem for solution and in computing time. On the other hand, a transient elastic to steady-state analysis gives rise, as we shall see, to initial value problems. These require less problem preparation and little extra computing time and are automatic – i.e. no initial guessed solution is necessary. As we have mentioned before, the methods of solution that we have developed for steady creep are of no use for transient creep. We must start afresh.

Three topics will be discussed in this chpater. First we will demonstrate to what type of equations the transient creep problem leads. Secondly, numerical methods for the solution of the resulting equations will be described. Finally, we will examine how these techniques are incorporated into the most common computer stress analyses – finite elements. Most of the examples that we present will be concerned with steady loading. Nevertheless the methods can be applied to variable loading; the only limitation is the size of the computer, time and resources. What it does not require, however, is imagination.

7.1 Example: Forward creep of a pin-jointed framework

We have already examined the forward creep of a bar structure in Section 3.3. It will not hurt to look at another. Consider the pin-jointed framework shown in

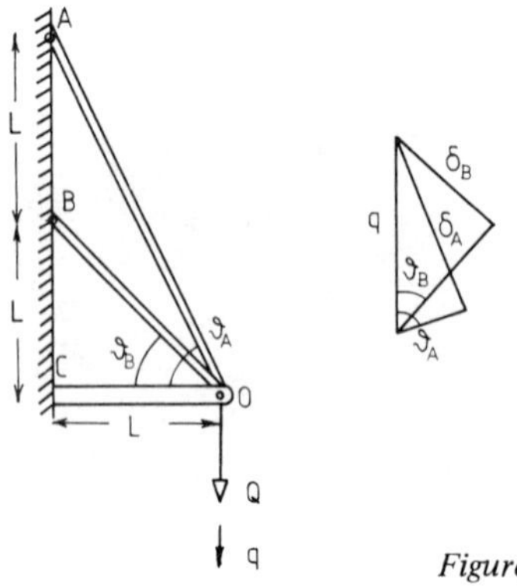

Figure 7.1 Geometry of pin-jointed framework

Figure 7.1. The bars of the framework are such that one end of each bar is attached to a rigid wall length L apart. The material of the bars OA and OB, which have the same cross-sectional area A, is subject to creep, while the bar OC is rigid. A vertical load Q is prescribed at O and the aim is to determine the resultant vertical displacement q. Let δ_A, δ_B be the extensions of the creeping bars (*Figure 7.1*). Then their strains are

$$\epsilon_A = \frac{\delta_A}{\sqrt{5}L} \qquad \epsilon_B = \frac{\delta_B}{\sqrt{2}L}$$

and from simple geometry the compatibility relation is

$$q = \frac{\sqrt{5}}{2}\delta_A = \sqrt{2}\delta_B$$

Finally if σ_A, σ_B are the stresses in the bars, the equilibrium equation is

$$\frac{2}{\sqrt{5}}\sigma_A + \frac{1}{\sqrt{2}}\sigma_B = \frac{Q}{A}$$

We will assume for simplicity that the constitutive relations take the form of a simple time hardening power law as in Section 2.3,

$$\dot{\epsilon}_A = \frac{\dot{\sigma}_A}{E} + g(t)\sigma_A^n \qquad \dot{\epsilon}_B = \frac{\dot{\sigma}_B}{E} + g(t)\sigma_B^n$$

where E is Young's modulus, and $g(t)$ and n are material parameters. For clarity, set

$$S_A = \frac{\sigma_A}{Q/A} \qquad S_B = \frac{\sigma_B}{Q/A}$$

$$G(t) = E(Q/A)^{n-1}g(t)$$

Then differential equations for S_A and S_B are obtained by substituting the constitutive relations into compatibility and using equilibrium, whence

$$\frac{dS_A}{dt} = \frac{G(t)}{5/4 + 2\sqrt{(2/5)}}\left(S_B^n - \frac{5}{4}S_A^n\right) \tag{7.1}$$

$$\frac{dS_B}{dt} = \frac{-2\sqrt{(2/5)}G(t)}{5/4 + 2\sqrt{(2/5)}}\left(S_B^n - \frac{5}{4}S_A^n\right) \tag{7.1}$$

Initially the fully linear elastic solution must prevail, which can be found as

$$S_A(0) = \frac{1}{2/\sqrt{5} + (5/4)/\sqrt{2}} \qquad S_B(0) = \frac{5/4}{2/\sqrt{5} + (5/4)/\sqrt{2}} \tag{7.2}$$

The first-order differential equation (7.1) together with the initial condition (7.2) constitute an *initial value problem* in time for the variables S_A and S_B. (The reader will no doubt notice that these can be reduced to a single equation in S_A using the equilibrium equations to eliminate S_B – this has no particular advantage.)

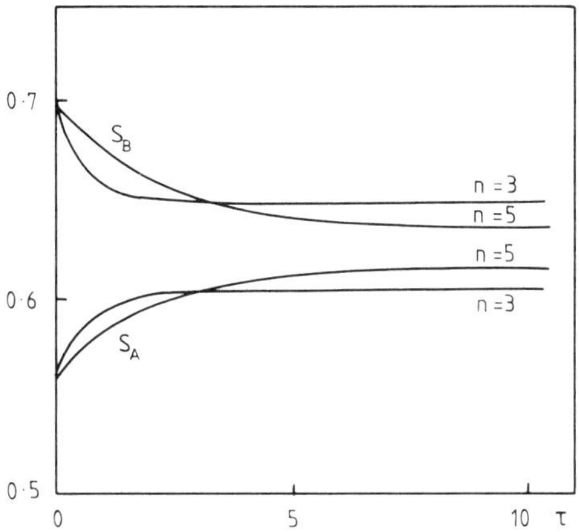

Figure 7.2 Redistribution of normalised stresses

These equations must be solved numerically – we will discuss this in detail in the following sections. For the time being, we can readily apply a standard scientific subroutine package from a computer library for the solution of initial value problems. In *Figure 7.2* we plot the variation of S_A and S_B for various values of the exponent n in the normalised time scale

$$\tau = \int G(t)\, dt$$

which leads to the equations

$$\frac{dS_A}{d\tau} = a\left(S_B^n - \frac{5}{4}S_A^n\right)$$

$$\frac{dS_B}{d\tau} = b\left(S_B^n - \frac{5}{4}S_A^n\right)$$

where we have set

$$a = \frac{1}{5/4 + 2\sqrt{(2/5)}} \qquad b = \frac{-2\sqrt{(2/5)}}{5/4 + 2\sqrt{(2/5)}}$$

It can be seen that the normalised stress S_A increases to a constant value, whilst S_B decreases to a constant, both from the elastic solution. These final constant values are, of course, the steady-state stresses

$$S_B = (\sqrt{5}/2)[(4/5)^{1/n} + \sqrt{10}/4]^{-1} \qquad S_A = (\sqrt{5}/2)(1 - S_B/\sqrt{2})$$

The variation in time is called stress redistribution.

7.2 Example: Forward creep of a thin tube in bending

As a second example, consider a long thin constant-thickness circular cross-section cylindrical shell of mean radius r, thickness h (*Figure 7.3*). It is supposed that $h << r$, so that the radial stress components are negligibly small,

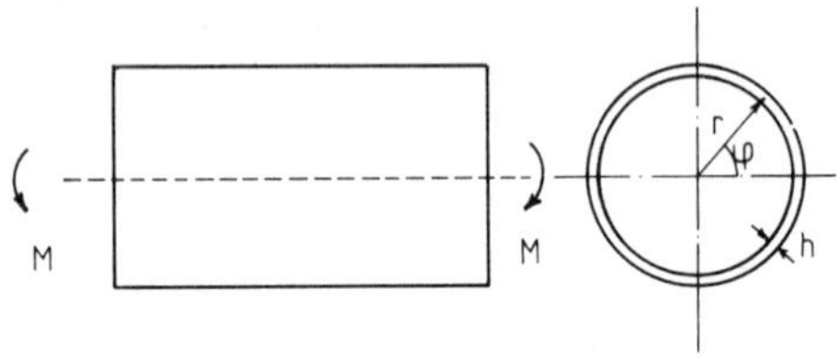

Figure 7.3 Geometry of thin tube under bending

with the remaining longitudinal stress $\sigma(\varphi, t)$ being constant through the thickness. We are therefore assuming simple beam bending. The pipe is subject to a bending moment M which is kept fixed as the pipe creeps. The basic field equations are as follows.

Equilibrium with external moment:

$$M = hr^2 \int_0^{2\pi} \sigma \sin\varphi \, d\varphi$$

Compatibility of longitudinal strain ϵ with curvature κ:

$$\epsilon = \kappa r \sin\varphi$$

Constitutive relations:

$$\epsilon = \epsilon_E + \epsilon_C$$

$$\epsilon_E = \sigma/E \qquad \dot{\epsilon}_C = g(t)\sigma^n$$

As in the previous example, we derive a differential equation for the stress. First differentiate all equations with respect to time

$$\int_0^{2\pi} \dot{\sigma} \sin\varphi \, d\varphi = 0 \qquad \dot{\epsilon} = \dot{\kappa}\, r \sin\varphi \qquad \dot{\epsilon} = \dot{\sigma}/E + \dot{\epsilon}_C$$

From the constitutive relation we have

$$\dot{\sigma} = E(\dot{\epsilon} - \dot{\epsilon}_C)$$

and on substituting into equilibrium and using strain–displacement

$$\dot{\sigma} = E\left(\frac{\sin\varphi}{\pi} \int_0^{2\pi} \dot{\epsilon}_C \sin\varphi - \dot{\epsilon}_C\right)$$

Finally, from the creep law there results the differential equation

$$\frac{d\sigma}{dt} = Eg(t)\left(\frac{\sin\varphi}{\pi}\int_0^{2\pi}\sigma^n \sin\varphi \, d\varphi - \sigma^n\right) \tag{7.3}$$

which, together with the initial elastic solution

$$\sigma(\varphi, 0) = \frac{M}{hr^2}\,\frac{\sin\varphi}{\pi} \tag{7.4}$$

again form an *initial value problem* for the stress. However, it is to be recalled that the stress is a function of the angle φ; then equation (7.3) is called an 'integro-differential' equation. It must be solved numerically. The best way to do this is to reduce the equation to a finite system of first-order differential equations by examining the stresses only at a finite number of points

$$\varphi_k,\ k = 1, 2, \ldots, M \qquad \varphi_1 = 0 \qquad \varphi_M = 2\pi$$

and replacing the integral by a finite sum using some approximate integration rule, e.g. Simpson's rule. Then equations (7.3) and (7.4) become

$$\frac{d\sigma_k}{dt} = Eg(t)\left(\frac{\sin\varphi_k}{\pi}\sum_{j=1}^{M} a_j\,\sigma_j^n - \sigma_k^n\right)$$

$$\sigma_k(0) = \frac{M}{hr^2}\,\frac{\sin\varphi_k}{\pi} \qquad k = 1, 2, \ldots, M \tag{7.5}$$

where $\sigma_k = \sigma(\varphi_k, t)$ and a_k, $k = 1, 2, \ldots, M$, are constants which specify the integration rule. For example, with Simpson's rule

$$\varphi_k = 2\pi\frac{k-1}{M-1} \qquad \Delta\varphi = \frac{2\pi}{M-1}$$

$$\left\{a_k\right\} = \frac{\Delta\varphi}{3}\left\{1 \quad 2 \quad 4 \quad 2 \quad 4 \quad 2 \quad \ldots \quad 4 \quad 2 \quad 1\right\}$$

The system of equations can now be resolved using some standard routine from a computer library. In *Figure 7.4* we plot the variation of the stress at several points around the circumference of the pipe with time for various values of n. For convenience only, the stresses have been normalised as $S = \sigma/\sigma_0$ where $\sigma_0 = M/\pi hr^2$ in the time scale

$$\tau = E\sigma_0^{n-1}\int g(t)\,dt$$

Once again, redistribution of the stresses from the initial elastic through to the steady state is observed. (The steady state was evaluated in Section 3.1.) One interesting phenomenon can be observed if we plot the distribution of stress around the cross-section for various instants of time (*Figure 7.5*). It can be seen that there exists a point on the cross-section where the stress distribution remains

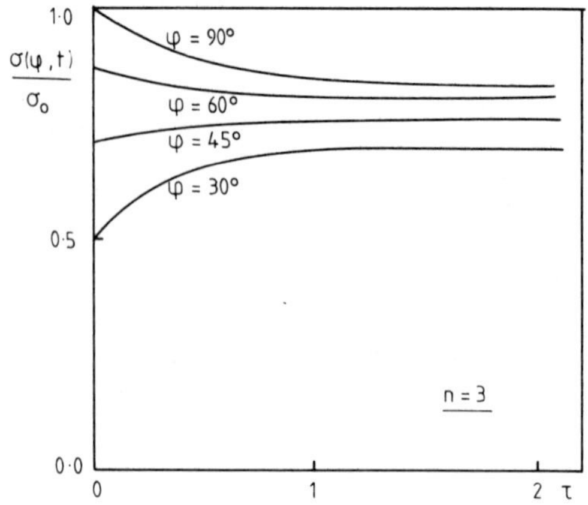

Figure 7.4 Redistribution of normalised longitudinal stresses

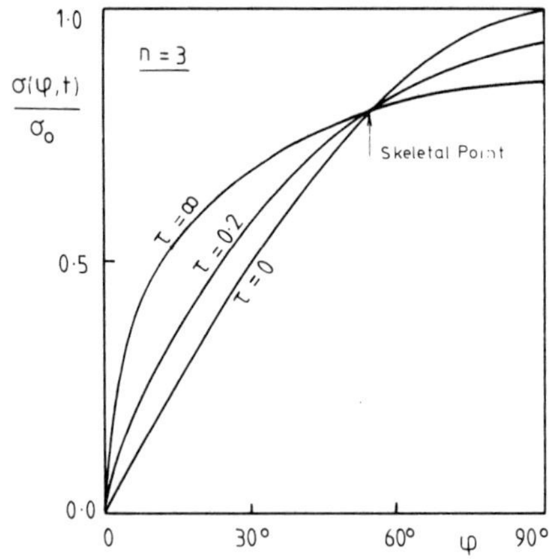

Figure 7.5 Variation of longitudinal stress distribution with time

virtually constant in time. It is given as that point where the elastic and steady-state stresses coincide and is obtained as

$$\varphi_S = \sin^{-1}\left[\left(\frac{\pi}{D_n}\right)^{n/(n-1)}\right]$$

This is known as the *skeletal point* and can be observed in many structures during stress redistribution.

7.3 The general equations of stress redistribution

We have seen in these two examples that transient creep can be represented by an initial value problem for the stresses. The numerical solution of transient creep problems is wholly concerned with this initial value problem. The question is – how do we construct the initial value problem in the general case?

7.3.1 The boundary value problem for transient creep

We begin by formulating the general boundary value problem for transient creep as in Section 4.7.

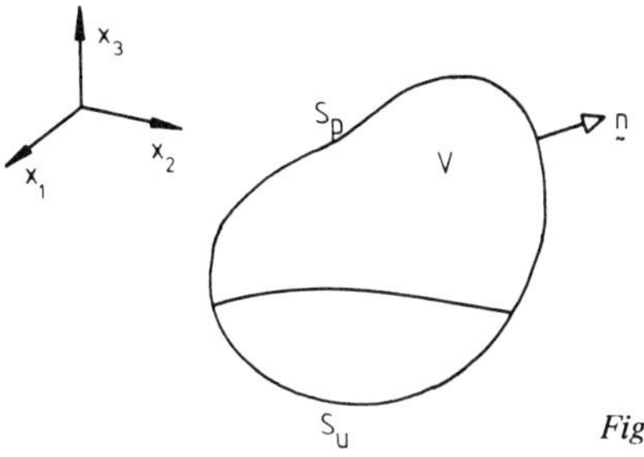

Figure 7.6 Geometry of a continuous body

Consider a solid body occupying a volume V bounded by a surface S (*Figure 7.6*). Denote the stress and strain fields in the body by the tensors σ_{ij}, ϵ_{ij} and the displacement field by u_i. The basic field equations are then

Equilibrium:

$$\sigma_{ij,j} + \bar{b}_i = 0$$

Strain-displacement:

$$\epsilon_{ij} = \frac{1}{2}(u_{i,j} + u_{j,i})$$

where $\bar{b}_i$ are applied body forces.

It is assumed that over part of the surface S_p tractions are prescribed, while over the remainder S_u displacements are imposed; the boundary conditions are then

$$u_i = \bar{u}_i \qquad \text{on } S_u$$

$$\sigma_{ij}n_j = p_i = \bar{p}_i \qquad \text{on } S_p$$

The field equations and boundary conditions are completed by the constitutive relations. For clarity, it is assumed that the strain ϵ_{ij} can be decomposed into an elastic ϵ_{ij}^E and a creep part ϵ_{ij}^C

$$\epsilon_{ij} = \epsilon_{ij}^E + \epsilon_{ij}^C + \epsilon_{ij}^T$$

where ϵ_{ij}^T is a thermal strain commonly given by

$$\epsilon_{ij}^T = \alpha_T \Delta T$$

where α_T is the coefficient of thermal conductivity and ΔT the increase in absolute temperature. The elastic strain is related to stress through Hooke's law

$$\epsilon_{ij}^E = C_{ijkl}\sigma_{kl}$$

in tensorial form, where C_{ijkl} is the fourth-order tensor of elastic compliance.

For simplicity, the creep strain rate is assumed to be derivable from a potential function of stress using a time hardening theory as a specialisation of Section 4.1 (equation (4.8))

$$\dot{\epsilon}_{ij}^{C} = g(t)\,\frac{\partial \Omega}{\partial \sigma_{ij}}$$

We will write this relationship in the form

$$\dot{\epsilon}_{ij}^{C} = f_{ij}\,(\sigma_{kl},\ t)$$

for convenience.

In general, the applied forces are functions of time, i.e.

$$\bar{b}_i = \bar{b}_i(t) \qquad \bar{p}_i = \bar{p}_i(t) \qquad \bar{u}_i = \bar{u}_i(t)$$

as may also be the temperature

$$T = T(t)$$

This being so, we say that the structure has a mechanical or thermal loading *history*. Otherwise, it is steadily loaded; we consider this situation first.

7.3.2 Derivation of an initial value problem for constant loads

Assume that the applied mechanical and thermal loadings are constant in time and consider the rate form of the field equations.

Equilibrium:

$$\dot{\sigma}_{ij,j} = 0$$

Strain–displacement:

$$\dot{\epsilon}_{ij} = \frac{1}{2}(\dot{u}_{i,j} + \dot{u}_{j,i})$$

Boundary conditions:

$$\dot{u}_i = 0 \qquad \text{on } S_{\mathrm{u}}$$

$$\dot{p}_i = 0 \qquad \text{on } S_{\mathrm{p}}$$

Constitutive:

$$\dot{\epsilon}_{ij} = C_{ijkl}\,\dot{\sigma}_{kl} + \dot{\epsilon}_{ij}^{C}$$

If we think of the creep strain rate $\dot{\epsilon}_{ij}^{C}$ as being given, then the above comprises an *elastic* problem for the field quantities $\dot{\sigma}_{ij}$, $\dot{\epsilon}_{ij}$ and $\dot{u}_i$. We may formally write

$$\dot{\sigma}_{ij} = R_{ij}(\dot{\epsilon}_{kl}^{C})$$

where the linear function R_{ij} is obtained by solving this elastic problem. Then we have immediately from the constitutive relation

$$\frac{\mathrm{d}\sigma_{ij}}{\mathrm{d}t} = R_{ij}(f_{kl}(\sigma_{mn},t)) \qquad (7.6)$$

which, together with the initial linear elastic stress field σ_{ij}^{o},

$$\sigma_{ij}(0) = \sigma_{ij}^{\mathrm{o}} \tag{7.7}$$

forms an *initial value problem* for the stress field σ_{ij}.

The equation which defines this initial value problem, equation (7.6), we will call the *equation of stress redistribution*. It is the general form for the two examples we considered, i.e. equation (7.1) for the two-bar structure and equation (7.3) for the thin tube in bending. In those examples, the relationship between the stress rate and the creep strain rate treated as an initial strain given by the linear function R_{ij} and defined by the solution of an elastic problem is readily apparent. Of course, in general, such closed-form expressions cannot be obtained – but the relationship can be constructed by simply solving the required elastic problem.

7.3.3 General form of the equation of stress redistribution

The preceding result can be extended to deal with histories of loading in the following manner. We define *two* elastic problems.

The *equivalent elastic problem* for the loading $\bar{b}_i$, $\bar{p}_i$, $\bar{u}_i$ and thermal strain $\epsilon_{ij}^{\mathrm{T}}$ is given by a solution of the equations

Equilibrium:

$$\sigma_{ij,j}^{\mathrm{o}} + \bar{b}_i = 0$$

Strain–displacement:

$$\epsilon_{ij}^{\mathrm{o}} = \frac{1}{2}(u_{i,j}^{\mathrm{o}} + u_{j,i}^{\mathrm{o}})$$

Boundary conditions:

$$u_i^{\mathrm{o}} = \bar{u}_i \qquad \text{on } S_{\mathrm{u}}$$

$$p_i^{\mathrm{o}} = \bar{p}_i \qquad \text{on } S_{\mathrm{p}}$$

Constitutive:

$$\epsilon_{ij}^{\mathrm{o}} = C_{ijkl}\,\sigma_{kl}^{\mathrm{o}} + \epsilon_{ij}^{\mathrm{T}}$$

The *residual elastic problem* for an 'initial' strain field ϵ_{ij}^{*} is given by the solution of the equations

Equilibrium:

$$\sigma_{i,j}^{\mathrm{R}} = 0$$

Strain–displacement:

$$\epsilon_{ij}^{\mathrm{R}} = \frac{1}{2}(u_{i,j}^{\mathrm{R}} + u_{j,i}^{\mathrm{R}})$$

Boundary conditions:

$$u_i^{\mathrm{R}} = 0 \qquad \text{on } S_{\mathrm{u}}$$

$$p_i^{\mathrm{R}} = 0 \qquad \text{on } S_{\mathrm{p}}$$

Constitutive:

$$\epsilon_{ij}^{R} = C_{ijkl}\,\sigma_{kl}^{R} + \epsilon_{ij}^{*}$$

and we *define* the relationship

$$\sigma_{ij}^{R} = R_{ij}(\epsilon_{kl}^{*})$$

Then from examination of the rate of change of the basic equations, we have the identity

$$\dot{\sigma}_{ij} = \dot{\sigma}_{ij}^{o} + R_{ij}\,(\dot{\epsilon}_{kl}^{C})$$

and consequently the initial value problem

$$\frac{d\sigma_{ij}}{dt} = R_{ij}\,(f_{kl}\,(\sigma_{mn},\,t)) + \dot{\sigma}_{ij}^{o} \tag{7.8}$$

together with the initial elastic solution

$$\sigma_{ij}(0) = \sigma_{ij}^{o}\,(0) \tag{7.9}$$

This reduces to the previous equations (7.6) and (7.7) for constant loads where the equivalent elastic problem is constant in time.

The linear tensor-valued function R_{ij} is the same as that developed for constant loads. It may be algebraic, Section 7.1, integral, Section 7.2, differential or functions of these. As before, it is defined by the solution of an elastic problem – which we call here the residual elastic problem. We adopt this terminology since, rather than derive an equation for the actual stress, we could have similarly derived an equation for the *residual stress* (cf. Section 3.4)

$$\rho_{ij} = \sigma_{ij} - \sigma_{ij}^{o}$$

so that equations (7.8) and (7.9) can be written

$$\frac{d\rho_{ij}}{dt} = R_{ij}(f_{kl}(\sigma_{mn}^{o} + \rho_{mn},\,t)) \tag{7.10}$$

$$\rho_{ij}(0) = 0 \tag{7.11}$$

7.3.4 Use of more accurate constitutive relations

Although the differential equation of stress redistribution has been developed in the above for a simple time hardening relation, this approach can also be applied to more complex constitutive relations. Of particular interest are the internal-variable theories described in Section 2.3. If α_j are a number of internal or hidden variables which appear in the constitutive relations

$$\dot{\epsilon}_{ij}^{C} = f_{ij}(\sigma_{kl},\,t,\,\alpha_k)$$

$$\dot{\alpha}_i = g_i(\sigma_{kl},\,t,\,\alpha_k)$$

then the equations (7.8) become

$$\frac{d\sigma_{ij}}{dt} = R_{ij}(f_{kl}(\sigma_{mn}, t, \alpha_m)) + \dot{\sigma}_{ij}^{o} \tag{7.12}$$

$$\frac{d\alpha_i}{dt} = g_i(\sigma_{kl}, t, \alpha_k)$$

which define the required initial value problem for the stress and internal variables.

7.3.5 An initial value problem for inelastic strain

The equations of stress redistribution are not the only means of mathematically describing the evolution of a creeping structure. In many circumstances, it may be more appropriate to use an alternative formulation in terms of the inelastic strain.

It can be fairly easily demonstrated, using arguments similar to those given above, that the inelastic strain ϵ_{ij}^{C} satisfies the equations

$$\frac{d\epsilon_{ij}^{C}}{dt} = f_{ij}(\sigma_{kl}^{o} + R_{kl}(\epsilon_{mn}^{C}), t) \tag{7.13}$$

for a time hardening material model. Together with the initial condition

$$\epsilon_{ij}^{C}(0) = 0 \tag{7.14}$$

these equations form an initial value problem for the inelastic strains.

What are the differences in the formulations of equations (7.8) and (7.13)? Mathematically they are equivalent; the only differences occur when they are solved numerically. In the equations of stress redistribution equation (7.8), only the stress rates satisfy equilibrium and consequently if the equations are being solved numerically the stresses will only satisfy equilibrium approximately. On the other hand, in the formulation for inelastic strain, equation (7.13), the stresses computed from the inelastic strain will always satisfy equilibrium. This difference can usually be ignored for most practical purposes but becomes important if the material is deteriorating, as we will see in a later chapter.

7.3.6 Discussion

We have attempted in this section to show the type of mathematical equations to which the field equations of transient creep give rise. That is, they lead to initial value problems in the stress, equations (7.1), (7.10), (7.12) and (7.13), defined by the solution of two elastic problems. The form of these equations tells us the sort of techniques we require in order to solve the transient creep problem. First we need some means of solving the elastic problems for the structure – the way in which this is approached is of course at the discretion of the analyst. Whatever means is used – energy methods, finite difference or finite element – the equations will reduce mathematically to a *finite system of first-order ordinary differential equations* of the form

$$\frac{d}{dt}\mathbf{y} = \mathbf{f}(\mathbf{y}, t) \qquad \mathbf{y}(0) = \mathbf{y}_0 \tag{7.15}$$

in vector notation. We have seen this already for the thin tube in bending and will discuss it in more detail for the finite element method. Thus, secondly, we need some numerical techniques for resolving systems of first-order equations forming an initial value problem. Such techniques are the subject of a subsequent section.

In fact, from a numerical and computational viewpoint, the transient creep problem is more straightforward than the steady creep problem since it only needs the solution of linear boundary value problems coupled with an initial value problem. However, if the loading varies in time, this may prove quite expensive for complex three-dimensional structures. For constant loading, this process is in many cases a more attractive alternative to steady-state analysis and may be used as a method of solving such problems.

7.4 Numerical solution of initial value problems

Methods for the numerical solution of systems of first-order ordinary differential equations which form an initial value problem have received much attention in the literature and the reader is referred to any number of texts on the subject. Here we will restrict ourselves to discussing techniques which have found application in transient creep analysis.

The fundamental object of solving equations like (7.15) is to find the solution $\mathbf{y}$ at a sequence of discrete instants of time $t_1, t_2, \ldots$ and so on. Methods which do this are called 'time stepping', 'marching' or 'step-by-step' methods and provide a rule for computing the approximation at step t_i in terms of the solution at t_{i-1} and possibly preceding steps. We will call this approximation $\mathbf{y}_i$ Such time-stepping methods (or 'algorithms') are the methods most frequently found in computer scientific subroutine libraries.

When we think of approximating a solution by the application of some numerical algorithm we are concerned with two things. First we should be interested in the *accuracy* of our approximate solution. In particular, we want to be sure that we can select the time step small enough to obtain any desired accuracy. Of course, two approximations will interfere with this. We are not solving the actual transient creep equations, only a discretised version of them (cf. the thin tube in bending or the use of finite elements); in addition, for a computer solution the calculated solution can only be represented to a finite precision. These inherent errors are called 'discretisation' (or 'truncation') errors and 'round-off' errors. Given the unavoidable nature of these, we want to be certain that our algorithm does not make them worse so as to accumulate without bound. Mathematically, we wish to be reassured that the algorithm is *stable*. The satisfaction of these conditions – accuracy and stability – are of particular importance in the analysis of complex structures where the number of unknowns may be high and the subsequent cost of computation expensive. We could make these concepts more precise – but will refrain from doing so here. It is sufficient to know that they are important.

7.4.1 An example – Euler's method

The simplest method – and indeed one which has been widely adopted for transient creep problems – is the Euler method. In this, the value of $\mathbf{y}_{i+1}$ is

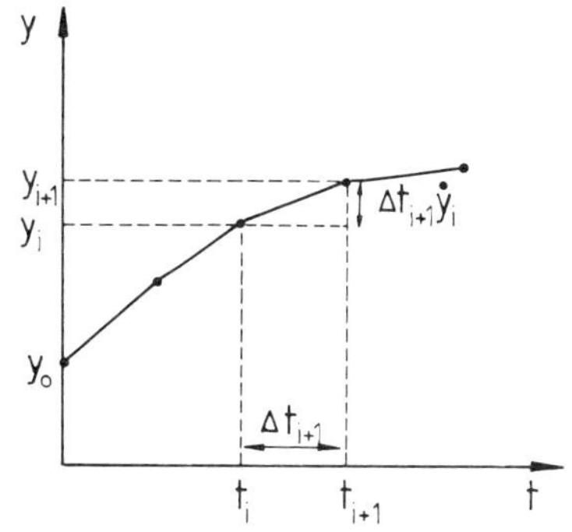

Figure 7.7 Graphical representation of Euler's method

calculated by a straight extrapolation from $\mathbf{y}_i$ (*Figure 7.7*). From a Taylor series expansion, we have

$$\mathbf{y}_{i+1} = \mathbf{y}_i + \Delta t_{i+1}\,\dot{\mathbf{y}}_i \qquad i = 0, 1, 2, \ldots$$

where $\Delta t_{i+1} = t_{i+1} - t_i$. Thus

$$\mathbf{y}_{i+1} = \mathbf{y}_i + \Delta t_{i+1}\,\mathbf{f}(t, \mathbf{y}_i) \tag{7.16}$$

We will examine the convergence and stability of this method for the simple pin-jointed framework (Section 7.1) assuming a uniform time step $\Delta\tau_{i+1} = \Delta\tau$ in the normalised time scale. Results for various increasing values of $\Delta\tau$ are shown in *Figure 7.8*. It can be seen that the method is convergent, i.e. the smaller the time

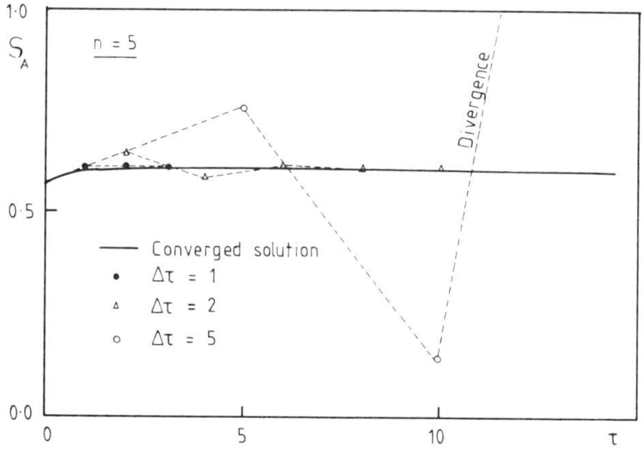

Figure 7.8 Effect of varying time step on Euler's method for the pin-jointed framework

interval the closer the approximate solutions become to the exact solution. But it is also seen that as the time steps become larger the method becomes *unstable*. The Euler method is then *conditionally stable* – its stability depends on the time step used.

It is possible to develop a stability criterion for this method – we will not go into detail here but simply quote the main result. For an arbitrary structure, stability of the Euler method is ensured if we keep the time step to the upper limit

$$\Delta t_{i+1} \leqslant \frac{4}{3} \frac{(1+\nu)}{Eg(t)} \frac{1}{n\,\bar{\sigma}_i^{n-1}} \tag{7.17}$$

where $\bar{\sigma}_i$ is the value of the effective stress at the time t_i. For our present example, we see that the method is stable provided, roughly for all time

$$\Delta\tau \leqslant 1$$

The stability criterion is only a necessary condition – the Euler method may be stable for higher time steps (e.g. in our example, it is stable for $\Delta\tau = 2$). Thus such criterions may specify a smaller time step than we could get away with in practice. (Of course, such considerations only become important if it is expensive to perform the analysis.)

In general, application of Euler's method is really too crude to represent the solution with accuracy for economically acceptable time steps. (But do not dismiss it – it is very easy to program if cost is not a problem!) A more accurate method which allows larger time steps is required.

7.4.2 Improved computational methods

There are several possibilities here. Such techniques are usually assigned to two types – those in which the function **f** is evaluated only at the points $(t_i, \mathbf{y}_i)$ and those which evaluate **f** at points other than $(t_i, \mathbf{y}_i)$. The former are commonly called 'numerical integration' or 'multistep' methods and include the popular 'predictor–corrector' methods. The latter are called 'Runge–Kutta' methods. However, reference to the literature shows that there are many time-stepping algorithms each with its own special properties. Here we will give a short discussion of suitable algorithms which have been found to have particular application to transient creep. Nevertheless, it should be borne in mind that if the problem is trivial – and of necessity most of the problems in this book can be classed as this – then any readily available method from a computer subroutine library can be used with confidence. Our discussion mainly refers to the non-trivial problems dealing with complex structures under histories of loading where efficiency (i.e. optimisation of computing cost with accuracy) is of paramount importance.

One important point in the choice of a suitable algorithm is that it often proves more efficient to have a method which allows the time step to be continually adjusted. For example, in the pin-jointed framework under constant loading, larger time steps can be used as the steady state is approached. Additionally, the loads may be discontinuous, e.g. a sudden step change in load. Ideally we require a 'one-step' method; this would rule out some of the popular multistep methods such as Adams–Bashforth or Adams–Moulton methods which make it difficult to change the step size. Thus, although they are generally slower to gain the same accuracy as the multistep methods and generally have weaker stability properties, the Runge–Kutta methods are more desirable.

The most popular Runge–Kutta methods are of fourth order (this means that four evaluations of the functions **f** per time step are required – as a matter of interest an equivalent multistep method would need only two such evaluations).

Euler's method is a Runge–Kutta method of first order. The particular method we will describe here is due to Gill and is achieved in four stages:

$$\left.\begin{aligned} \mathbf{k}^1 &= \Delta t_{i+1}\,\mathbf{f}(t_i, \mathbf{y}_i) \\ \mathbf{y}^1_{i+1} &= \mathbf{y}_i + \frac{1}{2}(\mathbf{k}^1 - 2\mathbf{q}^o) \\ \mathbf{q}^1 &= \mathbf{q}^o + \frac{3}{2}(\mathbf{k}^1 - 2\mathbf{q}^o) - \frac{1}{2}\mathbf{k}^1 \end{aligned}\right\} \quad (1)$$

$$\left.\begin{aligned} \mathbf{k}^2 &= \Delta t_{i+1}\,\mathbf{f}(t_i + \Delta t_{i+1}/2, \mathbf{y}^1_{i+1}) \\ \mathbf{y}^2_{i+1} &= \mathbf{y}^1_{i+1} + (1 - \sqrt{\tfrac{1}{2}}(\mathbf{k}^2 - \mathbf{q}^1)) \\ \mathbf{q}^2 &= \mathbf{q}^1 + 3(1 - \sqrt{\tfrac{1}{2}})(\mathbf{k}^2 - \mathbf{q}^1) - (1 - \sqrt{\tfrac{1}{2}})\,\mathbf{k}^2 \end{aligned}\right\} \quad (2)$$

$$\left.\begin{aligned} \mathbf{k}^3 &= \Delta t_{i+1}\,\mathbf{f}(t_i + \Delta t_{i+1}/2, \mathbf{y}^2_{i+1}) \\ \mathbf{y}^3_{i+1} &= \mathbf{y}^2_{i+1} + (1 + \sqrt{\tfrac{1}{2}})(\mathbf{k}^3 - \mathbf{q}^2) \\ \mathbf{q}^3 &= \mathbf{q}^2 + 3(1 + \sqrt{\tfrac{1}{2}})(\mathbf{k}^3 - \mathbf{q}^2) - (1 + \sqrt{\tfrac{1}{2}})\,\mathbf{k}^3 \end{aligned}\right\} \quad (3)$$

$$\left.\begin{aligned} \mathbf{k}^4 &= \Delta t_{i+1}\,\mathbf{f}(t_i + \Delta t_{i+1}, \mathbf{y}^3_{i+1}) \\ \mathbf{y}^4_{i+1} &= \mathbf{y}^3_{i+1} + \frac{1}{6}(\mathbf{k}^4 - 2\mathbf{q}^3) \\ \mathbf{q}^4 &= \mathbf{q}^3 + \frac{1}{2}(\mathbf{k}^4 - 2\mathbf{q}^3) - \frac{1}{2}\mathbf{k}^4 \end{aligned}\right\} \quad (4)$$

Then finally the solution at time step t_{i+1} is given by

$$\mathbf{y}_{i+1} = \mathbf{y}^4_{i+1}$$

and for the subsequent time step we set

$$\mathbf{q}^o = \mathbf{q}^4$$

with initially $t = 0$, $\mathbf{q}^o = 0$. (This may seem complicated, but the reader is invited to write a computer program to use this method in as few lines as possible, whence it can be seen that it is actually very simple to code.)

The Runge–Kutta–Gill method has the advantage that it minimises the number of 'extra' storage required, i.e. **y** together with two more storage locations of the same size **k** and **q** being successively overwritten in each stage. Also the coefficients are chosen so as to reduce the amount of round-off error. From stability considerations, it is important to have some means of controlling the time step at each stage in order to increase efficiency – we do not want it too small so as to waste valuable time, or too large so as to lose accuracy and to invite possible instability.

The most commonly used technique for error control is called 'step doubling'. Each basic time step of size Δt is done twice – once with two successive steps of size $\Delta t/2$ and once as a single step of size Δt. Let the result of the former be $\mathbf{y}^a$ and the latter be $\mathbf{y}^b$, and let ϵ be the desired accuracy. Finally, let $\mathbf{y}_{max}$ contain the maximum value of each computed component calculated so far. Then for each component the relative error is given by

$$\epsilon_{rel} \simeq \frac{|y_i^a - y_i^b|}{(y_{max})_i}$$

for a fourth-order Runge–Kutta. If $\epsilon_{rel} < \epsilon$ then we proceed to the next time step. Further, it can be shown that roughly

$$\epsilon_{rel} \simeq \frac{15}{16} \Delta t^5$$

and the next time step is chosen as

$$\Delta t = \left(\frac{16}{15} \epsilon_{rel}\right)^{1/5} \tag{7.18}$$

On the other hand, if $\epsilon_{rel} < \epsilon$ the present step is repeated with the step size recommended by equation (7.18).

Incidentally, for the Euler method the recommended time step with the doubling procedure is

$$\Delta t = \left(\frac{4}{3} \epsilon_{rel}\right)^{1/2} \tag{7.19}$$

As an alternative to the use of equation (7.18), the time step may be doubled if ϵ_{rel} is smaller than a specified minimum accuracy.

7.4.3 Stiff equations

It is often found that when integrating an initial value problem using a method with time step control, unusually small time steps are selected. This is usually an indication that the equations are *stiff*. It is not easy to give an *a priori* condition which identifies stiffness in a set of equations, but it frequently becomes apparent in differential equations which arise from physical processes which gave greatly differing time constants. In terms of creep mechanics, this means that the rate of stress redistribution greatly differs throughout the body. This behaviour is not usually met in the simpler problems involving a time hardening theory with constant loading where a normalised time scale can be introduced. Rather, it is more apparent when we employ some of the internal-variable theories and will become a particular problem when we come to deal with creep failure.

While the numerical integration methods that we have described so far can still be used, the step size has to be intolerably small before acceptable accuracy is obtained in practice. Then the accumulation of round-off error and excessive computation time become more critical. What are required are algorithms which do not restrict the step size for stability reasons – accuracy becomes the sole criterion. These are called *unconditionally stable* algorithms. Unfortunately, with these methods, there is a price to be paid for the use of larger time steps – usually more evaluations per step. Thus the stiff algorithms must allow a sufficiently large time step to offset this increased computational time per step.

We will briefly describe the simplest stiff algorithm which has found application in creep mechanics. It is the first-order implicit Euler scheme

$$\mathbf{y}_{i+1} = \mathbf{y}_i + \Delta t_{i+1}\,[(1-\alpha)\,\mathbf{f}_i + \alpha\mathbf{f}_{i+1}]$$

where $0 \leqslant \alpha \leqslant 1$. When $\alpha = 0$ this reduces to the Euler method; when $\alpha = ½$ it is the well known implicit trapezoidal scheme. It can be shown that for any $\alpha \geqslant ½$ this scheme is unconditionally stable.

A slight variation of the above is the algorithm

$$\mathbf{y}_{i+1} = \mathbf{y}_i + \Delta t_{i+1}\,\mathbf{f}((1-\alpha)t_i + \alpha t_{i+1}(1-\alpha)\mathbf{y}_i + \alpha\mathbf{y}_{i+1})$$

which again reduces to Euler's method if $a = 0$. This scheme has similar stability characteristics to the above.

The trouble with these algorithms is that they are *implicit*: for each time step, we have a nonlinear equation for the updated solution $\mathbf{y}_{i+1}$. This could be solved directly since it is in fixed-point form, but we can derive a simpler algorithm if we use quasilinearisation so that

$$\mathbf{f}_{i+1} \simeq \mathbf{f}_i + \left[\frac{d\mathbf{f}}{d\mathbf{y}_i}\right](\mathbf{y}_{i+1} - \mathbf{y}_i)$$

and the first-order implicit scheme becomes

$$\mathbf{y}_{i+1} = \left[\,[I] - \alpha\Delta t_{i+1}\left[\frac{d\mathbf{f}}{d\mathbf{y}_i}\right]\right]^{-1}\left\{\mathbf{y}_i + \Delta t_{i+1}\left(\mathbf{f}_i - \alpha\left[\frac{d\mathbf{f}}{d\mathbf{y}_i}\right]\mathbf{y}_i\right)\right\} \qquad (7.20)$$

In practice, it is difficult to identify and demonstrate a particular simple problem in transient creep as being stiff *before* numerical solution. As an example here, we demonstrate stiffness in a relative manner.

7.4.4 Example: Forward creep of a non-uniformly heated structure

As an example, we solve the forward creep problem for the non-uniformly heated three-bar structure of Section 3.2. Assuming a time hardening power law of creep, the equation of stress redistribution in the centre bar is fairly easily shown to be

$$\frac{d\sigma_1}{dt} = Eg(t)\,e^{-\gamma/T}\left(\frac{\sigma_1^n + 2e^{-\bar{\gamma}}\cos\vartheta[½(Q/A - \sigma_1)/\cos\vartheta]^n}{1 + 2\cos^3\vartheta} - \sigma_1^n\right)$$

The initial elastic solution is

$$\sigma_1(0) = \frac{Q/A + E\alpha_T(\Delta T_1 + 2\Delta T_2\cos\vartheta)}{1 + 2\cos^3\vartheta} - E\alpha_T\Delta T_1$$

where a_T is the coefficient of thermal expansion, $\bar{\gamma} = \gamma\,(1/T_2 - 1/T_1)$ and ΔT_1, ΔT_2 are the temperature rises in the bars above that in the unstressed state.

The solution of this problem can be obtained on defining a normalised stress $S_1 = \sigma_1/(Q/A)$ and a normalised time scale

$$\tau = E(Q/A)^{n-1} e^{-\bar{\gamma}/T_1} \int g(t)\, dt$$

As a numerical example, the following values are assumed for the various constants

$$\vartheta = 45^\circ \qquad n = 5 \qquad \Delta T_1 = 0 \qquad \frac{E\alpha_T \Delta T_2}{Q/A} = 0.5 \qquad \bar{\gamma} = 10$$

together with the isothermal case

$$\Delta T_1 = \Delta T_2 = 0 \qquad \bar{\gamma} = 0$$

These two examples are solved using Euler's method, the Runge–Kutta–Gill method and the implicit scheme, equation (7.20); results are shown in *Figures 7.9, 7.10* and *7.11*, respectively. The following observations are made: (1) The non-isothermal problem requires a far smaller time step than the equivalent isothermal problem to obtain a desired accuracy. It can be said to be stiffer. (2) Both the Runge–Kutta–Gill and implicit schemes can achieve the same accuracy as the simpler Euler scheme with larger time steps. For the isothermal problem, they are not significantly computationally cheaper. However, for the stiffer non-isothermal problem, they are both more efficient, with the implicit

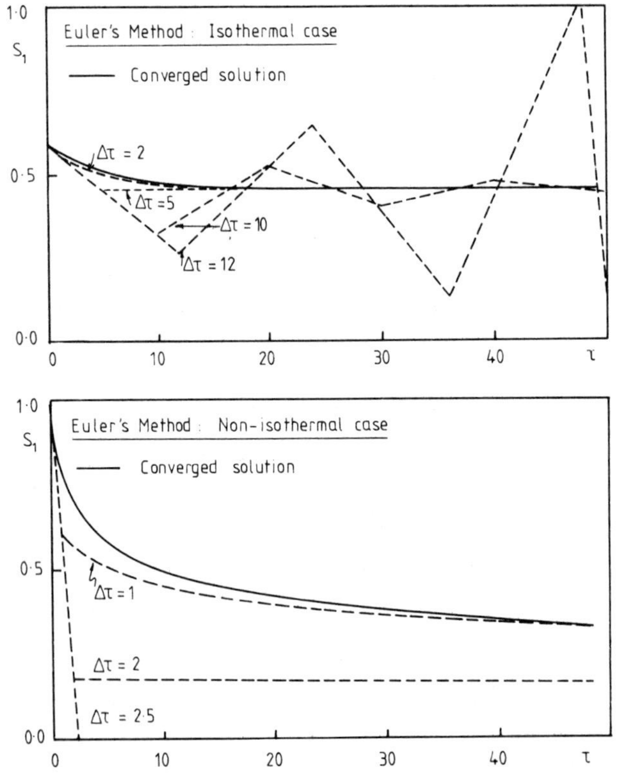

Figure 7.9 Illustration of convergence, stability and stiffness: Euler's method

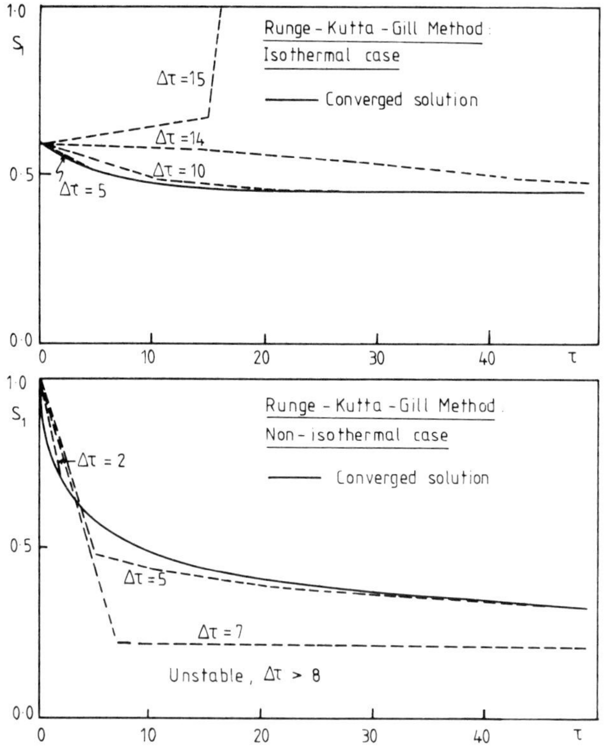

Figure 7.10 Illustration of convergence, stability and stiffness: Runge–Kutta–Gill method

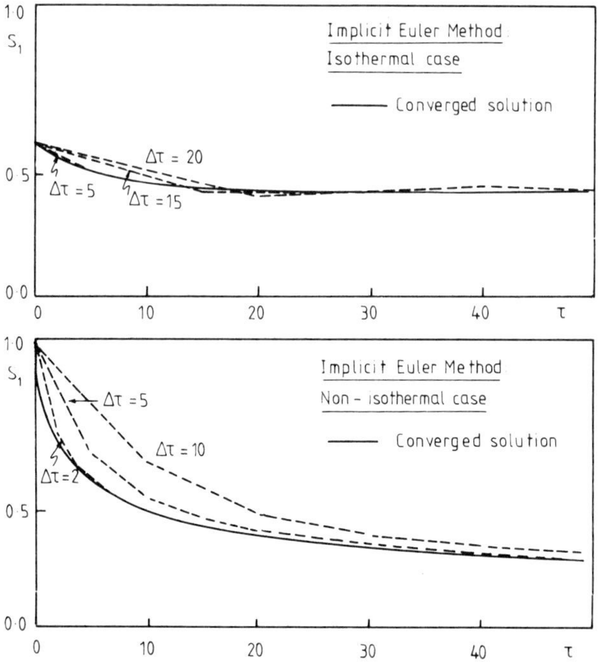

Figure 7.11 Illustration of convergence, stability and stiffness: implicit Euler method

scheme having a clear advantage. (3) Instability is observed for the Runge–Kutta–Gill and simple Euler schemes for increasing time steps. For the isothermal problem their stability range is about the same, but for the non-isothermal problem the Runge–Kutta–Gill method has a better range of stability – indeed, it is quite accurate for a time step outside the stability range of the Euler method. As expected, the implicit scheme is unconditionally stable.

7.4.5 Conclusions

In practice, it is difficult to recommend a single ideal method – its choice depends strongly on the problem in hand. The stiff algorithms allow an increase in step size but at the cost of an increased number of calculations per step and these must be carefully balanced; it must also be borne in mind that common examples given in the literature on differential equations deal with problems where the 'cost' of function evaluation is cheap (compared to those of creep mechanics!) Of more importance is recognising when the equations have the potential to be stiff – suitable techniques for doing this during a numerical time integration are available.

In this book – except where stated otherwise – we have used the Runge–Kutta–Gill method with error control. In the next chapter we will recompute some of the steady creep problems for transient creep in order to discuss approximate methods. However, to finish off this discussion of exact techniques, we will give two further applications. The first is the plane strain problem of a thick sphere, but using an internal-variable theory and assuming a temperature gradient.

7.5 Example: Creep of a pressurised thick sphere with a radial temperature gradient

Consider a thick sphere of internal radius a and external radius b subject to an internal pressure $p_a(t)$ and an external pressure $p_b(t)$ respectively (*Figure 7.12*). The inside temperature is T_a and the outside T_b, both being constant in time. This problem has certain similarities to the thick cylinder of Section 4.3, except that we have spherical symmetry described by the coordinates (r, ϑ, φ) (*Figure 7.12*), which can be taken as the principal directions. The only non-zero component of displacement is that in the radial direction $u = u(r, t)$; the principal strains are ϵ_r, $\epsilon_\vartheta = \epsilon_\varphi$ while the principal stresses are σ_r, $\sigma_\vartheta = \sigma_\varphi$, all being functions of r and t alone.

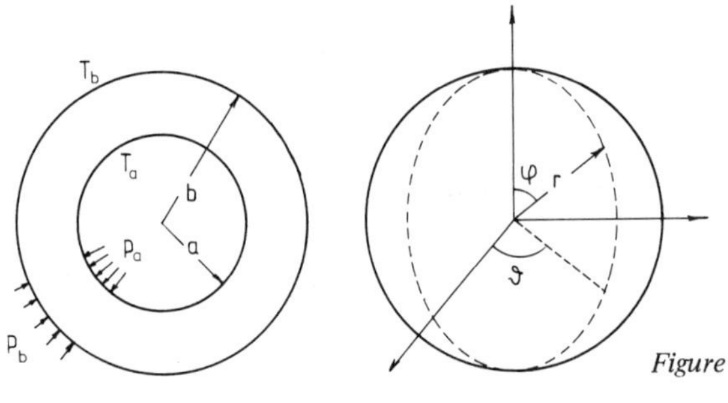

Figure 7.12 Geometry of thick sphere

The appropriate field equations which define this problem are

Equilibrium:

$$\frac{d\sigma_r}{dr} + \frac{2}{r}(\sigma_r - \sigma_\vartheta) = 0$$

Strain–displacement:

$$\epsilon_r = du/dr \qquad \epsilon_\vartheta = \epsilon_\varphi = u/r$$

together with the boundary conditions

$$\sigma_r(a,\ t) = -\,p_a(t) \qquad \sigma_r(b,\ t) = -\,p_b(t)$$

Assuming a radial temperature gradient, the temperature distribution is given by

$$T = T_a + (T_b - T_a)\,\frac{1 - a/r}{1 - b/r} \tag{7.21}$$

The basic equations are then completed by the constitutive relations, which, for spherical symmetry, take the form

$$\epsilon_r = \epsilon_r^{\mathrm{E}} + \epsilon_r^{\mathrm{C}} + \alpha_{\mathrm{T}}\Delta T$$

$$\epsilon_\vartheta = \epsilon_\vartheta^{\mathrm{E}} + \epsilon_\vartheta^{\mathrm{C}} + \alpha_{\mathrm{T}}\Delta T$$

where for the elastic strains

$$\epsilon_r^{\mathrm{E}} = \frac{1}{E}\,(\sigma_r - 2\nu\sigma_\vartheta) \qquad \epsilon_\vartheta^{\mathrm{E}} = \frac{1}{E}\,[(1 - \nu)\,\sigma_\vartheta - \nu\sigma_r]$$

where a_r is the coefficient of thermal expansion with ΔT the temperature rise above the stress-free temperature (assumed here to be $T_b \leqslant T_a$). The material model for the creep behaviour will be taken here as Hart's theory, Sections 2.3 and 4.2. For spherical symmetry, these are

$$\dot{\epsilon}_r^{\mathrm{C}} = \left(\frac{\dot{\bar{\epsilon}}^{\mathrm{C}}}{\bar{\sigma}_\alpha}\right)\,[(\sigma_r - \sigma_\vartheta) - \frac{3}{2}\,a_r\,] \tag{7.22}$$

$$\dot{\epsilon}_\vartheta^{\mathrm{C}} = \dot{\epsilon}_\varphi^{\mathrm{C}} = -\,\frac{1}{2}\,\dot{\epsilon}_r^{\mathrm{C}}$$

from incompressibility. Noting further that $a_\vartheta = a_\varphi = -\dfrac{1}{2}a_r$ we have the invariants

$$\bar{\sigma}_\alpha = [(\sigma_r - \sigma_\vartheta) - \frac{3}{2}a_r]$$

$$\bar{\alpha} = \frac{3}{2}\,a_r$$

The growth equation for a_r can be obtained as

$$\dot{a}_r = \frac{2}{3}c_2\,(\dot{\epsilon}_r^{\mathrm{C}} - \frac{3}{2}\,f_2 a_r) \tag{7.23}$$

Finally, the multiaxial relations, equations (7.22) and (7.23), are completed by the equivalent uniaxial relations

$$\dot{\bar{\epsilon}}^{C} = c_1\,(\bar{\sigma}_{\alpha})^{m} \qquad f_2 = (\dot{\epsilon}^*/\bar{\alpha})\,[\ln(\sigma^*/\bar{\alpha})]^{-1/\lambda} \tag{7.24}$$

where

$$\dot{\epsilon}^* = c_3(\sigma^*)^{n}\exp(-\Delta H/RT)$$

and

$$\dot{\sigma}^* = f_2(\bar{\alpha},\sigma^*)\,\sigma^*\bar{\alpha}\,\Gamma(\sigma^*,\bar{\alpha}) \tag{7.25}$$

The problem will be resolved using equations for the residual stress, equation (7.10), which here reduce to

$$\frac{\mathrm{d}\rho_r}{\mathrm{d}t} = R_r[\dot{\epsilon}_r^{C}(\sigma_r^{o} + \rho_r,\ \sigma_\vartheta^{o} + \rho_\vartheta)]$$
$$\frac{\mathrm{d}\rho_\vartheta}{\mathrm{d}t} = R_\vartheta\,[\dot{\epsilon}_r^{C}(\sigma_r^{o} + \rho_r,\ \sigma_\vartheta^{o} + \rho_\vartheta)] \tag{7.26}$$

where

$$\rho_r = \sigma_r - \sigma_r^{o} \qquad \rho_\vartheta = \sigma_\vartheta - \sigma_\vartheta^{o}$$

The stresses for the equivalent elastic problem can be obtained as

$$\sigma_r^{o} = -\frac{2\alpha_T E}{1-\nu}\,\frac{1}{r^3}\int_a^r \Delta T\,r^2\,\mathrm{d}r + \frac{EC_1}{1-2\nu} - \frac{2EC_2}{1+\nu}\,\frac{1}{r^3}$$

$$\sigma_\vartheta^{o} = \frac{\alpha_T E}{1-\nu}\,\frac{1}{r^3}\int_a^r \Delta T\,r^2\,\mathrm{d}r + \frac{EC_1}{1-2\nu} + \frac{EC_2}{1+\nu}\,\frac{1}{r^3} - \frac{\alpha_T E\Delta T}{1-\nu}$$

The constants C_1 and C_2 are obtained from the boundary conditions as

$$\frac{C_2}{a^3} = \frac{(p_a - p_b)/E + [2\alpha_T/(1-\nu)]\,(a/b)^3\int_a^r \Delta T\,(r/a)^3\,\mathrm{d}r}{[2/(1+\nu)]\,[1-(a/b)^3]}$$

$$C_1 = (1-2\nu)\left(\frac{2C_2/a^3}{1+\nu} - \frac{p_a}{E}\right)$$

Similarly, the radial displacement is

$$u^{o} = \frac{1+\nu}{1-\nu}\,\alpha_T\,\frac{1}{r^3}\int_a^r \Delta T r^2\,\mathrm{d}r + C_1 r + C_2/r^2$$

The functions R_r and R_ϑ are defined through a solution of the linear elastic boundary value problem for the initial strains ϵ^*_r, $\epsilon^*_\vartheta = -\tfrac{1}{2}\epsilon^*_r$

Equilibrium:

$$\frac{\mathrm{d}\sigma_r^{R}}{\mathrm{d}r} + 2\frac{(\sigma_r^{R} - \sigma_\vartheta^{R})}{r} = 0$$

Strain–displacement:

$$\epsilon_r^R = \frac{du^R}{dr} \qquad \epsilon_\vartheta^R = \epsilon_\varphi^R = \frac{u^R}{r}$$

Boundary conditions:

$$\sigma_r^R(a) = 0 \qquad \sigma_r^R(b) = 0$$

Constitutive relations:

$$\epsilon_r^R = \frac{1}{E}(\sigma_r^R - 2\nu \ \sigma_\vartheta^R) + \epsilon_r^*$$

$$\epsilon_\vartheta^R = \frac{1}{E}[(1-\nu)\sigma_\vartheta^R - \nu\sigma_r^R] + \epsilon^*$$

This problem can be solved to yield

$$R_r(\epsilon_r^*) = \sigma_r^R = \frac{E}{1-\nu}\left(\int_a^r \frac{\epsilon^*_r}{r} \, dr - \frac{r^3 - a^3}{b^3 - a^3} \frac{b^3}{r^3} \int_a^b \frac{\epsilon^*_r}{r} \, dr\right)$$

$$R_\vartheta(\epsilon_r^*) = \sigma_\vartheta^R = \frac{E}{1-\nu}\left(\int_a^r \frac{\epsilon_r^*}{r} \, dr - \frac{1}{2} \frac{2r^3 + a^3}{b^3 - a^3} \frac{b^3}{a^3} \int_a^b \frac{\epsilon_r^*}{r} dr + \frac{\epsilon_r^*}{2}\right)$$

The basic equations (7.26), together with the constitutive relations (7.22)–(7.25), are now in a suitable form for numerical solution, with the initial conditions

$$\rho_r(0) = 0 \qquad \rho_\vartheta(0) = 0 \tag{7.27}$$

The radial displacement can be obtained by including the equation

$$\frac{d}{dt}(u - u^o) = \frac{r}{E}[(1-\nu)R_\vartheta(\dot{\epsilon}_r^C) - \nu R_r(\dot{\epsilon}_r^C)] \tag{7.28}$$

in the solution process.

In the following numerical calculations, the equations have been conveniently normalised according to

$$S = \frac{\sigma}{\sigma_0} \qquad U = \frac{u}{a} \qquad \xi = \frac{r}{a} \qquad \eta = \frac{b}{a}$$

where σ is any quantity with the dimensions of stress, and σ_o is any suitable normalising stress.

The basic equations (7.26) and (7.28) must be reduced to a finite system of first-order differential equations (cf. equation (7.15)), as we did for the thin tube in bending in Section 7.2. This is done by considering the stresses and internal variables α_r and σ^* only at a finite number of points through the thickness of the sphere. The integrals are then replaced by finite sums – some necessary details of this are given below.

If the interval $[1, \eta]$ is subdivided into equal intervals of length denoted by

$$\xi_k = 1 + (k-1)h \qquad k = 1, 2, \ldots M$$

where $h = (\eta - 1)/(M - 1)$ and M is the number of points with $\xi_1 = 1, \xi_M = \eta$, it is required to evaluate the open integral

$$I_k = I(\xi_k) = \int_1^{\xi_k} \frac{\dot{\epsilon}_r^C}{\xi}\, d\xi$$

A suitable sequence of approximations is given as

$$I_1 = 0$$

$$I_2 = I_1 + \frac{1}{12} h(5f_1 + 8f_2 - f_3)$$

$$I_3 = I_2 + \frac{1}{12} h(5f_2 + 8f_3 - f_4)$$

$$I_k = I_{k-1} + \frac{1}{24} h\,(9f_k + 19f_{k-1} - 5f_{k-2} + f_{k-3})$$

denoting

$$f_k = \left.\frac{\dot{\epsilon}_r^C}{\xi}\right|_{\xi=\xi_k}$$

Numerical calculations have been carried out here for a type 304 stainless steel. The work hardening function in this case has the form

$$\Gamma(\sigma^*, \bar{a}) = (\mu/\sigma^*)^l (\bar{\alpha}/\sigma^*)^{\mu/\sigma^*}$$

derived from experiments, where μ and l are material constants. If the stress-free temperature is assumed to be 400° C, then the material constants are given by

$c_1 = 0.6937 \times 10^{25} \times (1/c_2)^m\,\text{s}^{-1}$ $\quad c_2 = 0.132 \times 10^8$ psi $\quad c_3 = 1.1813 \times 10^9 \times (1/G)^n\,\text{s}^{-1}$

$m = 7.8 \quad n = 5 \quad \lambda = 0.15 \quad \mu = 0.179 \times 10^6$ psi $\quad l = 1.33$

In the above, the stress and hardness are in units of psi and time is in seconds. Finally the elastic constants are

$E = 0.244 \times 10^8$ psi $\quad \nu = 0.298 \quad G = 45\,670$ psi

with an initial uniform hardness

$\sigma^*(0) = 17\,000$ psi

The thermal constants are

$\Delta H = 380\,753$ J mol^{-1} $\quad R = 8.314$ J mol^{-1} K^{-1} $\quad \alpha_T = 1 \times 10^{-5}$ °C^{-1}

In the particular numerical example considered here, the outside temperature is taken as 400° C with the inside temperature 500° C. The loading condition considered has the outside pressure $p_b = 0$ with the inside pressure p_a initially at 10 000 psi for 3 h followed by a step up to 15 000 psi.

One further point needs to be made concerning the numerical solution with Hart's equations. In equation (7.24) care must be taken to avoid the situation

$\sigma^* \simeq \bar{\alpha}$ (in general, $\sigma^* \geqslant \bar{\alpha}$). If this is the case it has been suggested that equation (7.24) takes the simpler form

$$f_2 = \frac{\bar{\dot{\epsilon}}^C/\bar{\alpha}}{1 + \sigma^*\Gamma/c_2}$$

which has been specialised here for spherical symmetry. This form can be used whenever $f_2\bar{\alpha}$ exceeds a certain critical value $\dot{\epsilon}^*_o(\sigma^*/\sigma^*_o)^n$. For the present material, it is found experimentally that

$$\dot{\epsilon}^*_o = 0.2123 \times 10^{-8}\ \mathrm{s}^{-1} \qquad \sigma^*_o = 10\,000\ \mathrm{psi}$$

In *Figure 7.13* results are presented for the normalised displacement U at $r = a$ against time. In *Figures 7.14* and *7.15* the spatial variation of the normalised

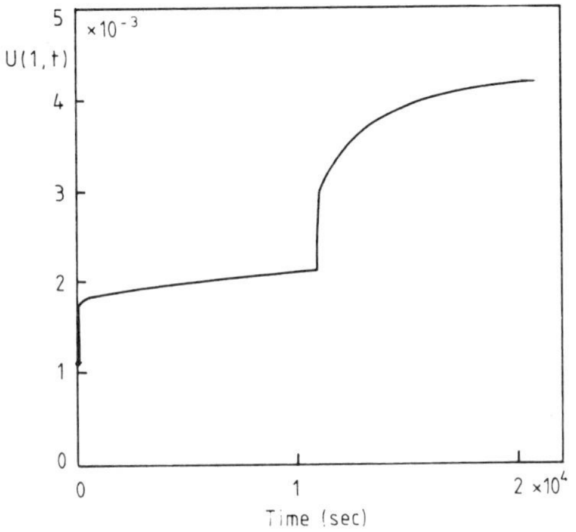

Figure 7.13 Variation of normalised displacement at inner surface with time

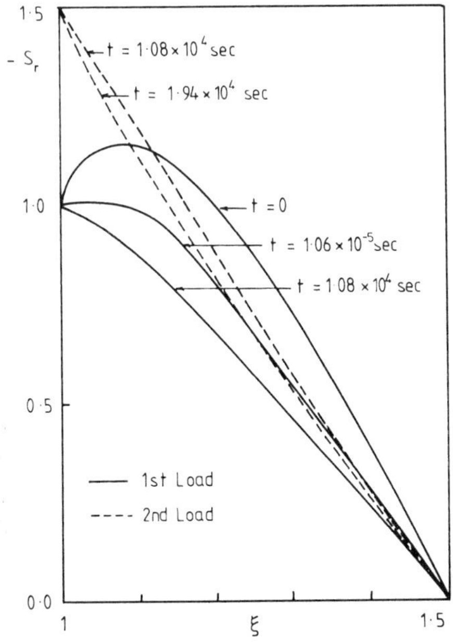

Figure 7.14 Variation of normalised radial stress distribution with time

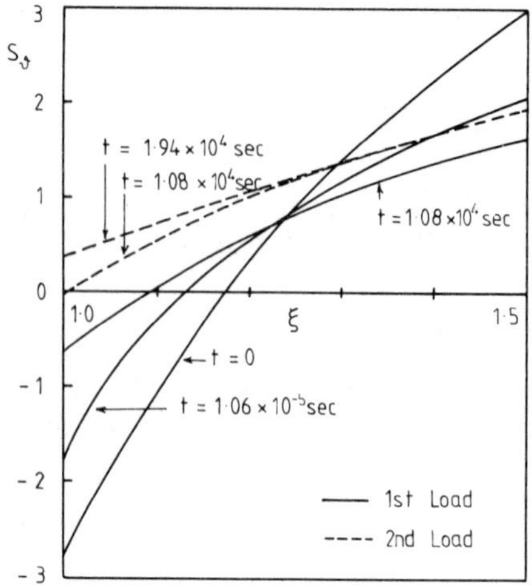

Figure 7.15 Variation of normalised circumferential stress distribution with time

stresses S_r and S_ϑ respectively are shown for various values of time, taking $\sigma_0 =$ 10 000 psi.

The quantitative results are not of much interest since we have arbitrarily chosen the material and loading. Nevertheless, a number of qualitative features of the material model can be seen. From *Figure 7.13* the deformation can be seen to consist of four distinct phases. An initial elastic deformation is followed by a rapid increase in the 'creep' strain occurring in a very short time – this is in fact the more familiar 'plastic' strain. This is then followed by a period of primary creep and finally a period of steady creep, which is interrupted by the load increase.

7.6 Numerical solution of transient creep problems using the finite element method

We here briefly discuss the practical application of these numerical algorithms in the finite element method. We can present here only the main features of the method; the notation we employ is, by and large, standard – the symmetric stress and strain field tensors are replaced by six-dimensional vectors $\boldsymbol{\sigma}$ and $\boldsymbol{\epsilon}$ while the displacement is represented by the three-dimensional vector $\mathbf{u}$. Thus, for example, compatibility of strains with displacements is given by the linear differential operator

$$\boldsymbol{\epsilon} = [L]\,\mathbf{u} \tag{7.29}$$

where

$$[L] = \begin{bmatrix} \dfrac{\partial}{\partial x_1} & 0 & 0 & 0 & \dfrac{\partial}{\partial x_3} & \dfrac{\partial}{\partial x_2} \\ 0 & \dfrac{\partial}{\partial x_2} & 0 & \dfrac{\partial}{\partial x_3} & 0 & \dfrac{\partial}{\partial x_1} \\ 0 & 0 & \dfrac{\partial}{\partial x_3} & \dfrac{\partial}{\partial x_2} & \dfrac{\partial}{\partial x_1} & 0 \end{bmatrix}$$

and the total strain is decomposed into an elastic ϵ_E and a creep ϵ_C component

$$\epsilon = \epsilon_E + \epsilon_C \tag{7.30}$$

such that the elastic relation is

$$\boldsymbol{\sigma} = [D]\epsilon \tag{7.31}$$

where $[D]$ is the 6 × 6 matrix of elastic coefficients. Thermal strains are neglected for simplicity.

The basic concepts of the finite element displacement method is the representation of a structure by a model structure consisting of a finite assemblage of *simple elements* of finite size (*Figure 7.16*). These elements are connected to

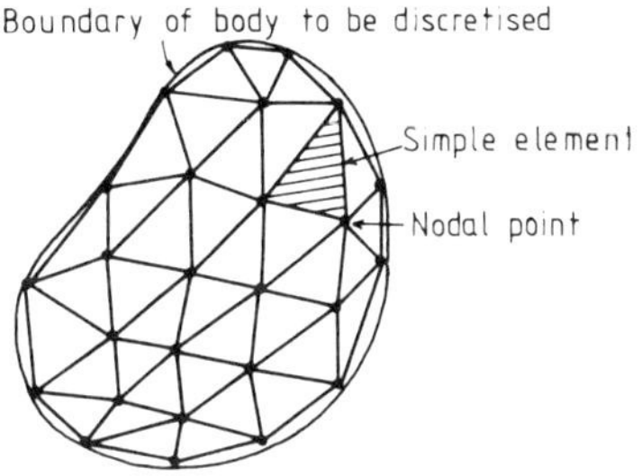

Figure 7.16 Discretisation of a continuous body using simple elements

adjacent elements at a finite number of points called *nodal points*. The displacement field **u** at any point within an element is expressed in terms of known *shape functions* $[N]_e$

$$\mathbf{u} = [N]_e \mathbf{a}_e$$

where the vector represents a series of parameters, originally associated with the displacements at the nodes, but it is not necessary to restrict them as such.

If we denote by **a** the aggregate of (unknown) parameters determining the approximate solution, then we can formally assemble the element trial expansions of **u** into a global form

$$\mathbf{u} = [N]\mathbf{a} \tag{7.32}$$

The dimensions of **a** will of course depend on the discretisation.

The aim of the displacement method is to derive suitable equations for the parameters **a** – however, this procedure must be adapted to deal with the transient effects of creep. Rather than derive the appropriate initial value problem by definition from Section 7.3, we will obtain them from first principles by way of illustration.

As is common in most finite element analyses, we will start from Green's theorem (i.e. virtual work). If the stress σ is in equilibrium with body force $\bar{\mathbf{b}}$ and surface forces $\bar{\mathbf{p}}$ on part of the boundary S_p and if the strain and displacement are compatible with prescribed surface displacements, then

$$\int_V \epsilon^T \sigma \; dV - \int_V \mathbf{u}^T \bar{\mathbf{b}} \; dV - \int_{S_p} \mathbf{u}^T \bar{\mathbf{p}} \; dS = 0$$

where superscript T denotes transpose (and thus $\boldsymbol{\epsilon}^T \sigma$ is the vector product; recall that we have adopted *matrix* notation).

Using equations (7.29) and (7.32), this reduces to

$$\mathbf{a}^{\mathrm{T}}\left(\int_V ([L]\,[N])^{\mathrm{T}}\,\sigma\,\mathrm{d}V - \int_V [N]^{\mathrm{T}}\,\bar{\mathbf{b}}\,\mathrm{d}V - \int_{S_{\mathrm{p}}} [N]^{\mathrm{T}}\bar{\mathbf{p}}\;\;\mathrm{d}S\right) = 0$$

and we equate the expression inside the brackets to zero so that

$$\int_V [B]^{\mathrm{T}}\sigma\,\mathrm{d}V - \mathbf{F} = 0$$

where

$$[B] = [L]\,[N] \qquad \mathbf{F} = \int_V [N]^{\mathrm{T}}\,\bar{\mathbf{b}}\,\mathrm{d}V + \int_{S_{\mathrm{p}}} [N]^{\mathrm{T}}\,\bar{\mathbf{p}}\,\mathrm{d}S$$

Then substituting the constitutive relations (7.30) and (7.31), there results

$$[K]\,\mathbf{a} - \mathbf{F}_{\mathrm{c}} - \mathbf{F} = 0 \tag{7.33}$$

on defining the *stiffness matrix*

$$[K] = \int_V [B]^{\mathrm{T}}\,[D]\,[B]\,\mathrm{d}V$$

and the creep force vector is

$$\mathbf{F}_{\mathrm{c}} = \int_V [B]^{\mathrm{T}}\,[D]\,\epsilon_{\mathrm{c}}\,\mathrm{d}V$$

The equations of stress redistribution are then obtained as follows. From equations (7.30), (7.31) and (7.32) in rate form

$$\dot{\sigma} = [D]\,[B]\,\dot{a} - [D]\,\dot{\epsilon}_{\mathrm{c}}$$

and from a solution of equation (7.33) we have

$$\dot{\sigma} = [D]\,[B]\,[K]^{-1}\,(\dot{\mathbf{F}}_{\mathrm{c}} + \dot{\mathbf{F}}) - [D]\,\dot{\epsilon}_{\mathrm{c}}$$

Thus from the creep constitutive relation, say, for simplicity's sake

$$\dot{\epsilon}_{\mathrm{c}} = \mathbf{f}(\sigma, t)$$

we have the initial value problem

$$\frac{\mathrm{d}}{\mathrm{d}t}(\sigma - \sigma^{\mathrm{o}}) = [D]\;[B]\,[K]^{-1}\left(\int_V [B]^{\mathrm{T}}\,[D]\,\mathbf{f}(\sigma, t)\,\mathrm{d}V\right) - [D]\,\mathbf{f}(\sigma, t) \tag{7.34}$$

where σ^{o} is the equivalent elastic stress

$$\sigma^{\mathrm{o}} = [D]\,[B]\,[K]^{-1}\,\mathbf{F}$$

Finally the initial condition is

$$\sigma(0) = \sigma^{\mathrm{o}}(0) \tag{7.35}$$

Attention is now focused on a finite set of values σ_i, $i = 1, 2, \ldots, G$ at G *integrating points* within V so that any integration over the volume becomes

$$\int_V f(\mathbf{x})\,\mathrm{d}V = \sum_{i=1}^{G} c_k f(\mathbf{x}_k)$$

where the numerical quadrature formula is left unspecified, but the weighting coefficients c_k are positive.

If we denote by σ_i, $[B_i]$, the values of σ and $[B]$ at these integrating points $i = 1, 2, \ldots, G$, then the equation (7.34) is replaced by the set of equations

$$\frac{d}{dt}\sigma_i = [D]\,[B_i]\,[K]^{-1}\left(\sum_{j=1}^{G} c_j [B_j]^{T}[D]\,\mathbf{f}(\sigma_j, t)\right) - [D]\,\mathbf{f}(\sigma_i, t) + \dot{\sigma}_j^{o} \qquad (7.36)$$

$i = 1, 2, \ldots, G$

A similar set of equations can be developed for the evolution of inelastic strain ϵ_c. Without following a detailed derivation, the resulting equations are

$$\frac{d}{dt}\epsilon_{ci} = \mathbf{f}\left([D]\,[B_i]\,[K]^{-1}\sum_{j=1}^{G} c_j [B_j]^{T}[D]\,\epsilon_{cj} - [D]\,\epsilon_{ci} + \sigma_i^{o}, t\right) \qquad (7.37)$$

$i = 1, 2, \ldots, G$

In practice, the equations (7.36) or (7.37) are not directly solved in this form; rather, certain computational algorithms are adopted. We describe two of these below.

7.6.1 Explicit Euler algorithm

This is perhaps the best established of all the computational schemes for a finite element implementation of an incremental solution to a creep problem. The algorithm proceeds fairly obviously as follows:

(1) Solve initial elastic problem by forming $\mathbf{F}$, $[K]^{-1}$. Set $\epsilon_{ci} = 0$ at each integrating point and evaluate ($t = 0$)

$$\sigma_i = [D]\,[B_i]\,[K]^{-1}\,\mathbf{F}$$

(2) Determine time increment Δt and calculate $\dot{\epsilon}_{ci}$ from the constitutive relation

$$\dot{\epsilon}_{ci} = \mathbf{f}(\sigma_i, t)$$

(3) Increment the creep strains

$$\epsilon_{ci} = \epsilon_{ci} + \Delta t\,\dot{\epsilon}_{ci}$$

$$t = t + \Delta t$$

(4) Form $\mathbf{F}$ and $\mathbf{F}_c$ according to

$$\mathbf{F}_c = \sum_{j=1}^{G} c_j\,[B_j]^{T}\,[D]\,\epsilon_{cj}$$

(5) Determine

$$\mathbf{a} = [K]^{-1}\,(\mathbf{F} + \mathbf{F}_c)$$

$$\sigma_i = [D]\,[B_i]\,\mathbf{a} - [D]\,\epsilon_{ci}$$

(6) Output desired σ_i and displacement $\mathbf{u} = [N]\,\mathbf{a}$, etc.

(7) If predetermined time reached, terminate computation. Otherwise return to step (2).

This algorithm is usually called the *initial strain* method, since we are in fact solving equation (7.37). If we had solved equation (7.36) by incrementing the stresses then we would have the *initial stress* method. Clearly, various refinements can be added to control the computation in order to make it automatic, for example in the choice of the current time increment in step (2). The algorithm is fairly simple in theory, but there are a number of computational details in the finite element implementation which need to be discussed.

The most important of these is in the formation of the stiffness matrix $[K]$ and the solution of the full system of simultaneous equations for **a**. In practice equation solvers are used and the stiffness matrix is evaluated and stored in some partially reduced form at the start of the calculation and subsequently recalled for each time increment. This results in a considerable saving in computer time.

7.6.2 Implicit Euler algorithm

We describe here an implicit scheme based on an application of the first order implicit scheme to equation (7.37). It is a variation of the initial strain method outlined above. The algorithm proceeds as follows for some chosen $\alpha \geqslant ½$:

(1) Solve initial elastic problem by forming **F**, $[K]^{-1}$; set $\epsilon_{ci} = 0$ at each integrating point and evaluate

$$\sigma_i(0) = [D][B_i][K]^{-1}\mathbf{F}$$

(2) Determine time increment Δt and calculate $\dot{\epsilon}_{ci}$ from the constitutive relation

$$\dot{\boldsymbol{\epsilon}}_{ci} = \mathbf{f}(\sigma_i, t)$$

(3) Set

$$\dot{\epsilon}_{ci}^{t+\Delta t} = \dot{\epsilon}_{ci} \qquad\qquad \boldsymbol{\epsilon}_{ci}^{\text{old}} = \epsilon_{ci}$$

(4) Evaluate

$$\boldsymbol{\epsilon}_{ci}^{t+\Delta t} = \epsilon_{ci} + \Delta t\,((1-\alpha)\,\dot{\epsilon}_{ci} + \alpha\,\dot{\epsilon}_{ci}^{t+\Delta t})$$

(5) If max $|\epsilon_{ci}^{t+\Delta t} - \epsilon_{ci}^{\text{old}}| < \epsilon$, some prescribed error tolerance, then proceed to step (10).

(6) Form $\mathbf{F}^{t+\Delta t}$ and $\mathbf{F}_{\text{c}}^{t+\Delta t}$ according to

$$\mathbf{F}_{\text{c}}^{t+\Delta t} = \sum_{j=1}^{G} c_j[B_j]^{\text{T}}[D]\,\epsilon_{cj}^{t+\Delta t}$$

(7) Determine

$$a^{t+\Delta t} = [K]^{-1}(\mathbf{F}^{t+\Delta t} + \mathbf{F}_{\text{c}}^{t+\Delta t})$$

$$\sigma_j^{t+\Delta t} = [D][B_i]\mathbf{a}^{t+\Delta t} - [D]\,\epsilon_{ci}^{t+\Delta t}$$

(8) Set

$$\dot{\epsilon}_{ci}^{t+\Delta t} = f(\sigma_i^{t+\Delta t}, t+\Delta t) \qquad \epsilon_{ci}^{\text{old}} = \epsilon_{ci}^{t+\Delta t}$$

(9) Return to step (4).

(10) Set

$$\sigma_i = \sigma_i^{t+\Delta t} \qquad \mathbf{a} = \mathbf{a}^{t+\Delta t} \qquad t = t + \Delta t$$

(11) Output desired σ_i and displacement $\mathbf{u} = [N]\,\mathbf{a}$, etc.

(12) If prescribed time reached, then terminate; otherwise return to step (2).

This is the most straightforward implicit scheme which can be envisaged. Unfortunately, convergence of the iteration loop in steps (3) – (9) can be slow. An alternative would be to employ quasilinearisation, i.e. equation (7.20), but this is rarely done in practice. There are numerous suggestions in the literature for accelerating this convergence by suitable redefinition of the problem. We cannot go into these here, but the reader is referred to the references at the end.

Further reading

Hoff, N.J., Approximate analysis of structures in the presence of moderately large creep deformations, *Q. J. Appl. Math.*, **12**, 49–55, 1954.

Goodey, W.J., Creep deflexion and stress distribution in a beam, *Aircraft Eng.*, **30**, 170–1, 1958.

Mendelsohn, A., *et al.*, A general approach to the practical solution of creep problems, *Trans. ASME, J. Basic Eng.*, **81**, 585–9, 1959.

Lin, T.H., *Theory of Inelastic Structures*, Wiley, 1968.

Penny, R.K. and Marriott, D.L., *Design for Creep*, McGraw-Hill, 1971.

Boyle, J.T., A rational approach to creep mechanics, *Arch. Mech.*, **29**, 229–49, 1977.

Gear, C.W., *Numerical Initial Value Problems in Ordinary Differential Equations*, Prentice-Hall, 1971.

Hall, G., and Watt, J.M., (Eds), *Modern Numerical Methods for Ordinary Differential Equations*, Clarendon Press, 1976.

Cormeau, I., Numerical stability in quasi-static elasto-visco-plasticity, *Int. J. Numer. Meth. Eng.*, **9**, 109–27, 1975.

Shih, C.F. *et al.*, A stable computational scheme for stiff time dependent constitutive equations, *Proc. 4th Int. Conf. on Structural Mechanics in Reactor Technology, San Francisco, 1977*, Paper L2/2.

Krieg, R.D., Numerical integration of some new unified plasticity–creep formulations, *Proc. 4th Int. Conf. on Structural Mechanics in Reactor Technology, San Francisco, 1977*, Paper M64.

Kumar, V. *et al.*, Numerical integration of some stiff constitutive models of inelastic deformation, *Trans. ASME, J. Eng. Mat. Techn.*, **102**, 92–6, 1980.

Hayhurst, D.R. and Krzeczkowski, A.J., Numerical solution of creep problems, *Comp. Meth. Appl. Mech. Eng.*, **20**, 151–71, 1979.

Mukherjee, S. *et al.*, Elevated temperature inelastic analysis of metallic media under time varying loads using state variable theories, *Int. J. Solids Struct.*, **14**, 663–79, 1978.

Chang, K.J. *et al.*, Inelastic bending of beams under time varying moments a state variable approach, *Trans. ASME, J. Pressure Vessel Techn.*, **101**, 305–10, 1979.

Delph, T.J., Creep, relaxation and cyclic behaviour of a beam using a state-variable constitutive model, *Nucl. Eng. Des.*, **65**, 411–21, 1981.

Zienkiewkz, O.C., *The Finite Element Method*, McGraw-Hill, 3rd Edn, 1977.

Greenbaum, G.A. and Rubenstein, M.F., Creep analysis of axisymmetric bodies using finite elements, *Nucl. Eng. Des.*, **7**, 378–97, 1968.

Carpenter, W.C. and Gill, P.A.T., Automated solutions of time dependent problems, in *The Mathematics of Finite Elements and Applications,* Ed. J.R. Whiteman, Academic Press, 1973.
Donea, J., The application of computer methods to creep analysis, in *Creep of Engineering Materials and Structures,* Eds. G. Bernasconi and G. Piatti, Applied Science Pub., 1979.
Zienkiewicz, O.C. and Cormeau, I.C., Visco-plasticity – plasticity and creep in elastic solids – a unified numerical solution approach, *Int. J. Numer. Meth. Eng.,* 8, 821–45, 1974.
Hughes, T.J.R., and Taylor, R.L., Unconditionally stable algorithms for quasi-static elasto-visco-plastic finite element analysis, *Comp. Struct.,* 8, 169–72, 1978.
Argyris, J.H. *et al.*, Improved solution methods for inelastic rate problems, *Comp. Meth. Appl. Mech. Eng.,* **16**, 231–77, 1978.
Synder, M.D. and Bathe, K.J., A solution procedure for thermo-elasto-plastic and creep problems, *Nucl. Eng. Des.,* **64**, 49–80, 1981.
Owen, D.R.J. and Hinton, E., *Finite Elements in Plasticity: Theory and Practice,* Pineridge Press, 1980.

Chapter 8

Approximate solution of transient creep problems

The numerical solution techniques described in Chapter 7 have general applicability, yet for large structures and complex loading histories they can be expensive, if not beyond available computing capacity. Even for more tractable problems, it is always good engineering practice to have some simple check on the calculations. While such a check is usually straightforward in elasticity, for material creep it may not be quite so obvious. But no matter what the complexity of the problem, it is often the case that the accuracy and detail of the solution can be well in excess of our confidence in the material model of the basic creep data – and bearing in mind that knowledge of the virgin state of the actual structure is hazy, to say the least. There is a place for *simplified* methods of stress analysis for transient creep – particularly for variable loading – which avoids a detailed assessment of the history of the structure.

In creep mechanics, simplification of the problem can be achieved on several levels according to the particular physical or mechanical principle that is being relaxed. It is worth describing these in some more detail, although there are basically two interrelated approaches.

(1) *Simplification of the problem* This of course is done in any stress analysis – the discretisation of a structure for numerical solution is an obvious example. However, in transient creep, three possible simplifications can be achieved:

(i) The model of the *material* behaviour can be simplified. Since we deal almost exclusively with phenomenological theories of creep, this is always done to some extent. Nevertheless, the most obvious simplification is the use of a time-hardening model over a more complex internal-variable model. Ignoring elastic strains is another example, and the use of the simple power law another. In simplifying the material behaviour, it is of course important to use a model which at least qualitatively characterises the particular response which is expected, for example, fast transients or recovery.

(ii) The *geometry* can be simplified. This can be done directly by ignoring parts of the structure – for example, assuming a thin tube is infinitely long rather than of finite length with end connections – or indirectly. We have already met some indirect simplifications in the generalised models of Chapter 4 and 5 which allow the number of spatial dimensions being considered to be reduced.

This would also involve some simplification of the material behaviour – for example, in a thin shell assuming all points through the thickness respond in the same way at the same time.

(iii) The *loading* can be simplified. We can assume constant loads throughout the life of a structure even if this means ignoring start-up and possible short-term load or heat loss. For example, it is common in creep to ignore the transient heat transfer problem. Another obvious simplification is to assume long-term cyclic loading – even if this is often in the short term. These are simplifications of the loading history – it is also, of course, possible to simplify the spatial character of the loading, for example, to assume that it is symmetrical on a symmetric structure. This is linked to the geometrical simplification as well.

None of the above simplifications to the problem necessarily allows us to avoid a transient analysis. This can be done as follows:

(2) *Characterisation of the solution in terms of known solutions* The essence of this approach is to characterise the transient solution in terms of non-transient (stationary) solutions. Such known solutions are commonly the equivalent elastic or steady state creep solutions, although additional stress analyses can also be done – for example, for one cycle of a loading history, or to calculate the deformation rates on loading. It is also important to *bound* the actual solution and, ideally, to have some measure of the likely error, although the latter might not always be possible.

Obviously both of these approaches can be used at the same time – indeed it will be shown that we can only get meaningful bounds for particular material idealisations.

8.1 Constant load – isothermal creep

In this, we will use the notation introduced in Section 5.4.1 and initially we will assume a time hardening theory of creep so that the constitutive relation is given by

$$\epsilon = \epsilon_E + \epsilon_C$$

$$\boldsymbol{\epsilon}_E = \mathbf{C}\sigma \qquad \dot{\epsilon}_C = g(t)\,\frac{\mathrm{d}\Omega}{\mathrm{d}\sigma} \tag{8.1}$$

where $\mathbf{C}$ is the matrix of elastic constants. We will make particular use of the monotonicity property of the creep law

$$(\dot{\epsilon}_C^B - \dot{\epsilon}_C^A).(\sigma^B - \sigma^A) \geqslant 0 \tag{8.2}$$

together with the implied convexity of the potential Ω, equation (5.32).

Since we have adopted a time hardening theory of creep, we can introduce the normalised time scale

$$\tau = \int g(t)\,\mathrm{d}t$$

However, *for notational simplicity, we will replace* τ *by* t without confusion.

The transient creep response of a structure to constant load we have seen briefly in the previous chapter: *Figure 8.1* shows the type of behaviour observed. Some typical deflection, once it has reached the steady state, is composed of

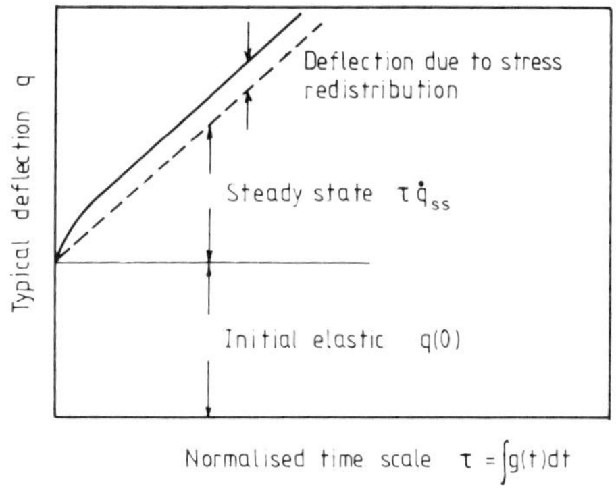

Figure 8.1 Time variation of a typical displacement in a creeping structure under constant load

three parts – the initial elastic deflection, the accumulated steady-state deflection rate and a deflection due to the transient phase of redistribution. Intuitively, the simple addition of the elastic and steady state should provide a lower estimate of the accumulated deformation. Indeed, this is the essence of the method of superposition of deformation states introduced in Section 3.3. However, it would be safer also to have an upper estimate – an upper bound on the deflection due to the stress redistribution. In the following we will develop these intuitive concepts for a general structure.

Firstly, although it may seem obvious, we need to know that the structure under constant load will approach the steady state. The structural problem we consider is that of a structure of volume V bounded by a surface S supported on part of its surface and subject to generalised point loads Q_k, $k = 1, 2, \ldots$ (*Figure 8.2*). We will examine the generalised deflections q_k, $k = 1, 2, \ldots,$ in the direction

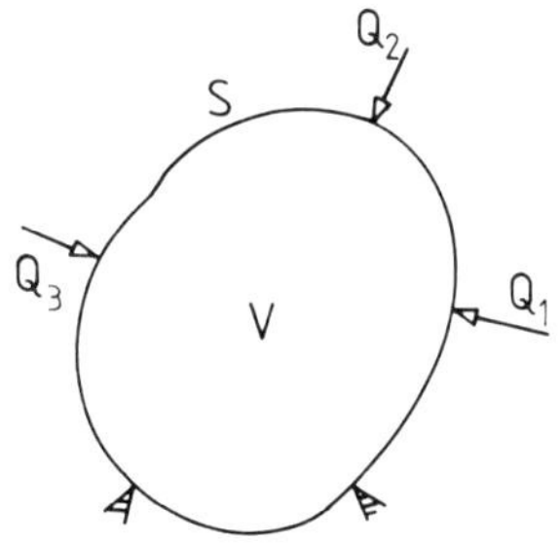

Figure 8.2 Geometry of continuous body under the action of generalised point loads

of these loads. (The following may be adapted to deal with body forces and surface tractions – but we avoid these for clarity.) We will make use of Green's theorem in the rate form: If σ^A is in equilibrium with generalised loads $\mathbf{Q}^A$ and if $\dot{\epsilon}^B$ is compatible with the corresponding generalised deflection $\dot{\mathbf{q}}^B$ and the support conditions, then

$$\mathbf{Q}^A . \dot{\mathbf{q}}^B = \int_V \sigma^A . \dot{\epsilon}^B \, dV \tag{8.3}$$

We can establish the following result.

Result 1 In the structural problem for constant loads, the stresses approach the steady-state stresses.

Let σ, ϵ be the actual stress and strain and σ_{ss}, $\dot{\epsilon}_{ss}$ the steady-state stress and strain rate. Then the difference $\sigma - \sigma_{ss}$ is self-equilibrating with zero loads, and application of Green's theorem gives

$$\int_V (\sigma - \sigma_{ss}) \,.\, \dot{\epsilon} \; \mathrm{d}V = 0$$

Then from the constitutive relation, equations (8.1) and (8.2), we have

$$\frac{\mathrm{d}}{\mathrm{d}t} \in(\sigma - \sigma_{ss}) = - \int_V (\sigma - \sigma_{ss}).\, \dot{\epsilon}_C \; \mathrm{d}V$$

where we have introduced the *elastic energy* associated with a stress field σ denoted by

$$\in(\sigma) = \frac{1}{2} \int_V C\sigma \,.\, \sigma \, \mathrm{d}V$$

The term on the right-hand side can be rearranged as

$$\int_V (\sigma - \sigma_{ss}) \,.\, \dot{\epsilon}_C \; \mathrm{d}V = \int_V (\dot{\epsilon}_C - \dot{\epsilon}_{ss}) \,.\, (\sigma - \sigma_{ss}) \, \mathrm{d}V + \int_V \dot{\epsilon}_{ss} \cdot (\sigma - \sigma_{ss}) \, \mathrm{d}V \quad (8.4)$$

From monotonicity, equation (8.2), the first term is positive. Also from an application of Green's theorem, since $\sigma - \sigma_{ss}$ is self-equilibrating, the second term is zero. Therefore

$$\frac{\mathrm{d}}{\mathrm{d}t} \; \mathrm{E}(\sigma - \sigma_{ss}) \leqslant 0$$

This means that the elastic energy of the difference stress state is decreasing, and it is also bounded below (since it is positive); it must therefore tend to a limit. To prove the result it remains to demonstrate that this limit is zero.

Since $\mathrm{E}(\sigma - \sigma_{ss})$ tends to a limit,

$$t \to \infty \qquad \frac{\mathrm{d}}{\mathrm{d}t} \; \mathrm{E}(\sigma - \sigma_{ss}) \to 0$$

and from equation (8.4)

$$\int_V (\dot{\epsilon}_C - \dot{\epsilon}_{ss}) \,.\, (\sigma - \sigma_{ss}) \, \mathrm{d}V \to 0$$

Now, equality in the monotonicity condition, equation (8.2), is presumed to hold if and only if $\sigma^B = \sigma^A$. Therefore, in the limiting state, we must have $\sigma = \sigma_{ss}$. Hence

$$\mathrm{E}(\sigma - \sigma_{ss}) \to 0$$

The result is therefore established. The elastic energy of the difference between the instantaneous stress and the steady state gives us a measure of how close they are.

We can be sure, for the structural problem that we have posed, that a steady state will eventually be reached. The next problem is to characterise the transient solution in terms of the elastic and steady-state solutions in a more rigorous fashion. In particular, we want to know if we can generalise our intuitive concepts regarding superposition of deformation states.

From Green's theorem, equation (8.3), for the actual stress and strain rate

$$\mathbf{Q}\,.\,\dot{\mathbf{q}} = \int_V \sigma\,.\,\dot{\epsilon}\,\mathrm{d}V = \int_V \sigma\,.\,\dot{\epsilon}_C\,\mathrm{d}V + \frac{\mathrm{d}}{\mathrm{d}t}\,\mathrm{E}(\sigma)$$

Into this expression we can introduce the steady-state solutions σ_{ss}, $\dot{\epsilon}_{ss}$ and $\dot{\mathbf{q}}_{ss}$ related by

$$\mathbf{Q}\,.\,\dot{\mathbf{q}}_{ss} = \int_V \sigma_{ss}\,.\,\dot{\epsilon}_{ss}\;\mathrm{d}V$$

and on integrating with respect to time

$$\mathbf{Q}\,.\,\mathbf{q} = \mathbf{Q}\,.\,(\mathbf{q}_0 + \dot{\mathbf{q}}_{ss}t + \Delta\mathbf{q}) \tag{8.5}$$

where we have defined

$$\mathbf{Q}\,.\,\Delta\mathbf{q} = \int_0^t \int_V (\sigma\cdot\dot{\epsilon}_C - \sigma_{ss}\,.\,\dot{\epsilon}_{ss})\,\mathrm{d}V\,\mathrm{d}t + \mathrm{E}(\sigma(t)) - \mathrm{E}(\sigma_0)$$

with σ_0, $\mathbf{q}_0$ the initial elastic solutions.

The relation given by equation (8.5) is an expression of the concept of superposition of states in terms of work. Indeed, we can write

$$\mathbf{q}(t) = \mathbf{q}_0 + \dot{\mathbf{q}}_{ss}t + \Delta\mathbf{q} \tag{8.6}$$

although, except for single loading, we cannot easily characterise $\Delta\mathbf{q}$ – merely the work term $\mathbf{Q}\,.\,\Delta\mathbf{q}$ which we may use as a measure of the likely error. We would like to estimate this error, but surprisingly this is difficult in general. Nevertheless, we can derive particular results for the *power law* since we have

$$\int_V \sigma\,.\,\dot{\epsilon}_C\,\mathrm{d}V = (n+1)U_c(\sigma) \qquad\qquad U_c(\sigma) = \int_V \Omega\,\mathrm{d}V$$

so that

$$\mathbf{Q}\,.\,\Delta\mathbf{q} = (n+1)\int_0^t [U_c(\sigma) - U_c(\sigma_{ss})]\,\mathrm{d}t + \mathrm{E}(\sigma(t)) - \mathrm{E}(\sigma_0) \tag{8.7}$$

We derive the following bounds on the accumulated error.

Result 2 For the power law, once a structure is in the steady state, the following bounds on the accumulated error hold

$$\mathrm{E}(\sigma_{ss}) - \mathrm{E}(\sigma_0) \leqslant \lim_{t\to\infty} \mathbf{Q}\,.\,\Delta\mathbf{q} \leqslant (n+2)\,[\mathrm{E}(\sigma_{ss}) - \mathrm{E}(\sigma_0)]$$

The lower bound is immediately obvious on applying the theorem of minimum complementary dissipation in steady creep, Section 5.4.3, to the first term in equation (8.7): since σ is statically admissible

$$U_c(\sigma) \geqslant U_c(\sigma_{ss})$$

Therefore, for all time,

$$\mathbf{Q}\,.\Delta\mathbf{q} \geqslant \mathrm{E}(\sigma(t)) - \mathrm{E}(\sigma_0)$$

This lower bound is clearly positive by application of the theorem of minimum complementary energy in elasticity. The lower bound results on taking the limit as $t\to\infty$ where $\sigma\to\sigma_{ss}$.

The upper bound is less obvious. We make use of the convexity condition, equation (5.32).

$$\Omega(\sigma^A) - \Omega(\sigma^B) \leqslant (\sigma^A - \sigma^B) \, . \, \dot{\epsilon}_C^A$$

identifying $\sigma^A = \sigma$, $\sigma^B = \sigma_{ss}$ and integrating over the volume so that

$$U_c(\sigma) - U_c(\sigma_{ss}) \leqslant \int_V (\sigma - \sigma_{ss}) \, . \, \dot{\epsilon}_C \, dV$$

$$= \int_V (\sigma - \sigma_{ss}) \, . \, \dot{\epsilon} \, dV - \int_V (\sigma - \sigma_{ss}) \, . \, C\dot{\sigma} \, dV$$

The first term is zero by Green's theorem since $\sigma - \sigma_{ss}$ is self-equilibrating. Hence

$$\int_0^t U_c(\sigma) - U_c(\sigma_{ss}) \, dt \leqslant \int_V \sigma_{ss} C(\sigma - \sigma_0) \, dV - [E(\sigma) - E(\sigma_0)]$$

But again by Green's theorem, since both σ_{ss} and σ_o are statically admissible,

$$\int_V \sigma_{ss} \, . \, \epsilon_0 \, dV = \int_V \sigma_0 \, . \, \epsilon_0 \, dV = 2E(\sigma_0)$$

and since $\sigma \to \sigma_{ss}$, $t \to \infty$

$$\lim_{t \to \infty} \int_V \sigma_{ss} . C\sigma \, dV = 2E(\sigma_{ss})$$

Therefore

$$\lim_{t \to \infty} \int_0^t [U_c(\sigma) - U_c(\sigma_{ss})] \, dt \leqslant E(\sigma_{ss}) - E(\sigma_0)$$

and the result is proved.

8.2 Some examples

We demonstrate these results for two elementary problems: the simple hinged-bar structure of Section 3.3 and the rectangular beam in bending of Section 3.1.

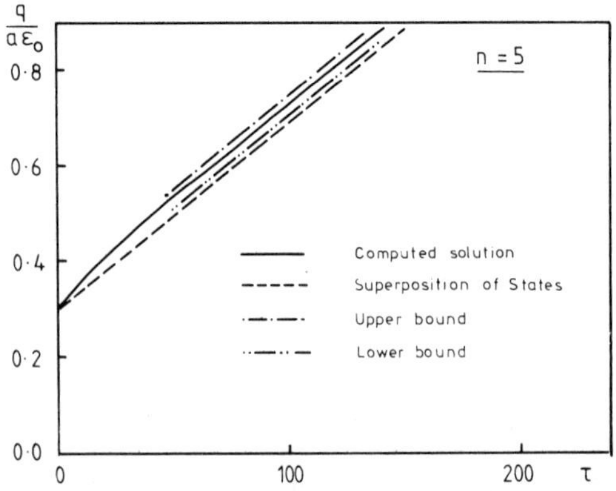

Figure 8.3 Comparison of computed and bounding displacements for hinged bar structure

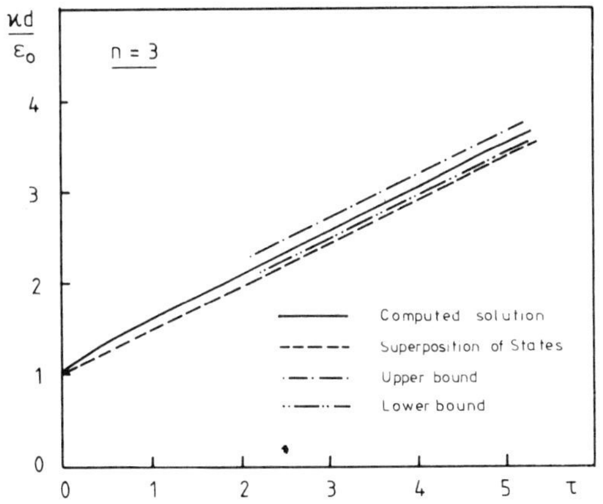

Figure 8.4 Comparison of computed and bounding displacements for beam in bending

In both cases the actual behaviour has been obtained using the numerical techniques of Chapter 7. These computed results are compared in *Figures 8.3* and *8.4* with the superposition of states approximation, equation (8.6), including the upper and lower bounds from Result 2. The latter can be evaluated as follows.

In the case of the hinged-bar structure a normalised time scale

$$\tau = E\int g(t)\,\mathrm{d}t\ \sigma_0^{n-1}$$

has been defined for convenience with $\sigma_0 = 3Q/2A$. If q is the vertical displacement of the point of application of the load Q (*Figure 3.3*), then the superposition of states approximation can be expressed as

$$q = q(0) + \tau\frac{\mathrm{d}}{\mathrm{d}\tau}q_{ss} + \Delta q$$

where the initial elastic solution is

$$q(0)/a\epsilon_0 = \frac{3}{10}$$

with $\epsilon_0 = \sigma_0/E$, and the steady-state solution is

$$\frac{\mathrm{d}}{\mathrm{d}\tau}\frac{q_{ss}}{a\epsilon_0} = \frac{3/2}{(1+2^{1+1/n})^n}$$

The error term Δq can be bounded from Result 2 as

$$\delta \leqslant \Delta q/a\epsilon_0 \leqslant (n+2)\delta$$

where the normalised elastic energy of the difference between the steady and elastic stresses is given by

$$\delta = \frac{1}{2}\left[\left(\frac{2^{1/n}}{1+2^{1+1/n}}\right)^2 + \left(\frac{1}{1+2^{1+1/n}}\right)^2 - \left(\frac{2}{5}\right)^2 - \left(\frac{1}{5}\right)^2\right]$$

This is evaluated below for several values of the exponent n:

n	3	5	7	9	11
δ	0.0044	0.0067	0.0078	0.0085	0.0089

In the case of the beam bending with a rectangular cross-section, Section 3.1, a normalised time scale

$$\tau = E\sigma_0^{n-1}\int g(t)\,\mathrm{d}t$$

has also been used, with $\sigma_0 = 3M/2bd^2$, $\epsilon_0 = \sigma_0/E$. The superposition of states approximation, equation (8.6), applied to the curvature κ is

$$\kappa = \kappa(0) + \tau\frac{\mathrm{d}}{\mathrm{d}\tau}\kappa + \Delta\kappa$$

where the initial and steady-state solutions are

$$\frac{\kappa(0)d}{\epsilon_0} = 1 \qquad \frac{\mathrm{d}}{\mathrm{d}\tau}\left(\frac{\kappa_{ss}d}{\epsilon_0}\right) = \left[\frac{2}{3}\left(1 + \frac{1}{2n}\right)\right]^n$$

The error term $\Delta\kappa$ can easily be bounded from Result 2 as

$$\delta \leqslant \Delta\kappa d/\epsilon_0 \leqslant (n+2)\delta$$

with

$$\delta = \left[\frac{n}{n+2}\left(1 + \frac{1}{2n}\right)^2 - \frac{3}{4}\right]\frac{2}{3}$$

which is evaluated below:

n	3	5	7	9	11
δ	0.0444	0.0762	0.0953	0.1077	0.1165

In both these examples the superposition of states approximation, itself a lower bound, is a fairly good representation of the actual behaviour, since the effect of the transient period is relatively small. The upper and lower bounds on the accumulated error in the steady state diverge to a significant degree as the exponent n increases. This is perhaps the price to be paid for obtaining such simple bounds in the first place.

8.3 Constant load – non-isothermal creep

The preceding results may be extended to deal with non-isothermal creep. Suppose there is a thermal gradient in the structure which induces a thermal strain ϵ_T so that equation (8.1) becomes

$$\epsilon = \epsilon_E + \epsilon_C + \epsilon_T$$

and the constitutive relation for $\dot{\epsilon}_C$ (and ϵ_E if necessary) is temperature-dependent. It is presumed that the thermal strain is constant in time. Then it can be verified that Result 1 is retained so that the structure will tend to the steady state. Again we are interested in the superposition of states approximation, equation (8.6)

$$q = q_0 + \dot{q}_{ss}t + \Delta q$$

However, for non-isothermal creep, the upper bound on the error must be modified as

$$Q \cdot \Delta q \leqslant (n+2)\,[E(\sigma_{ss}) - E(\sigma_0)] + (n+1)\int_V \sigma_{ss} \cdot \epsilon_T \, dV$$

Moreover, we cannot assert that the lower bound is necessarily positive, since from the theorem of minimum complementary energy in thermoelasticity we merely have

$$E(\sigma(t)) - E(\sigma_0) \geqslant \int_V (\sigma_0 - \sigma) \cdot \epsilon_T dV \rightarrow \int_V (\sigma_0 - \sigma_{ss}) \cdot \epsilon_T \, dV$$

at large times $t \rightarrow \infty$.

By way of illustration, the superposition of states approximation is applied to the three-bar structure of Sections 3.2 and 7.4.4 (*Figure 8.5*). It can be seen that the approximation is not so good in this case. Further, if the bounds on the

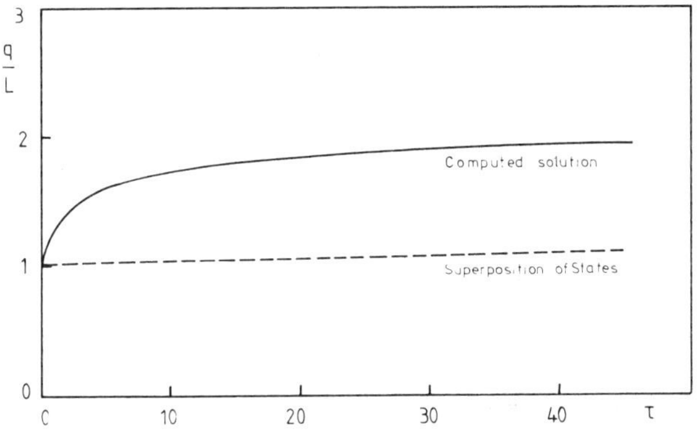

Figure 8.5 Comparison of computed displacement with superposition of states approximation for non-uniformly heated three-bar structure

error are evaluated (this is left to the reader as an exercise), it is found that they are poorer than in the isothermal case.

8.4 Variable loading

The bounding results that we have developed for constant loading are rather special since we knew beforehand what the ultimate state of the structure would be. This knowledge allowed us to characterise the gross effects of the transient period, before the steady state was reached, in terms of the initial and ultimate state of the structure – that is, in terms of the elastic and steady-state stresses. In addition, the proofs took particular advantage of the fact that the loading did not vary in time. Unfortunately, if we consider variable loading, we cannot use these techniques – indeed, in general, we can claim that the structure is always in the transient state (although we can do something better for cyclic loading as we shall see later) and will not assume some well defined asymptotic state.

Nevertheless, some weaker results can be developed which do allow us to express the solution in terms of known solutions.

Again we presume the special case of time hardening creep with a power law and make particular use of Martin's inequality in a slightly different form: from equation (5.42) identify

$$\sigma^{A} = \frac{n}{n+1}\,\sigma^* \qquad \sigma^{B} = \sigma$$

Then after rearrangement

$$(\sigma^* - \sigma) \,.\, \dot{\epsilon}_C \leqslant \left(\frac{n}{n+1}\right)^n \Omega(\sigma^*) \tag{8.8}$$

We may also rearrange the convexity condition as

$$(\sigma^* - \sigma) \,.\, \dot{\epsilon}_C \leqslant \Omega(\sigma^*) - \Omega(\sigma) \tag{8.9}$$

where $\dot{\epsilon}_C$ is the creep strain corresponding to $\boldsymbol{\sigma}$.

The structural problem that we consider is as follows: a body of volume V supported on part of its surface S is subject to a history of thermal strains $\epsilon_T(t)$ and a history of generalised loads $\mathbf{Q}(t)$; there are no body forces. The resultant actual stress and strain we will denote by $\boldsymbol{\sigma}, \boldsymbol{\epsilon}$ and the actual generalised deflections by $\mathbf{q}(t)$. First of all we establish the following result which is fundamental.

Result 3 (The Comparison Theorem) Let σ^* be the stress history associated with an equivalent *elastic* problem with the same thermal loading but with mechanical loading $\mathbf{Q}^*$ not necessarily equal to $\mathbf{Q}$. Let $\mathbf{q}^*$ be the associated elastic generalised deflection. Then the following inequalities hold, for a time interval (t_1, t_2)

$$\int_{t_1}^{t_2} (\mathbf{Q}^* - \mathbf{Q}) \,.\, (\dot{\mathbf{q}} - \dot{\mathbf{q}}^*)\, \mathrm{d}t \leqslant \mathrm{E}^{**}(t_1) - \mathrm{E}^{**}(t_2) + \left(\frac{n}{n+1}\right)^n \int_{t_1}^{t_2} U_c(\sigma^{**})\, \mathrm{d}t$$

$$\int_{t_1}^{t_2} (\mathbf{Q}^* - \mathbf{Q}) \,.\, (\dot{\mathbf{q}} - \dot{\mathbf{q}}^*)\, \mathrm{d}t \leqslant \mathrm{E}^{**}(t_1) - \mathrm{E}^{**}(t_2) + \int_{t_1}^{t_2} [U_c(\sigma^{**}) - U_c(\sigma)]\, \mathrm{d}t$$

defining $\boldsymbol{\sigma}^{**} = \boldsymbol{\sigma}^* + \boldsymbol{\rho}^*$ where $\boldsymbol{\rho}^*$ is any constant self-equilibrating stress field and

$$\mathrm{E}^{**}(t) = \mathrm{E}(\sigma^{**} - \sigma)$$

The proof of this result is quite straightforward; from Green's theorem

$$(\mathbf{Q}^* - \mathbf{Q}) \,.\, (\dot{\mathbf{q}} - \dot{\mathbf{q}}^*) = \int_V (\sigma^{**} - \sigma) \,.\, (\dot{\epsilon} - \dot{\epsilon}^{**})\, \mathrm{d}V$$

where $\dot{\boldsymbol{\epsilon}}^{**} = \mathbf{C}\dot{\boldsymbol{\sigma}}^*$ and noting that $\boldsymbol{\rho}^*$ is constant in time and self-equilibrating then

$$(\mathbf{Q}^* - \mathbf{Q}) \,.\, (\dot{\mathbf{q}} - \dot{\mathbf{q}}^*) = -\frac{\mathrm{d}}{\mathrm{d}t}\mathrm{E}^{**} + \int_V (\sigma^* - \sigma) \,.\, \dot{\epsilon}_C\, \mathrm{d}V$$

The results then follow from the inequalities (8.8) and (8.9) on integrating over the time interval (t_1, t_2).

The comparison theorem is quite weak – it is perhaps surprising that we can obtain any sort of characterisation for variable loading – although we can optimise the bounds in terms of $\mathbf{Q}^*$ and ρ^*. The structural response is seen to be characterised not in terms of its own known behaviour but in terms of *another* structural problem. If we take the first inequality over the interval $(0, t)$, then we can bound the accumulated rate of work by ignoring the unknown quantity $\mathrm{E}^{**}(t)$, which is positive; this does not affect the bound. Similarly, for the second inequality over (t_1, t_2), if we take $\mathbf{Q}^* = \mathbf{Q}$ then $\sigma^* = \sigma_0$, the equivalent elastic solution, therefore we have our next result.

Result 4 Let σ_0 be the equivalent elastic solution and ρ^* an arbitrary self-equilibrating stress. Then the following upper bound holds on the accumulated creep complementary dissipation rate

$$\int_{t_1}^{t_2} U_c(\sigma)\, dt \leqslant \int_{t_1}^{t_2} U_c(\sigma^{**})\, dt + \mathrm{E}^{**}(t_1) - \mathrm{E}^{**}(t_2)$$

We may ignore the contribution of $\mathrm{E}^{**}(t_2)$ without violating the inequality.

Finally, it should be noted that we have so far only characterised the solution of the transient creep problem in terms of *elastic* solutions. Can we characterise the solution in terms of the *pure creep* problem?* We can obtain a rather obvious bound on the accumulated creep energy dissipation, namely the following result.

Result 5 Let σ_s be the equivalent pure creep solution. Then

$$\int_{t_1}^{t_2} U_c(\sigma_s)\, dt \leqslant \int_{t_1}^{t_2} U_c(\sigma)\, dt$$

This follows fairly obviously from the theorem of minimum complementary dissipation for pure creep.

From Results 4 and 5 we can bound the creep energy dissipation. Furthermore, if we apply the superposition of states approximation, equation (8.6)

$$\mathbf{q} = \mathbf{q}_0 + \int_0^t \dot{\mathbf{q}}_s\, dt + \Delta\mathbf{q}$$

then for *isothermal* problems we have

$$\int_0^t \mathbf{Q} \,.\, \Delta\dot{\mathbf{q}}\, dt \geqslant 0$$

which gives a measure of the error involved, although, unlike in the constant-load case, it is more difficult to interpret.

*The problem which results if elastic strains are ignored for variable loads, analogous to the steady creep problem.

8.5 Cyclic loading

We turn now to the special case of *cyclic loading*. By this, we mean mechanical or thermal loading which satisfies the condition

$$\mathbf{Q}(t+T) = \mathbf{Q}(t)$$

$$\epsilon_T(t+T) = \epsilon_T(t)$$

where T is called the period of the cycle, which we assume is the same for both loadings. Such loading has some importance for real structures which, for example, experience cyclic behaviour in routine operation. The importance of cyclic loading is that the stresses in a structure will eventually reach a cyclic state. This can be viewed as being the analogy of the steady state for constant loading. The proof of this is quite simple; however, we first establish the following basic result which is true for an arbitrarily loaded structure.

Result 6 (Theorem of convergent internal stresses). Consider two structures composed of the same material but suffering different loading conditions $\mathbf{Q}_1(t)$, $\epsilon_{T1}(t)$, and $\mathbf{Q}_2(t)$, $\epsilon_{T2}(t)$. Then, if after some finite time the loads become identical, ultimately the stresses in the structures, $\sigma_1(t)$ and $\sigma_2(t)$ shall also become identical.

From Green's theorem

$$(\mathbf{Q}_1 - \mathbf{Q}_2).(\dot{\mathbf{q}}_1 - \dot{\mathbf{q}}_2) = \int_V (\sigma_1 - \sigma_2).(\dot{\epsilon}_1 - \dot{\epsilon}_2)\,\mathrm{d}V$$

$$= \frac{\mathrm{d}}{\mathrm{d}t}\mathrm{E}(\sigma_1 - \sigma_2) + \int_V (\dot{\epsilon}_{C1} - \dot{\epsilon}_{C2}).(\sigma_1 - \sigma_2)\mathrm{d}V$$

$$+ \int_V (\sigma_1 - \sigma_2).(\dot{\epsilon}_{T1} - \dot{\epsilon}_{T2})\,\mathrm{d}V$$

From the monotonicity condition, equation (8.2), the second term is positive and after some finite time, by assumption

$$\mathbf{Q}_1 = \mathbf{Q}_2 \qquad \epsilon_{T1} = \epsilon_{T2}$$

It follows that

$$\frac{\mathrm{d}}{\mathrm{d}t}\mathrm{E}(\sigma_1 - \sigma_2) \leqslant 0$$

Therefore $\mathrm{E}(\sigma_1 - \sigma_2)$ is monotonic decreasing and positive. Following an argument similar to that of Result 1, it can be demonstrated that $\sigma_1 \to \sigma_2$ in such a way that $\mathrm{E}(\sigma_1 - \sigma_2) \to 0$. The result is therefore proved.

This fundamental result effectively tells us that, if we are interested in the stress distribution in a structure at large times, we can ignore its initial state – that is, any unknown residual stress imposed on the structure during manufacture can be ignored. However, it does not tell us anything about the deformations. We can apply this result to cyclic loading in the following way.

Let σ_1 correspond to the actual stress $\sigma(t)$ resulting from a periodic load applied at $t = 0$, with period T. Then let σ_2 correspond to the stress for the same structural problem but with loading first applied at $t = T$. From the theorem of convergent internal stresses eventually

$$\sigma_1(t) \to \sigma_2(t)$$

But by definition

$$\sigma_1(t) = \sigma(t) \qquad \sigma_2(t) = \sigma(t+T)$$

Hence

$$\sigma(t) \to \sigma(t+T)$$

That is, *the stresses are ultimately periodic with a period the same as the applied loading*. This asymptotic cyclic stress state is called the *cyclic stationary stress* and we will henceforth denote it by σ_{cs}. As it happens, the only way of determining the cyclic stationary stress is actually to follow the whole loading history! Unlike for the constant loading case, where we know that the asymptotic state is the steady-state creep solution, *we cannot characterise the cyclic stationary state in terms of some stationary boundary value problem*. It is therefore of paramount importance to develop some simple methods of analysis. Some use could be made of the inequalities in the comparison theorem. For example, from the first inequality integrated over a single cycle in the cyclic stationary state, we have

$$\int_t^{t+T} (\mathbf{Q}^* - \mathbf{Q}) \,.\, \dot{\mathbf{q}} \, \mathrm{d}t \leqslant \int_t^{t+T} (\mathbf{Q}^* - \mathbf{Q}) \,.\, \dot{\mathbf{q}}^* \, \mathrm{d}t + \left(\frac{n}{n+1}\right)^n \int_t^{t+T} U_c(\sigma^{**}) \, \mathrm{d}t$$

since $E^{**}(t) = E^{**}(t+T)$. Then if we wish to bound an accumulated generalised deflection q_k, say, over a cycle, we can put, as in the displacement bound approach, $\mathbf{Q}^* = \mathbf{Q}$ except for $Q_k^* = Q_k + R_k$ where R_k is an arbitrary constant load applied in the direction of Q_k. Then

$$R_k[q_k(t+T) - q_k(t)] \leqslant R_k[q_k^*(t+T) - q_k^*(t)] + \left(\frac{n}{n+1}\right)^n \int_t^{t+T} U_c(\sigma^{**}) \, \mathrm{d}t$$

This bound can then be optimised with respect to R_k and ρ^*.

Of particular interest is the behaviour of structures whose cycle time is short compared to some internal physical time scale – for example, the redistribution time associated with constant loading.

There are two time scales associated with a cyclically loaded structure. A global, discrete time scale γ which describes the start and end of loading cycles, e.g.

$$\mathbf{Q}(\gamma_0) = \mathbf{Q}(\gamma_1) = \ldots = \mathbf{Q}(\gamma_i) = \ldots$$

where $\gamma_0 = 0$, $\gamma_{j+1} - \gamma_j = T$. In addition to this, there is a local time scale, s, $0 \leqslant s \leqslant 1$, independent of the cycle period, such that $s = 0$ corresponds to the start of a cycle and $s = 1$ the end of a cycle. Clearly within the jth cycle (γ_j, γ_{j+1})

$$t = \gamma_j + sT$$

Thus any cyclic behaviour, e.g. the loading, or a cyclic stress field can be fully described by its distribution with s and the current cycle number (γ_j, γ_{j+1}).

The stress σ in a cyclically loaded structure can be written in terms of the equivalent elastic solution $\sigma_0(t)$ and a residual stress field $\rho(t)$

$$\sigma(t) = \sigma_0(t) + \rho(t)$$

and we could describe the behaviour of the structure using an equation for $\rho(t)$ (Section 7.3.3). In order to describe the actual stress fully, we must know the current cycle (γ_j, γ_{j+1}) the cyclic stress distribution $\sigma_0(s)$ and the residual stress $\rho(t)$ – which is not necessarily cyclic, although we do know from Result 6 that it is asymptotically cyclic.

We consider a particular statically admissible stress distribution which has the form $\sigma_{rs}(t)$

$$\sigma_{rs}(t) = \sigma(s) + \rho(\gamma_j)$$

for the cycle (γ_j, γ_{j+1}). That is, it is the sum of the equivalent elastic stress and the residual stress at the beginning of the cycle assumed constant over the cycle. In general, this stress distribution is discontinuous, since $\rho(\gamma_j)$ is discontinuous.

First, we can establish the following result.

Result 7 As the period $T \to 0$, the actual stress $\boldsymbol{\sigma}(t) \to \sigma_{rs}(t)$.

This can readily be proved if we rewrite the constitutive equation for a cycle as in the form

$$\frac{d}{ds}\epsilon = \mathbf{C}\frac{d}{ds}\sigma + \frac{d}{ds}\epsilon_C + \frac{d}{ds}\epsilon_T$$

where

$$\frac{d}{ds}\epsilon_C = T\frac{d\Omega}{d\sigma}$$

Obviously as $T \to 0$, $d\epsilon_C/ds \to 0$ and $d\sigma/ds \to d\sigma_0/ds$. Therefore

$$\sigma(t) \to \sigma_0(s) + \rho(\gamma_j) = \sigma_{rs}(t)$$

Because of this result, we call $\boldsymbol{\sigma}_{rs}(t)$ the *rapid cycle stress.* Its significance becomes clear if we note that once the cyclic stationary state $\sigma_{cs}(t)$ has been reached, the residual stress becomes cyclic, i.e.

$$j \to \infty \qquad \rho(\gamma_{j+1}) \to \rho(\gamma_j)$$

that is, $\rho(\gamma_i)$ *must become constant* in the time scale γ; we denote this asymptotic value by ρ_{rs}. However, the problem still remains of characterising ρ_{rs}. To do so, we use Result 4 which, for a cycle (γ_j, γ_{j+1}) in the stationary cyclic state, becomes

$$\int_0^1 U_c(\sigma_{cs})\, ds \leqslant \int_0^1 U_c(\sigma_0 + \rho^*)\, ds$$

as $t \to \infty$. The creep energy dissipation accumulated in the cyclic state is therefore bounded by that associated with the cyclic stress distribution $\sigma_0 + \rho^*$ where ρ^* is an arbitrary self-equilibrating stress. Obviously for the rapid cycle solution

ρ_{rs} must provide the minimum amongst all ρ^*. Therefore we find ρ_{rs} by evaluating

$$\min_{\rho^*} \int_0^1 U_c(\sigma_0 + \rho^*) \, ds \tag{8.10}$$

It is useful to summarise the above: *For a cyclically loaded structure whose period* T *is short compared to its redistribution time under constant load, the asymptotic cyclic stationary state can be approximated by the superposition of the equivalent elastic solution with a constant self-equilibrating residual stress found by satisfying the condition given by equation (8.10).*

Of course, if the condition of rapid cycles cannot be met, then the approximation is doubtful.

8.6 Some examples

The fundamental behaviour of creeping structures under cyclic load has already been demonstrated in Section 3.4 for the hinged bar structure. There, it was shown that a cyclic residual stress state was eventually reached; we have now proven this to be generally true for any structure with a material obeying a time-hardening, power law of creep.

The analysis of Section 3.4 is repeated here for the load cycle shown in *Figure 8.6* using various periods T. The normalised residual stress in bar 1 is shown in *Figures 8.7, 8.8* and *8.9* for a decreasing period; as expected in each case a cyclic state of residual stress is eventually achieved. As the period decreases, this cyclic residual stress becomes more and more constant in observance of the rapid cycle solution. This can be determined, without recourse to a solution of the equations of stress redistribution, by satisfying the condition of equation (8.10). A suitable self-equilibrating stress is, with $\sigma_0 = 3Q_0/2A$,

$$\rho_1 = \sigma_0 \lambda \qquad \rho_2 = -\sigma_0 \lambda$$

for some scalar λ. Then we should minimise the functional

$$\frac{AL}{n+1} \sigma_0^{n+1} \int_0^1 \left[\left(\frac{1}{5} \frac{Q(s)}{Q_0} + \lambda \right)^{n+1} + \left(\frac{2}{5} \frac{Q(s)}{Q_0} - \frac{\lambda}{2} \right)^{n+1} \right] g(s) \, ds$$

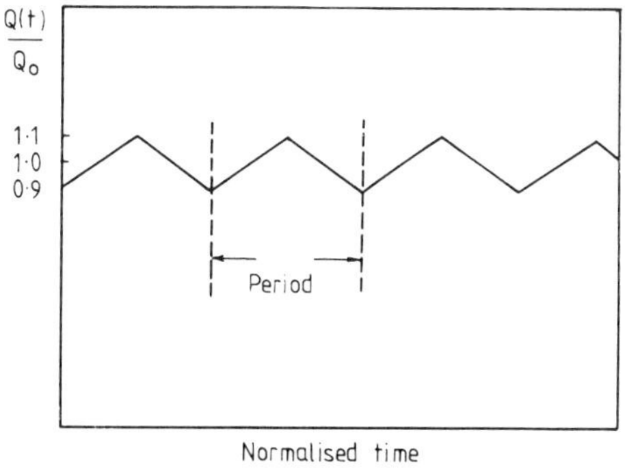

Figure 8.6 Load cycle for hinged-bar structure

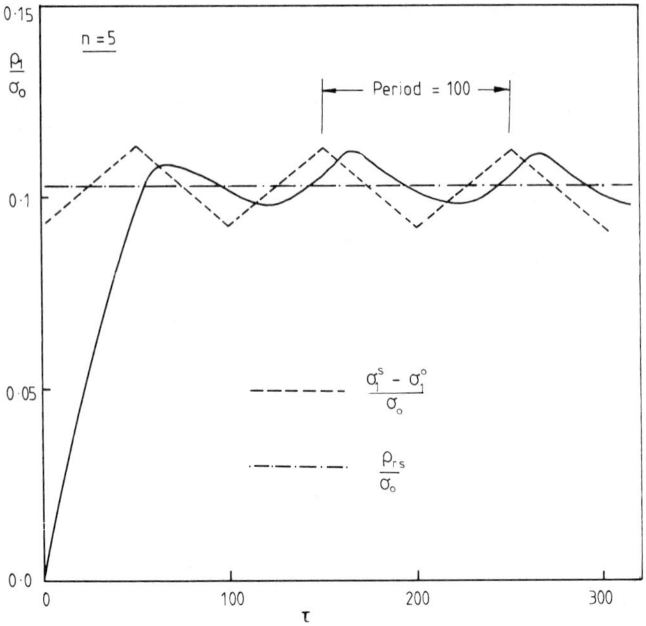

Figure 8.7 Variation of normalised residual stress in time: period = 100

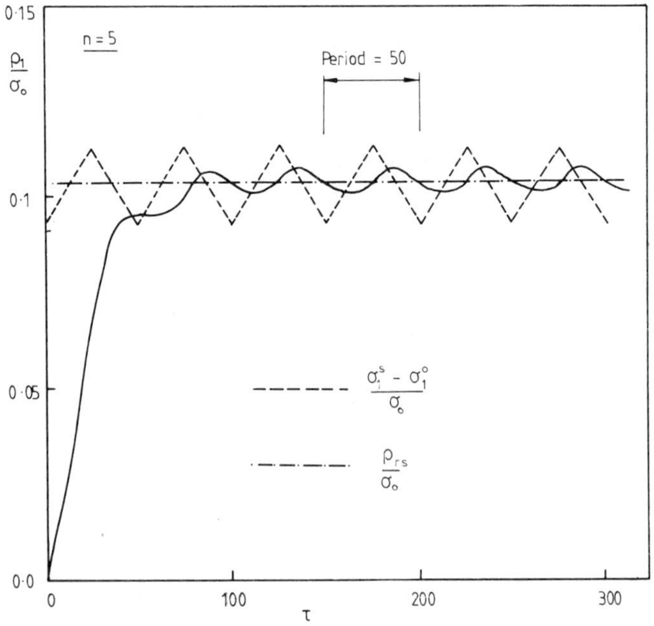

Figure 8.8 Variation of normalised residual stress in time: period = 50

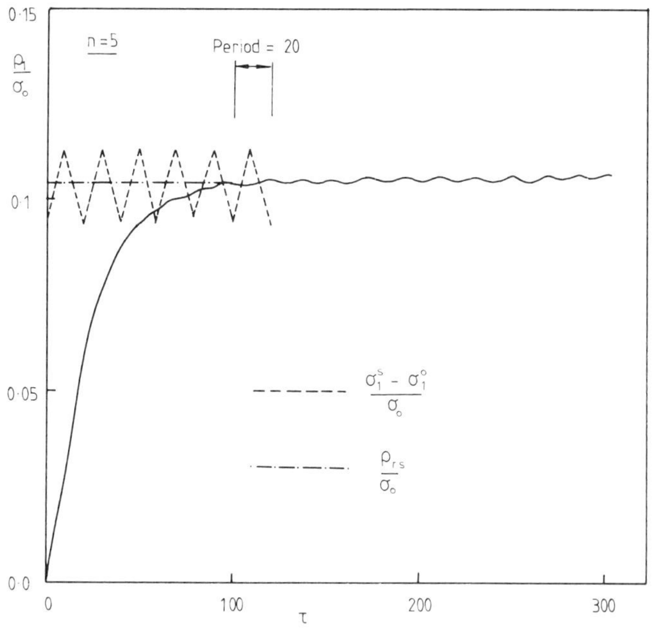

Figure 8.9 Variation of normalised residual stress in time: period = 20

with respect to λ. Then

$$\rho_{rs} = \sigma_0 \lambda_{min}$$

The actual residual stress is also compared with the cyclic residual stress state

$$\sigma_1^s - \sigma_1^o = \frac{3Q(t)}{2A}\left(\frac{1}{1 + 2^{1+1/n}} - \frac{1}{5}\right)$$

It can readily be seen that, as the period of the cycle increases, the peak residual stress approaches the peak of this fully cyclic state, although there is a phase shift observed.

As a second, more involved example we examine the cyclic creep of a rectangular beam in bending. The basic equations for this problem, assuming steady creep, have been set out in Section 3.1. The present problem can be analysed for transient creep using the techniques of Chapter 7; the numerical solution is in fact very similar to that outlined in Section 7.2 for a thin tube in bending. The cyclic history of the applied bending moment used in this example is shown in *Figure 8.10.*

The analysis is carried out here for a number of different periods of cycle. The variation in the residual stress at the outer fibre ($y = +d$) with time is shown in *Figure 8.11* using a normalised time scale defined by

$$\tau = E\sigma_0^{n-1} \int g(t)\, dt$$

where $\sigma_0 = 3M_0/2bd^2$. In each case it can be seen that a cyclic state of residual stress is eventually reached and that as the period of the cycle decreases a constant stress state is approached. The distribution of the limiting rapid cycle solution is shown in *Figure 8.12.*

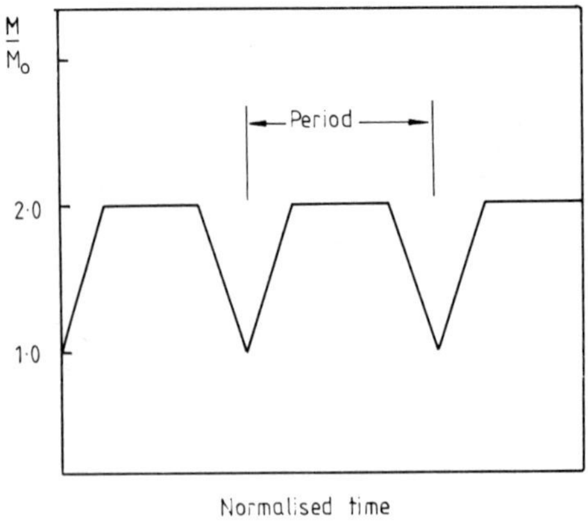

Figure 8.10 Load cycle for beam in bending

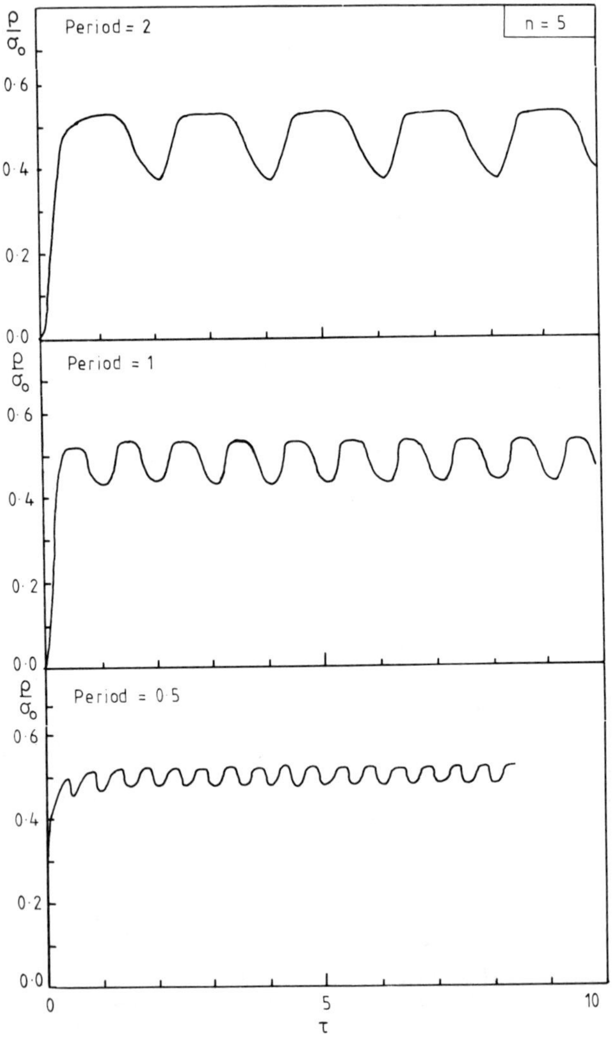

Figure 8.11 Variation of normalised residual stress at outer fibre in time

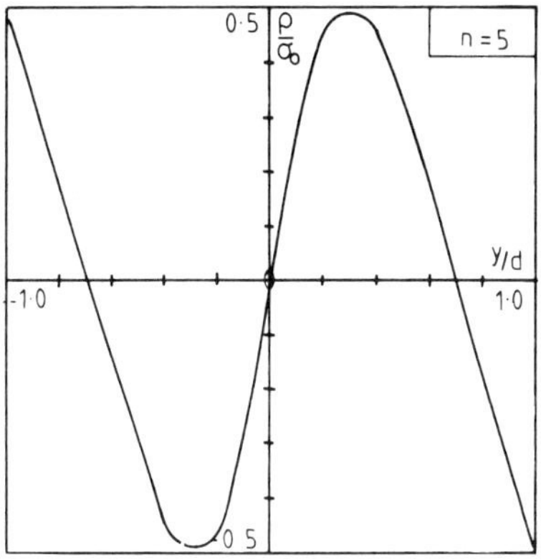

Figure 8.12 Distribution of normalised residual stress in cycle state

The rapid cycle solution can be obtained directly through minimisation of the functional

$$\frac{1}{n+1}\int_0^1\left[\int_{-1}^1\left(\rho^*(z)+\frac{3M(s)}{2bd^2}\,z\right)^{n+1}\mathrm{d}z\right]g(s)\,\mathrm{d}s$$

amongst all constant self-equilibrating stress distributions ρ^*; these must satisfy the constraint

$$\int_{-1}^{1}\rho^*(z)z\,\mathrm{d}z = 0$$

A suitable distribution of ρ^* is given by

$$\rho^*(z) = \sum_{j=1}^{\infty} a_j\left[\sin\left(\frac{(2j+1)\,\pi z}{2}\right) - \frac{(-1)^j}{(2j+1)^2}\sin\left(\frac{\pi z}{2}\right)\right]$$

The coefficients a_k which provide the required minimum can be found using a Rayleigh–Ritz technique as in Section 5.4.7.

The value of the rapid cycle solution thus obtained at the outer fibre is shown in *Figure 8.13* along with the cyclic residual stress derived from the stationary solution

$$\frac{|\sigma^s(d,t)-\sigma^o(d,t)|}{\sigma_0} = \left|\frac{2+1/n}{3}-1\right|\frac{M(t)}{M_0}$$

Once more, it can be seen that the actual residual stress is enveloped by the stationary solution, and that it tends to the rapid cycle solution as the period decreases.

In each of these examples, we have not used the comparison theorem other than to derive the rapid cycle solution. The reader is referred to the references at the end of this chapter for examples of bounds on the creep dissipation rate (Results 4 and 5). We have avoided this task here.

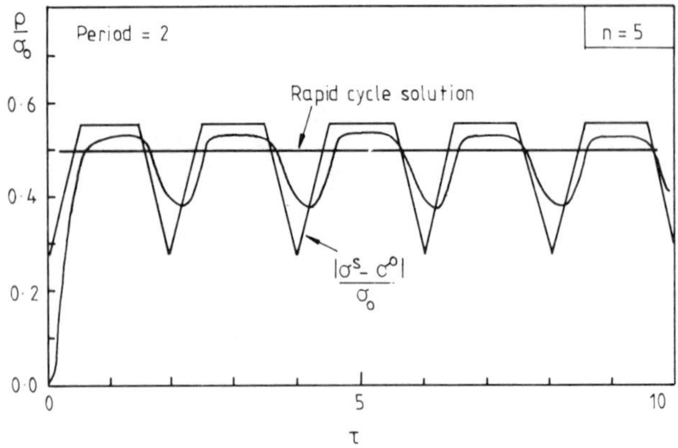

Figure 8.13 Comparison with rapid cycle and stationary solutions

8.7 Further developments

The simple characterisation results that we have described so far suffer from one major restriction, in that they have been developed for the simplest of constitutive relations – time hardening creep with a power law. While these constitutive relations may be expected to be sufficient for constant loading, they are unreliable for variable loading (and for this reason variable loading has been dealt with only briefly in the preceding sections). The question is, can we develop simple techniques for more complex constitutive relations – for example, the internal-variable theories which are capable of providing a unified description of creep and plasticity? As it happens, we can establish the comparison theorem if we replace equation (8.8) by the constitutive inequality

$$(\sigma^* - \sigma) \,.\, \dot{\epsilon}_C \leqslant \omega(\sigma^*, t) \tag{8.11}$$

where $\dot{\epsilon}_C$ is the creep strain rate associated with $\boldsymbol{\sigma}$ and σ^* is any stress. Then the bound

$$\int_{t_1}^{t_2} (\mathrm{Q}^* - \mathrm{Q}) \,.\, (\dot{\mathrm{q}} - \dot{\mathrm{q}}^*)\, \mathrm{d}t \leqslant \mathrm{E}^{**}(t_1) - \mathrm{E}^{**}(t) + \int_{t_1}^{t_2} \omega(\sigma^{**}, t)\, \mathrm{d}t$$

holds.

Therefore, our previous development remains valid *if* we can find a suitable functional ω. We have established this functional in the case of time hardening power-law creep

$$\omega(\sigma^*) = \left(\frac{n}{n+1}\right)^n \Omega(\sigma^*)$$

But, can we find such a functional for other constitutive equations – in particular, for internal-variable theories? In fact, for any time hardening model

$$\dot{\epsilon}_C = \mathrm{d}\Omega/\mathrm{d}\boldsymbol{\sigma}$$

such that Ω is convex, we can establish the inequality

$$\frac{1}{2}\,\sigma^{A}.\,\dot{\epsilon}_{C}^{A} + \sigma^{B}.\,\dot{\epsilon}_{C}^{B} \geqslant \sigma^{A}.\,\dot{\epsilon}_{C}^{B}$$

similar to Martin's inequality and as in the derivation of equation (8.8)

$$(\sigma^{*} - \sigma)\,.\,\dot{\epsilon}_{C} \leqslant \frac{1}{2}\sigma^{*}\,.\,\dot{\epsilon}_{C}^{*} = \omega(\boldsymbol{\sigma}^{*})$$

If we evaluate this for the power law

$$\omega(\sigma^{*}) = \frac{n+1}{2}\,\Omega(\sigma^{*})$$

Obviously

$$\frac{n+1}{2} > \left(\frac{n}{n+1}\right)^{n}$$

for all n and the bounds will be poorer. What we have been able to do with the power law is to obtain a better bound. Whether or not this can be done for other functions is open to question. The problem in fact reduces to that of solving

$$\omega(\sigma^{*}) = \max\,(\sigma^{*} - \sigma)\,.\,\dot{\epsilon}_{C}$$

In general, a suitable closed-form solution cannot be found.

In practice, the levels of the loading and thermal strain in a creeping structure are likely also to give rise to plastic straining. For a more complex constitutive model, such as Hart's theory which provides a unified description of creep and plasticity, it is unlikely that a simple expression like equation (8.11) can be established. Any such attempt to do so would in all probability be quite contrived. The only result that we can easily recover is the theorem of convergent internal stresses, but this will not be discussed further here. Nevertheless, if we use a simpler model of plasticity combined with a time-hardening creep law, the comparison theorem can still be used, but with a slight modification.

The constitutive relation (8.1) takes the form

$$\epsilon = \epsilon_{E} + \boldsymbol{\epsilon}_{P} + \epsilon_{C} + \epsilon_{T}$$

where the elastic ϵ_E, creep ϵ_C and thermal strains are as previously defined. The plastic strain rate is given by

$$\dot{\epsilon}_{P} = \dot{\lambda}\,\frac{d\psi}{d\sigma}$$

where $\psi(\sigma)$ is a convex yield surface with interior $\psi(\sigma) < 0$ and exterior $\psi(\sigma) > 0$ and $\dot{\lambda}$ is the plastic multiplier such that

$$\dot{\lambda} = 0 \qquad \psi < 0 \quad \text{or} \quad \dot{\sigma}\,.\frac{d\psi}{d\sigma} < 0$$

$$\dot{\lambda} \geqslant 0 \qquad \psi = 0 \quad \text{or} \quad \dot{\boldsymbol{\sigma}}\,.\frac{d\psi}{d\sigma} \geqslant 0$$

Plastic strains are normal to the yield surface and satisfy monotonicity, i.e.

$$(\dot{\epsilon}_P^* - \dot{\epsilon}_P) \cdot (\sigma^* - \sigma) \geqslant 0$$

If σ^* is within or on the yield, then $\psi(\sigma^*) \leqslant 0$, $\dot{\epsilon}_P^* = \mathbf{0}$

$$(\sigma^* - \sigma) \cdot \dot{\epsilon}_P \leqslant 0$$

which can be compared to the inequalities of equations (8.8), (8.9) and (8.11). This model of plasticity is the simplest, that of *perfect plasticity*, and the reader is referred to the standard texts for more detail.

Immediately we note from the monotonicity condition that the theorem of convergent internal stresses remains valid and therefore under cyclic loading a cyclic state is ultimately reached. In order to characterise the solution, we again turn to the comparison theorems. These can be stated as follows.

Result 3 (*contd*) Suppose that plastic strains are admitted. Then provided there exists a constant residual stress field ρ^* such that

$$\psi(\sigma^{**}) \leqslant 0$$

the inequalities remain valid. If this cannot be done, then redefine σ^* to be the solution of the equivalent *elasto–plastic* problem for the same thermal loads but mechanical loads $\mathbf{Q}^*$. Then we have for a time interval (t_1, t_2)

$$\int_{t_1}^{t_2} (\mathbf{Q}^* - \mathbf{Q}) \cdot (\dot{\mathbf{q}} - \dot{\mathbf{q}}^*) \, dt \leqslant \mathrm{E}^*(t_1) - \mathrm{E}^*(t_2) + \left(\frac{n}{n+1}\right)^n \int_{t_1}^{t_2} U_c(\sigma^*) \, dt$$

$$\int_{t_1}^{t_2} (\mathbf{Q}^* - \mathbf{Q}) \cdot (\dot{\mathbf{q}} - \dot{\mathbf{q}}^*) \, dt \leqslant \mathrm{E}^*(t_1) - \mathrm{E}^*(t_2) + \int_{t_1}^{t_2} [U_c(\sigma^*) - U_c(\sigma)] dt$$

where

$$\mathrm{E}^*(t) = \mathrm{E}(\sigma^* - \sigma)$$

The proof of this follows exactly the previous proof and is left to the reader to pursue. Again, it is noted that the second inequality leads to the energy bounds

$$\int_{t_1}^{t_2} U_c(\sigma) \, dt \leqslant \mathrm{E}^*(t_1) - \mathrm{E}^*(t_2) + \int_{t_1}^{t_2} U_c(\sigma^*) \, dt$$

and for cyclic loading $\mathrm{E}^*(t_1) = \mathrm{E}^*(t_2)$ over a cycle in the cyclic stationary state.

Finally Result 5 *et seq*. is valid if and only if

$$\psi(\sigma_s) \leqslant 0$$

It can be seen that provided the condition $\psi(\sigma^{**}) \leqslant 0$ can be met, the discussion of Section 8.5 is applicable in this case. In the terminology of plasticity theory, we say that the load and temperature variations are below *shakedown*. For loading above the shakedown limit, we can only bound the creep solution if we can evaluate an equivalent elastic–plastic problem. In the case of cyclic loading, this requires the calculation of the stationary cyclic plasticity solution. Unfortunately, this may necessitate a numerical solution of an equivalent order of magnitude to the creep problem in real time!

8.8 Constant displacements – the relaxation problem

The results of the previous section have dealt with the problem of applied mechanical or thermal loading. However, many structures have prescribed deformations, usually constant in time, imposed during construction; for example, piping systems and frameworks. We will deal here with the following structural problem: a structure occupying a volume V and bounded by a surface S is supported in some manner. Over a small part of the surface S_0 a constant generalised displacement q_0 is applied. It is assumed that there are no body forces present. The structural relaxation problem is then to determine the resultant relaxation of generalised load $Q(t)$ in the direction of q_0.

It is presumed that the structural relaxation problem is well posed and furthermore that the applied conditions are interchangeable, i.e. a prescribed q_0 results in a unique $Q(t)$ and vice-versa. Then we define the following associated stress distributions.

(a) The dual elastic problem: consider an identical body subject to a load $Q(t)$ (which of course is unknown) but composed of a purely *elastic* material – the stress distribution associated with this problem we denote by $\sigma_e(t)$.

(b) The dual pure creep problem: consider an identical body subject to a load $Q(t)$ but composed of a purely creeping material – the stress distribution associated with this problem we denote by $\sigma_s(t)$.

It follows from the linearity of the dual elastic problem that

$$\sigma_e(t) = Q(t)\, \sigma_e^* \tag{8.12}$$

where σ_e^* is the spatial distribution of σ_e independent of time. Obviously the initial elastic solution, $t = 0$, is given by

$$\sigma(0) = \sigma_e(0) = Q(0)\sigma_e^*$$

Similarly from the stationarity of the purely viscous problem

$$\sigma_s(t) = Q(t)\, \sigma^* \tag{8.13}$$

where σ_s^* is the spatial distribution of σ_s, independent of time.

Again we assume a *time hardening power law of creep*. Then it follows from Green's theorem, since $\dot{q}_0 = 0$, that

$$\frac{d}{dt} \mathrm{E}(\sigma) + (n + 1)\, U_c(\sigma) = 0 \tag{8.14}$$

Since $U_c(\sigma) \geqslant 0$ the elastic energy $\mathrm{E}(\sigma)$ is monotonic decreasing in time, and following an argument similar to that of Result 1 the stress must relax to zero.

We may now establish the following inequalities.

Result 8 For the structural relaxation problem the following inequalities hold

$$\frac{\mathrm{d}}{\mathrm{d}t} Q^2 + 2\lambda_U g(t)\, Q^{n+1} \leqslant - \frac{\mathrm{d}\xi}{\mathrm{d}t}$$

$$\frac{\mathrm{d}}{\mathrm{d}t} Q^2 + 2\lambda_L g(t)\, Q^{n+1} \geqslant n \frac{\mathrm{d}\xi}{\mathrm{d}t}$$

where

$$\lambda_U = \frac{(n+1)}{2} \frac{\bar{U}_c(\sigma_s^*)}{\mathrm{E}(\sigma_e^*)} \qquad \lambda_L = \frac{(n+1)}{2} \frac{\bar{U}_c(\sigma_e^*)}{\mathrm{E}(\sigma_e^*)}$$

$$U_c = g(t)\bar{U}_c \qquad \xi(t) = \frac{\mathrm{E}(\sigma) - \mathrm{E}(\sigma_e)}{\mathrm{E}(\sigma_e^*)}$$

The first inequality follows simply from equation (8.14) on using equations (8.12) and (8.13) and the theorem of minimum complementary dissipation for pure creep, i.e.

$$U_c(\sigma) \geqslant U_c(\sigma_s)$$

The second inequality results from summing a repeated application of Green's theorem

$$0 = \int_V \sigma_e \,.\, \dot{\epsilon}\; \mathrm{d}V$$

$$0 = \int_V (\dot{\sigma} - \dot{\sigma}_e) \,.\, \dot{\epsilon}_e\; \mathrm{d}V$$

$$0 = \int_V \sigma \,.\, \dot{\epsilon}\; \mathrm{d}V$$

to form

$$\frac{\mathrm{d}}{\mathrm{d}t}[\mathrm{E}(\sigma) - \mathrm{E}(\sigma_e)] = \int_V (\sigma_e - \sigma) \,.\, \dot{\epsilon}_C\; \mathrm{d}V \leqslant U_c(\sigma_e) - U_c(\sigma) \tag{8.15}$$

applying the convexity condition (8.9). Then combining with equation (8.14) and rearranging with equations (8.12) and (8.13), the inequality is established.

Let us examine the quantity

$$e(t) = \frac{\mathrm{d}}{\mathrm{d}t} \xi(t)$$

We know from an application of the theorem of minimum complementary energy in elasticity that

$$\mathrm{E}(\sigma) - \mathrm{E}(\sigma_e) \geqslant 0$$

and from initial and final conditions

$$t = 0 \qquad \mathrm{E}(\sigma) - \mathrm{E}(\sigma_e) = 0$$

$$t \to \infty \qquad \mathrm{E}(\sigma) - \mathrm{E}(\sigma_e) \to 0$$

while from the equality in equation (8.15)

$$t = 0 \qquad \frac{\mathrm{d}}{\mathrm{d}t}[\mathrm{E}(\sigma) - \mathrm{E}(\sigma_e)] = 0$$

$$t \to \infty \qquad \frac{\mathrm{d}}{\mathrm{d}t}[\mathrm{E}(\sigma) - \mathrm{E}(\sigma_e)] \nrightarrow 0$$

From these observations it must follow that in *some* initial time interval $(0, T)$ the quantity $\mathrm{E}(\sigma) - \mathrm{E}(\sigma_e)$ is *increasing,* * i.e.

$$e(t) \geqslant 0$$

Ultimately, however, the function $\mathrm{E}(\sigma) - \mathrm{E}(\sigma_e)$ must be *decreasing*. Thus *during the initial period of service of the structure* the inequalities

$$\frac{\mathrm{d}}{\mathrm{d}t}Q + \lambda_U g(t)\, Q^n \leqslant 0 \qquad \frac{\mathrm{d}}{\mathrm{d}t}Q + \lambda_L g(t) Q^n \geqslant 0$$

from Result 8 must hold. That is, applying the differential inequalities, *the relaxation of the load* Q(t) *is bounded in the short term by*

$$\left(1 + (n-1)\lambda_L Q(0)^{n-1}\int_0^t g(t)\mathrm{d}t\right)^{-1/(n-1)} \leqslant \frac{Q(t)}{Q(0)}$$
$$\leqslant \left(1 + (n-1)\lambda_U Q(0)^{n-1}\int_0^t g(t)\,\mathrm{d}t\right)^{-1/(n-1)} \qquad (8.16)$$

The lower bound is equivalent to making the approximation

$$\sigma(t) \simeq \sigma_e(t) = Q(t)\,\sigma^*_e \qquad (8.17)$$

that is, the stress is approximated by that of the dual elastic problem. The upper bound is equivalent to making the approximation

$$\dot{q}_0 \simeq \dot{q}_e(t) + \dot{q}_s(t) \qquad (8.18)$$

where $\dot{q}_e(t)$ is the generalised displacement rate from the dual elastic problem and $\dot{q}_s(t)$ the generalised displacement rate from the dual pure creep problem. It can be seen that the approximation (8.18) is an application of 'superposition of states' to the relaxation problem.

The short-term bounds of equation (8.16) are seen to be of the same form except for the factors λ_L and λ_U. Moreover, they have the form of a uniaxial test started at a stress σ_0 (Section 3.5)

$$\sigma(t)/\sigma_0 = \left(1 + (n-1)\mathrm{E}\sigma_0^{n-1}\int g(t)\,\mathrm{d}t\right)^{-1/(n-1)}$$

where for the upper bound

$$\sigma_0 = \lambda_U/\mathrm{E}^{1/(n-1)}Q(0)$$

*It should be recognised that this argument lacks rigour.

and for the lower bound

$$\sigma_0 = \lambda_L / E^{1/(n-1)} \; Q(0)$$

The performance of these bounds is best seen by way of example.

8.9 Some examples

As a rather elementary example, we consider the relaxation of the parallel two-bar structure introduced in Section 5.1 (*Figure 5.1*). The vertical deflection q_o is fixed and the problem is to determine the subsequent relaxation of the vertical load $Q(t)$.

The initial elastic solution is

$$\sigma_1(0) = \frac{L_2}{L_1 + L_2} \frac{Q(0)}{A} \qquad \sigma_2(0) = \frac{L_1}{L_1 + L_2} \frac{Q(0)}{A}$$

and the relaxation of $Q(t)$ is simply obtained as

$$\frac{Q(t)}{A} = \sigma_1(0)\left(1 + (n-1)E\,\sigma_1(0)^{n-1}\int g(t)\,\mathrm{d}t\right)^{-1/(n-1)} + \sigma_2(0)\left(1 + (n-1)E\,\sigma_2(0)^{n-1}\int g(t)\,\mathrm{d}t\right)^{-1/(n-1)}$$

The short-term lower bound can be evaluated as

$$\frac{Q(t)}{Q(0)} = \left(1 + (n-1)\,\mathrm{E}[\,Q(0)/A]^{n-1}\int g(t)\,\mathrm{d}t\,\Lambda_L\right)^{-1/(n-1)}$$

where

$$\Lambda_L = \frac{L_1^n + L_2^n}{(L_1 + L_2)^n}$$

and the short-term upper bound is

$$\frac{Q(t)}{Q(0)} = \left(1 + (n-1)\,\mathrm{E}[Q(0)/A]^{n-1}\int g(t)\,\mathrm{d}t\,\Lambda_U\right)^{-1/(n-1)},$$

where

$$\Lambda_U = \frac{L_1 + L_2}{(L_1^{1/n} + L_2^{1/n})^n}$$

These solutions are shown in *Figure 8.14*. It can be seen that the two estimates are initially bounds, but that eventually the short-term upper bound reduces to less than the true solution. In *Figure 8.15* we plot the quantity

$$\zeta(t) = \frac{\mathrm{E}(\sigma) - \mathrm{E}(\sigma_e)}{\mathrm{E}(\sigma_e(0))}$$

where the elastic energy is in this case

$$\mathrm{E}(\sigma) = \frac{1}{2E}\,(L_1\sigma_1^2 + L_2\sigma_2^2)$$

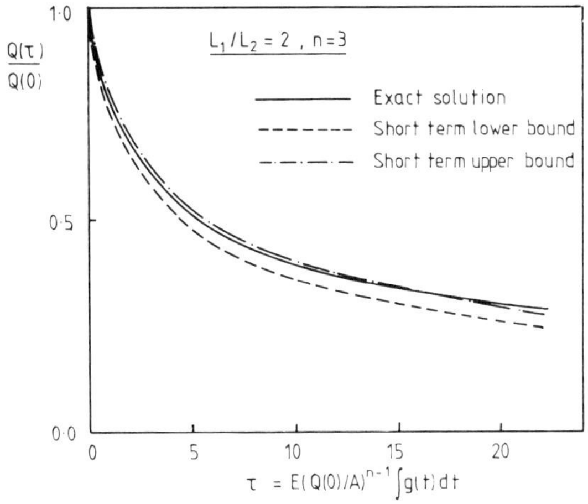

Figure 8.14 Relaxation of vertical load for two-bar structure: comparison of exact solution with short-term bounds

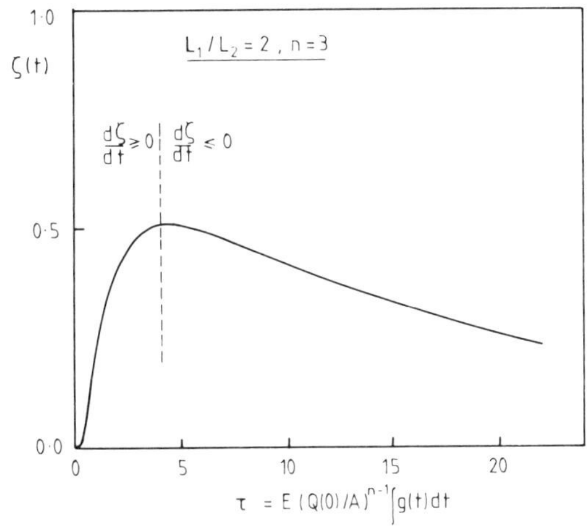

Figure 8.15 Variation of function $\zeta(t)$ with time

and the dual elastic solution is

$$\sigma_{1\,\mathrm{e}}(t) = \frac{L_2}{L_1 + L_2}\;\frac{Q(t)}{A} \qquad \sigma_{2\,\mathrm{e}}(t) = \frac{L_1}{L_1 + L_2}\;\frac{Q(t)}{A}$$

with $Q(t)$ from the exact solution.

As a second example, we may consider the beam in bending initially examined in Section 3.5. The relaxation of the bending moment $M(t)$ for a fixed curvature is shown for a rectangular cross-section in *Figure 3.16* together with the superposition of states approximation. This is reproduced in *Figure 8.16* with the addition of the lower bound

$$\frac{M(t)}{M_0} = \left(1 + (n-1)E\int g(t)\,\mathrm{d}t\,\Lambda_{\mathrm{L}}\,M_0^{n-1}\right)^{-1/(n-1)}$$

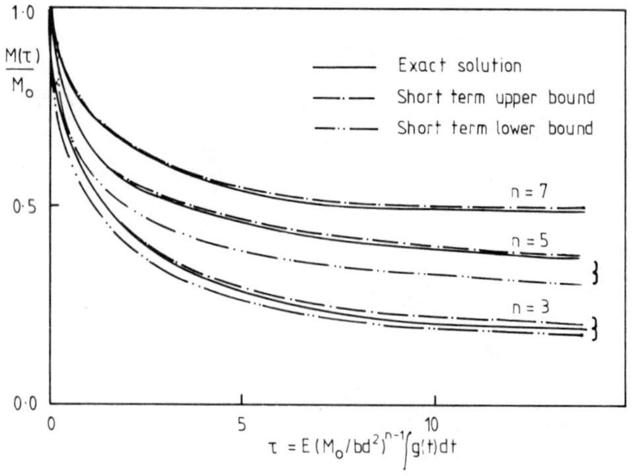

Figure 8.16 Relaxation of bending moment in beam: comparison of exact solution with short-term bounds

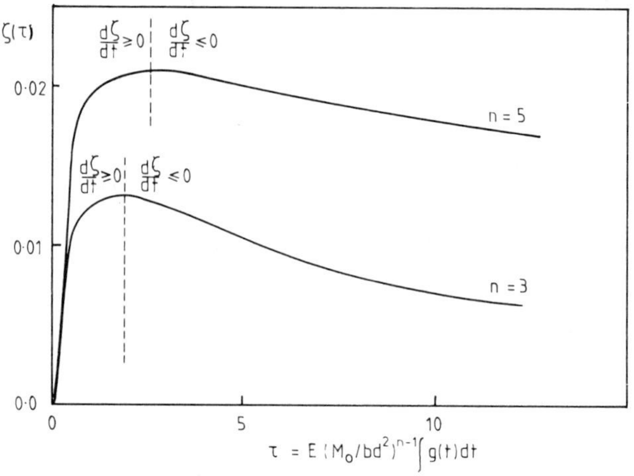

Figure 8.17 Variation of function $\zeta(t)$ with time

where

$$\Lambda_L = \frac{1}{I^n}\int_A y^{n+1}\, dA$$

In *Figure 8.17* we plot the quantity

$$\zeta(t) = \frac{E(\sigma) - E(\sigma_e)}{E(\sigma_e(0))}$$

where the elastic energy is

$$E(\sigma) = \frac{1}{2E}\int_A \sigma^2\, dA$$

and the dual elastic solution is

$$\sigma_e(t) = M(t)y/I$$

In both of these examples, the described behaviour of the quantity $\zeta(t)$ is clearly seen.

8.10 Generalised models in transient creep

It has become apparent that large-scale, detailed solutions of problems arising in structural analysis for creep are inconsistent with the needs of routine design. For this reason, interest has focused on the type of simplified bounding and comparison theorems that we have been discussing. But it is questionable if these are appropriate to the analysis of complex structural systems, such as frameworks or piping, or thin-shell structures. In elasticity, and to a lesser extent in plasticity, it is common to resolve these types of problems using generalised structural models which avoid costly three-dimensional analysis by reduction to a problem which is mechanically simpler. Examples of such (well known) models in elasticity are the one-dimensional theories of beams or rods, and the two-dimensional theories of plates and shells. We have already discussed these, for example, in the steady creep of shells, Section 5.7.4. As we saw there, the crucial aspect in the use of this approach is the practical development of 'generalised constitutive equations' which relate the chosen generalised measures of stress and strain. Below, we outline the stages in the development of a generalised model for transient creep. The procedure will be a familiar one; it closely parallels the development of the time hardening theory of creep from a power law in the steady state. There are three stages in the procedure:

(a) Initially, suitable generalised strain and stress measures should be selected. The generalised strain, which we denote by the vector **e** is chosen with reference to the deformation mechanisms in a particular structure. The components of **e** should be non-dimensional. The generalised stress s is then by definition the stress-type variables which must be associated with the generalised strain in order that the work per unit volume is given by

$$W = \mathbf{e} \cdot \mathbf{s}$$

(b) It is assumed, as an approximation, that the generalised strain can be partitioned into an elastic and a creep part

$$\mathbf{e} = \mathbf{e}_E + \mathbf{e}_C \tag{8.19}$$

such that the elastic component is exactly derived from the equivalent elastic model, i.e.

$$\mathbf{e}_E = \mathrm{d}\Pi/\mathrm{d}\mathbf{s} \tag{8.20}$$

where $\Pi(\mathbf{s})$ is the elastic energy expressed in terms of the generalised stress.

(c) Finally, the dependence of the creep component of generalised strain on generalised stress is assumed to be the same as that obtained from the structure in steady creep, that is

$$\dot{e}_C = d\Omega/ds \tag{8.21}$$

where $\Omega(t, \mathbf{s})$ is the creep energy dissipation, generally given as

$$\Omega = \frac{g(t)}{n+1}\,\overline{s}^{\,n+1}$$

where $\overline{s}$ represents the 'generalised effective stress' (Section 5.7, equation (5.58)), suitably normalised.

It cannot be overemphasised that the decomposition inherent in equation (8.19) is only approximate; the elastic and creep components of generalised strain cannot be related in any way to the elastic and creep components of the actual strain.

8.10.1 A generalised model for beam systems

We consider here the simple problem of a beam under tension and bending which was previously examined in Section 5.7.3. There, a generalised model for the steady creep of this component was developed using the theorem of nesting surfaces, which we will use here.

If κ is the curvature of the centre line and δ the extension, then we choose the following non-dimensional quantities as generalised strain

$$\mathbf{e} = (e_1 \quad e_2) = (\kappa\, y_0 \quad \delta)$$

(The reader is henceforth referred to *Figure 5.20* for further notation.) The generalised stress is then, by definition

$$\mathbf{s} = (s_1 \quad s_2) = \left(\frac{M}{y_0 A} \quad \frac{N}{A}\right)$$

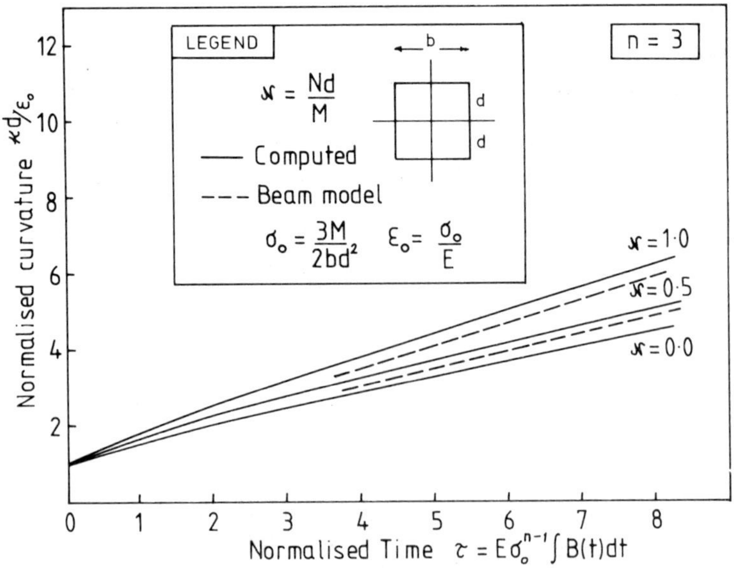

Figure 8.18 Comparison of exact and generalised beam models

Hence the generalised constitutive equation for an elastic time hardening creep response is given by equations (8.19)–(8.21) with

$$\Pi = \frac{1}{2E}(\mu_1^2 s_1^2 + s_2^2) \qquad \Omega \simeq \frac{g(t)}{n+1}(\mu_n^2 s_1^2 + s_2^2)^{(n+1)/2}$$

As an example of the use of this model, we consider a rectangular cross-section beam of depth $2d$ and breadth b subject to constant bending and axial load. Comparison is given in *Figure 8.18* between an exact analysis, obtained using the techniques of Chapter 7, and the above generalised model. It should be noted that for constant loads the approximation of equation (8.19) reduces to that of superposition of states.

8.10.2 A generalised model for thin shells

A generalised model for the steady creep of thin shells was developed in some detail in Section 5.7.4. Here we extend this to deal with transient creep.

A generalised strain for a rotationally symmetric shell is defined in beams of the deformation of the mid-surface

$$\mathbf{e} = (\epsilon_1 \quad \epsilon_2 \quad \kappa_1 h \quad \kappa_2 h)$$

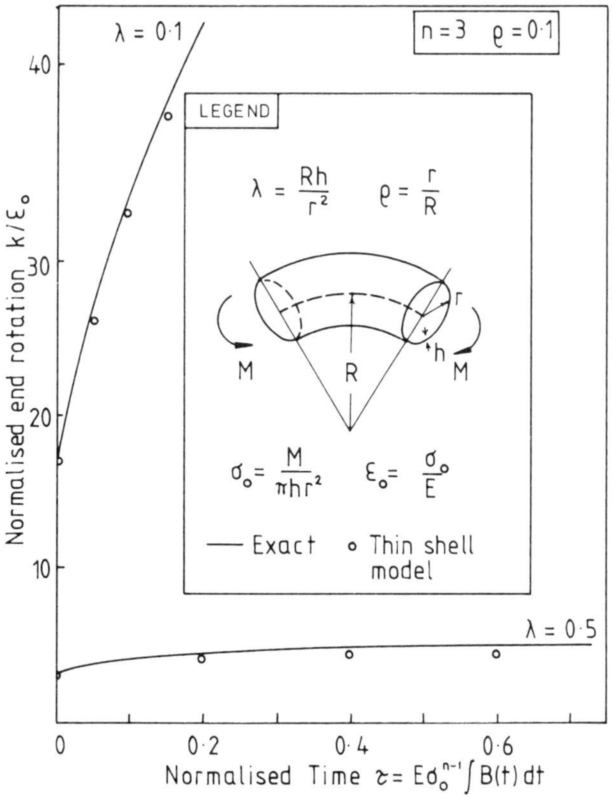

Figure 8.19 Comparison of exact and generalised thin-shell model for a curved pipe under in-plane bending

The corresponding generalised stress is again obtained on consideration of the work per unit volume as

$$\mathbf{s} = \left(\frac{N_1}{h} \quad \frac{N_2}{h} \quad \frac{2M_1}{h^2} \quad \frac{2M_2}{h^2}\right)$$

A generalised constitutive equation for the elastic creep response of the shell is then given by equations (8.19) – (8.21) with

$$\Pi = \frac{1}{2E}\left[(s_1^2 - 2\nu s_1 s_2 + s_2^2) + \mu_1^2(s_3^2 - 2\nu s_3 s_4 + s_4^2)\right]$$

$$\Omega \simeq \frac{g(t)}{n+1}\left[(s_1^2 - s_1 s_2 + s_2^2) + \mu_n^2(s_3^2 - s_3 s_4 + s_4^2)\right]$$

where ν is Poisson's ratio.

As an example of this model, we consider the forward creep of a thin curved pipe under constant in-plane bending examined for steady creep in Section 5.8. Comparison is given between the generalised model and a more exact analysis in *Figure 8.19*; agreement is clearly quite good.

Further reading

Marriott, D.L., Approximate analysis of transient creep deformation, *J. Strain Anal.*, **3**, 288–96, 1968.
Leckie, F.A. and Martin, J.B., Deformation bounds for bodies in a state of creep, *Trans. ASME, J. Appl. Mech.*, **34**, 411–17, 1967.
Ponter, A.R.S., On the stress analysis of creeping structures subject to variable loading. *Trans. ASME, J. Appl. Mech.*, **40**, 589–94, 1973.
Ponter, A.R.S., and Williams, J.J., Work bounds and associated deformation of cyclically loaded creeping structures, *Trans. ASME, J. Appl. Mech.*, **40**, 921–7, 1973.
Ponter, A.R.S., General bounding theorems for the quasi-static deformation of a body of inelastic material, with applications to metallic creep, *Trans. ASME, J. Appl. Mech.*, **41**, 947–52, 1974.
Ponter, A.R.S., The analysis of cyclically loaded creeping structures for short cycle times, *Int. J. Solids. Struct.*, **12**, 809–25, 1976.
Ainsworth, R.A., Approximate solutions for creeping structures subjected to periodic loading, *Int. J. Mech. Sci*, **18**, 149–59, 1976.
Ainsworth, R.A., Bounding structural deformation due to combined creep and plasticity, *Proc. IUTAM Symp. on Creep in Structures, Leicester, 1980*, Eds A.R.S. Ponter and D.R. Hayhurst, Springer-Verlag, 1981.
Leckie, F.A. and Ponter, A.R.S., Deformation bounds for bodies which creep in the plastic range, *Trans. ASME, J. Appl. Mech.*, **37**, 426–30, 1970.
Ponter, A.R.S., Deformation bounds for the Bailey–Orowan theory of creep, *Trans. ASME, J. Appl. Mech.*, **42**, 619–24, 1975.
Ponter, A.R.S., Convexity and associated continuum properties of a class of constitutive relationships, *J. Mecanique,* **15**, 527–42, 1976.
Zarka, J. *et al.*, Inelastic analysis of structures with applications to cyclic loadings, *Proc. 5th Int. Conf. on Structural Mechanics in Reactor Technology, Berlin, 1979*, Paper L3/2.
Edelstein, W.S. and Reichel, P.G., On bounds and approximate solutions of a class of transient creep problems, *Int. J. Solids. Struct.*, **16**, 107–18, 1980.
Spence, J. and Hult, J., Simple approximations for creep relaxation, *Int. J. Mech. Sci.*, **15**, 741–55, 1973.
Kachanov, L.M., *Theory of Creep,* 1960 (Transl. Natl Lending Library for Science and Technology, 1967).

Boyle, J.T. and Spence, J., Generalised structural models in creep mechanics, *Proc. IUTAM Symp. on Creep in Structures, Leicester, 1980*, Eds A.R.S. Ponter and D.R. Hayhurst, Springer-Verlag, 1981.
Nickell, R.E., *A Survey of Simplified Inelastic Analysis Methods,* Bulletin No. 253, Welding Research Council, 1979.

Chapter 9

Creep rupture

So far, we have only dealt with the calculation of stresses and distortions in a creeping structure (using either direct numerical techniques or simplified methods). Nevertheless, the ultimate aim in design for creep is the avoidance of structural failure during the expected life of a component. The question is – can we use the type of information that we already have to predict the lifetime of a structure? In order to attempt to resolve this question, it is necessary to understand the fundamental failure mechanisms which can occur due to creep. It is the purpose of this chapter to examine in detail one such mechanism, *creep rupture*, which can lead to a fracture of the structure causing a loss in load-bearing capacity.

As we have seen in Section 2.4, the creep process in a material will ultimately lead to creep rupture either through a ductile mechanism induced by large strains or through material embrittlement. Arguably the most insidious failure mechanism is the latter, which can occur at low strains. This will be examined in more detail here.

9.1 Constitutive equations for creep rupture

The material deterioration in a metal at elevated temperature which causes embrittlement proceeds mainly due to the nucleation and growth of fissures and voids on a microscopic level. The presence of such defects is loosely termed *damage* in the material. For a proper understanding of creep damage and its generalisation to multiaxial stress states, some description of the physical processes involved is necessary. This has been avoided so far with the adoption of a phenomenological approach. However, both a phenomenological and physical interpretation of damage is possible: these are outlined below.

9.1.1 Phenomenological approach

The notion of damage in a phenomenological context is only very vague. It can be introduced into a mathematical model of the creep process in one of two ways. First we can use the concept of a *net stress*, σ^*, such that if σ is the applied stress in a tensile test and ω the damage (related to a reduction in area, Section 2.4), then

$$\sigma^* = \frac{\sigma}{1-\omega}$$

Then in order to describe the tertiary stage of the basic creep curve, the net stress is introduced into constitutive equations developed for the primary and secondary stages.

Alternatively, the damage may be introduced as a macroscopic internal variable. Suppose that, in the secondary phase of creep, the creep strain rate, which is constant, has the following functional dependence on stress:

$$\dot{\epsilon}_C = f(\sigma)$$

In order to account for the increase in strain rate in the tertiary phase, a new internal variable, ω, is introduced such that as the damage increases the strain rate also increases

$$\dot{\epsilon}_C = f(\sigma, \omega)$$

The functional dependence of the strain rate on the damage is assessed from the form of the basic creep curves assuming an initial value of damage as zero and a final value at rupture of unity.

In either of these approaches, an equation relating the growth of damage to applied stress is required. This can take the form

$$\dot{\omega} = g(\sigma, \omega)$$

and can generally be determined from the rupture curves.

In general, both of these approaches lead to uniaxial creep equations of the form

$$\dot{\epsilon}_C = B\frac{\sigma^{n_1}}{(1-\omega)^{n_2}}$$

$$\dot{\omega} = D\frac{\sigma^{k_1}}{(1-\omega)^{k_2}} \tag{9.1}$$

where B, D, n_1, n_2, k_1, k_2 are material constants determined from the basic creep curves. If the concept of net stress is used, then we should have $n_1 = n_2 = n$ and $k_1 = k_2 = k$; it is found that as a rough relation $k \simeq 0.7n$. This form of the equation, and its multiaxial generalisation, will be used almost exclusively here.

The disadvantages of the phenomenological approach are obvious. While we can derive a good representation of existing data (usually for constant-load tests), the only means of assessment of its applicability to structures under a multiaxial state of stress is by analysis, experimentation and comparison. Thus, we cannot be entirely sure of the range of validity of the model. This takes on particular importance in modelling the tertiary phase of creep since we meet the problem of extrapolation to low stress levels. The only way in which we can gain more confidence in the model is through a physical understanding of the damage process.

9.1.2 Physical approach

Here we can only provide a rather sketchy description of deterioration in the tensile test.

Microscopic and metallographic observations indicate that deterioration proceeds through the formation of microcracks and voids at certain locations in the crystalline structure of a metal and their subsequent growth. There are then essentially two microscopic mechanisms; we call the former *nucleation* (the rate of formation of voids) and the latter *growth* (the rate of growth of void size). Thus we may define the nucleation rate, $\dot{n}$, and the rate of growth of cross-sectional areas of the voids $\dot{a}$. The important point here is that the latter depends on how long the voids have been in existence.

The total area A of the voids is a measure of the damage in the material; to construct a constitutive equation, we are interested in evaluating its rate of growth.

Let the time interval up to some instant t be split into a large number of discrete instants $\tau_1, \tau_2, \tau_3, \ldots, \tau_N$ such that $\tau_1 = 0$, $\tau_N = t$ where $\tau_{i+1} - \tau_i = \Delta\tau$, $i = 1, 2, \ldots, N$. The number of voids created in the interval (τ_i, τ_{i+1}) is $\dot{n}(\tau_i)\Delta\tau$. At time t the rate of growth of voids initiated at time τ_i is $\dot{a}(t, \tau_i)$. Then over a subsequent small interval Δt the change in area for all voids created in the interval (τ_i, τ_{i+1}) is $\dot{n}(\tau_i)\,\dot{a}(t, \tau_i)\Delta t\Delta\tau$. Taking into account each (small) time interval, the total change in area ΔA over the interval Δt is

$$\Delta A = \sum_{i=1}^{N} \dot{n}(\tau_i)\,\dot{a}(t, \tau_i)\Delta t\,\Delta\tau$$

i.e.

$$\frac{\Delta A}{\Delta t} = \sum_{i=1}^{N} \dot{n}(\tau_i)\,\dot{a}(t, \tau_i)\Delta\tau$$

Taking the limit as $\Delta t \to 0$, $\Delta\tau \to 0$ we have

$$\frac{dA}{dt} = \int_0^t \dot{n}(\tau)\,\dot{a}(t, \tau)\,d\tau$$

Immediately, a major drawback can be seen: the rate of change of damage in the material depends on the whole past history of the material, not just on its current state. This is conceptually equivalent to having an infinite number of state variables! Clearly this type of approach is not of much use in stress analysis (although neither is it intended to be). In addition, we have not managed to obtain an expression for the effect of damage on creep rate.

In practice, it is possible to construct an approximate constitutive equation for tertiary creep using nucleation growth of voids as a measure of damage. But this would involve direct measurement of the number of voids and their total volume on a microscopic level. The practical difficulties associated with this in developing a reliable constitutive equation should not need to be spelled out here.

Finally, it should also be noted that there have been some attempts at relating damage to some other macroscopically measurable quantity in a creep test – for example, changes in elastic modulus, plastic strain range or residual lifetime (or even the speed of sound in the material). While there is some argument as to the reliability of these measures in relation to damage, it is possible that this intermediate approach may ultimately prove the most fruitful.

9.2 Multiaxial constitutive equations for creep rupture

For the purposes of stress analysis, we need to extend the uniaxial relations, equations (9.1), to multiaxial stress states. Here we will continue to use the phenomenological approach, generalising the form of equations (9.1) based on observed behaviour. Unfortunately, when dealing with rupture, experimental evidence often only serves to complicate matters, suggesting that the simplified approach that we will describe below is generally inadequate. Nevertheless, it is probably good enough for most engineering purposes.

The results of experiment indicate that, during material deterioration in the tertiary phase of creep, the ratio of the strain-rate components remains virtually constant and equal to the value in the preceding secondary phase. The inference from this observation is that there exists a scalar quantity $\gamma(t)$ such that the multiaxial constitutive equations take the same form as those derived for secondary creep, Section 4.1,

$$\dot{\epsilon} = B\gamma(t)\frac{f(\bar{\sigma})}{\bar{\sigma}}\,\mathrm{s}$$

(in vector notation in the principal directions).

The scalar $\gamma(t)$ increases monotonically with time under constant load, reflecting the effect of damage: it may be related to the damage parameter, which is treated as an internal variable, by comparison with equations (9.1). Then we have

$$\dot{\epsilon} = B\,\frac{\bar{\sigma}^{n_1}}{(1-\omega)^{n_2}}\,\mathrm{s} \tag{9.2}$$

It remains to specify the growth law for the damage under a multiaxial stress state. This may be ascertained experimentally by identifying those combinations of stress which give the same rupture time; geometrically, these may be represented by surfaces in stress space, which we call *isochronous surfaces*. A typical surface is shown in *Figure 9.1* for a state of plane stress; in general, there are two extreme surfaces, that where the rupture time depends on the maximum principal stress

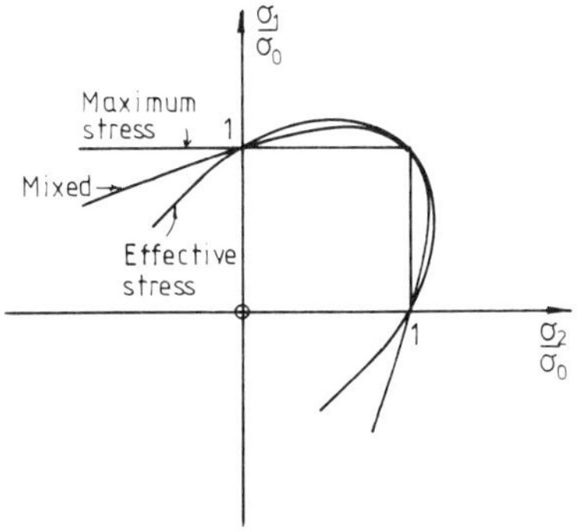

Figure 9.1 Isochronous rupture surfaces

and that where the rupture time depends on the effective stress imposed. Commercially pure copper and some aluminium alloys are found to give isochronous rupture surfaces close to one of these extremes. Mathematically, the isochronous rupture surface can be represented by a scalar invariant of stress $\Phi(\sigma_1, \sigma_2, \sigma_3)$. For a material whose rupture is governed by the maximum principal stress, $\Phi = \sigma_1$; for a material governed by the effective stress, $\Phi = \bar{\sigma}$. It has been suggested that the actual isochronous surface may be represented empirically by the linear interpolation

$$\Phi = a\sigma_1 + b\bar{\sigma} + 3c\sigma_V$$

where a, b, c are constants such that $a + b + c = 1$, determined from the experimental surfaces.

The importance of this is that it is observed that the stress states affecting the growth of $\gamma(t)$, and subsequently the damage ω, are the same as those found in the isochronous rupture surfaces. For example, in commercially pure copper which has an isochronous surface formed by the maximum principal stress, the voids and fissures are observed to form predominantly on the plane perpendicular to the (tensile) principal stress. We may then express the growth law for the damage in the form

$$\dot{\omega} = D \frac{\Phi(\sigma)^{k_1}}{(1-\omega)^{k_2}} \tag{9.3}$$

These multiaxial constitutive equations, equations (9.2) and (9.3), are based on experimental observations from constant-load, isothermal creep tests. The damage has little physical justification, it cannot be realistically related to a reduction in area and it appears simply as an internal variable. The extension of these equations to variable loading or variable temperature remains a matter of some contention. These deficiencies notwithstanding, the relations (9.2) and (9.3) can be criticised further on several grounds based on additional experimental observations. We now point out the principal objections.

In *Figure 9.1* the isochronous rupture surfaces have not been extended into the compressive region. In the analysis of engineering structures, some parts may be in compression and a suitable form of the constitutive equations is necessary. The precise form is open to debate. One can assume the form of equations (9.3) to hold for all regions, in which case the material continues to deteriorate in compression, or, as an example, the growth of damage may be assumed to stop if the material is in pure compression, i.e. $\sigma_v < 0$. Both of these are questionable, since it is observed that some metals do exhibit a limited 'healing' under compression.

In most metals, the formation of the voids and fissures occurs in preferred directions. This means that damage should be characterised not only quantitatively in terms of the total volume of voids and their rate of formation but also should incorporate some directional feature. Simply, in practice, damage is anisotropic. Moreover, this anisotropy cannot simply be expressed in terms of the principal stress directions as we have examined above. This suggests that the scalar measure of damage that we have used above is inappropriate and a tensorial measure should be introduced. A number of formal descriptions of damage are available, but these do not attempt to clarify the mechanics of multiaxial damage formation. One possibility which has emerged is that the original concept of net stress can be extended to multiaxial stress states. In order best to see how this can be done, we will digress for a moment and return to the tensorial description of stress introduced in Section 4.7.

Let us choose an arbitrary system of Cartesian coordinates (x_1, x_2, x_3) in a body and consider the infinitesimal tetrahedron formed by three surfaces parallel to the coordinate planes and one, of area ΔS, whose normal is the unit

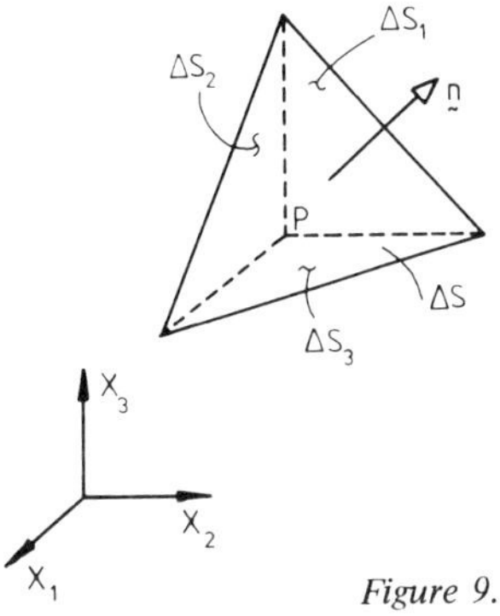

Figure 9.2 Geometry of infinitesimal tetrahedron

vector n_i (*Figure 9.2*). If p_i is the vector of surface tractions on ΔS, then the total force acting on this small area is

$$\Delta S p_i = \sigma_{ij} . \Delta S_j$$

where

$$\Delta S_j = \Delta S n_j$$

is called the vector element of area. This reduces to the well known Cauchy formula (Section 4.7).

$$p_i = \sigma_{ij} n_j$$

and can be interpreted as a generalisation of the familiar definition of stress $(\sigma = P/A_0)$ in the tensile test. In the latter, for a deteriorating material, damage ω is introduced by way of the net area $(1 - \omega)A_0$ and the constitutive relations are written in terms of the net stress $\sigma/(1 - \omega)$. This can be extended to multiaxial stress states on assuming that the area vector ΔS_i reduces to a *net area vector* $\Delta S_i{}^*$ through the linear transformation

$$\Delta S_i^* = (\delta_{ij} - \omega_{ij}) \, \Delta S_i$$

The (second-order) tensor ω_{ij} thus defined is called the *damage tensor*.

From Cauchy's formula, the total force acting on the net area is $\sigma_{ij}^* \; \Delta S_j^*$, where $\sigma_{ij}{}^*$ is introduced as the *net stress*, and this should remain unchanged during deterioration. That is

$$\sigma_{ij} \, \Delta S_j = \sigma_{ik}(\Delta_{kj} - \omega_{kj})^{-1} \Delta S_j^* = \sigma_{ij}^* \; \Delta S_j^*$$

and the net stress is defined by

$$\sigma_{ij}^* = \sigma_{ik}(\delta_{kj} - \omega_{kj})^{-1}$$

In general, this is non-symmetric and, instead of the tensor σ_{ij}^*, it is more practical to use a symmetric tensor

$$\sigma_{ij}^* = \frac{1}{2}(\sigma_{ik}\,\Phi_{kj} + \Phi_{ik}\sigma_{kj})$$

where $\Phi_{ij} = (\delta_{ij} - \omega_{ij})^{-1}$ is the damage effect tensor. It should be noted that we now have a three-dimensional version of the phenomenological concept of net stress, Section 9.1.1. With this theory, the criterion of creep rupture ($\omega = 1$) is generalised to the condition that the tensor $\delta_{ij} - \omega_{ij}$, treated as a matrix, is not invertible (singular).

It should be clear that the development of constitutive equations for the tertiary phase of creep is still in an early stage. In particular, it is not difficult to see that suitable constitutive equations based on a tensorial damage measure could prove extremely difficult to identify in practice. Here we will exclusively use the simpler equation (9.3) which should prove sufficient for isothermal creep of structures with a constant (or proportional) loading. Its use will allow us to illustrate the particular problems associated with the stress analysis of deteriorating materials and suitable methods of numerical solution.

The complete equations that we will use are given below in terms of the principal directions

$$\epsilon = \epsilon_E + \boldsymbol{\epsilon}_C$$

where, for simplicity,

$$\dot{\boldsymbol{\epsilon}}_C = B\,\frac{\bar{\sigma}^{n-1}}{(1-\omega)^n}\,\mathbf{s}$$
$$\dot{\omega} = D\,\frac{\Phi(\sigma)^k}{(1-\omega)^k} \qquad (9.4)$$

In particular, we will presume that the material constants are related by the condition

$$k = 0.7n$$

The above is an idealisation, but it serves our purpose admirably.

9.3 Example: Creep rupture of a multi-bar structure

We begin a brief study of stress analysis in the presence of material deterioration with a simple example which requires only the uniaxial constitutive relations. Nevertheless, it will serve to illustrate some of the difficulties which will be met in the analysis of more complex structures.

Consider the multi-bar structure shown in *Figure 9.3* consisting of M parallel bars of the same length L and cross-sectional area A attached an equal distance apart to a fixed horizontal surface at their upper ends and to a rigid horizontal bar at their lower ends. The lower rigid bar is hinged at one end and suffers a vertical downward load Q as shown.

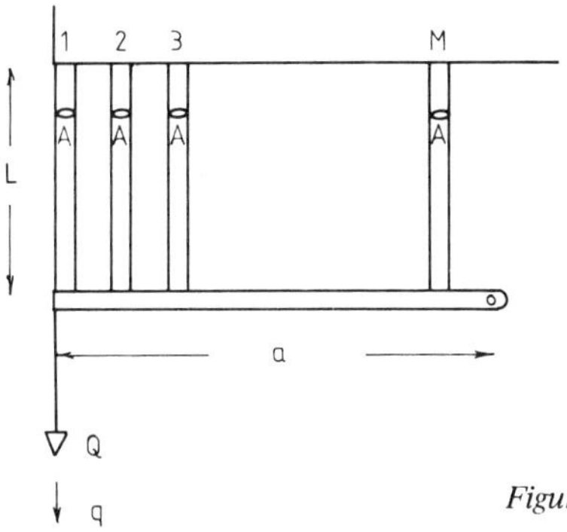

Figure 9.3 Geometry of a multi-bar structure

If σ_i, ϵ_i are the stress and strain in the ith bar and δ_i the vertical displacement of the ith bar, then the basic equations are equilibrium

$$\sum_{j=1}^{M} \frac{M+1-j}{M} \sigma_j = \frac{Q}{A}$$

and strain–displacement

$$\epsilon_i = \delta_i / L$$

together with the compatibility condition

$$\delta_i = \frac{M+1-i}{M} q$$

where q is the vertical displacement at the point of application of Q (that is $q = \delta_1$).

The constitutive equations are, from (9.4),

$$\epsilon_i = \sigma_i / E + \epsilon^c$$

$$\dot{\epsilon}_i^c = B\left(\frac{\sigma_i}{1-\omega_i}\right)^n \qquad \dot{\omega}_i = D\left(\frac{\sigma_i}{1-\omega_i}\right)^k$$

$$\epsilon_i^c(0) = 0 \qquad \omega_i(0) = 0$$

At the instant of loading ($t = 0$), the initial elastic solution is obtained as

$$\frac{q(0)}{L} = \frac{Q/AE}{I_1(M)}$$

$$\sigma_i(0) = \frac{M+1-i}{M} \frac{Q}{A} \frac{1}{I_1(M)}$$

on defining

$$I_1(M) = \sum_{j=1}^{M} \left(\frac{M+1-j}{M}\right)^2$$

These equations can now be combined into a form suitable for numerical solution. Previously we have dealt almost exclusively with the solution of the equations of stress redistribution, Section 7.3. The use of these equations in a rupture calculation is inappropriate on practical grounds. Let us consider the actual behaviour of the structure. As time progresses, the bars will fail in numerical order as they each reach the rupture criterion, $\omega_i = 1$, when the stresses reduce to zero. In practice, this condition obviously cannot be achieved exactly and the stress in a failed bar is not reduced to zero at rupture. This is important since as each bar fails the structural problem must be reformulated in terms of the remaining bars. Since the stress in each failed bar has not been fully reduced to zero, we cannot use the stresses at failure in the remaining bars as the initial stresses for the reformulated problem – these will *not* be in equilibrium. As an alternative, we can use the equations of evolution for creep strain Section 7.3.5. For this problem, these take the form

$$\frac{d\epsilon_i^c}{dt} = B\left\{\frac{M+1-i}{M}\left[\frac{Q}{A} + \sum_{j=1}^{M}\left(\frac{M+1-j}{M}\right)\epsilon_j^c\right]\Big/ I_1(M) - \epsilon_i^c\right\}^n \Big/ (1-\omega_i)^n \qquad (9.5)$$

$$i = 1, 2, \ldots, M$$

Using these equations, the creep strains at failure are taken as the initial creep strains in the remaining bars for each reformulated problem. The stresses, which are evaluated in terms of the creep strains, are always in equilibrium.

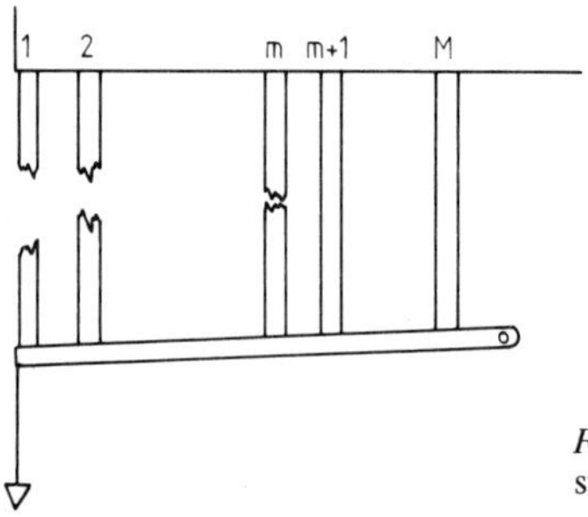

Figure 9.4 Propagation of failure in a multi-bar structure

Suppose that the mth bar has failed at time t_m and that the creep strains in the remaining bars at this instant (*Figure 9.4*) are denoted by

$$\epsilon_i^c(t_m) = \epsilon_{im}^c \qquad i = m+1, m+2, \ldots, M$$

The equations should now be reformulated in terms of the remaining bars. This results in the system

$$\frac{d\epsilon_i^c}{dt} = B\left\{\frac{M+1-i}{M}\left[\frac{Q}{A} + \sum_{j=m+1}^{M}\left(\frac{M+1-j}{M}\right)\epsilon_j^c\right]\Big/ I_m(M) - \epsilon_i^c\right\}^n \Big/ (1-\omega_i)^n \qquad (9.6)$$

$$i = m+1, m+2, \ldots, M$$

subject to the 'initial' condition

$$\epsilon_i^c(t_m) = \epsilon_{im}^c \qquad i = m+1, m+2, \ldots, M$$

where now

$$I_m(M) = \sum_{j=m+1}^{M} \left(\frac{M+1-j}{M}\right)^2$$

Before the first bar has failed, $m = 0$, and the system (9.6) is identical to (9.5). As rupture progresses, the failure times $t_1, t_2, \ldots, t_m, \ldots$ are noted; the structure finally fails completely at time t_M. The failure of this structure can be conveniently divided into three phases. Up until time t_1, no bar has failed – we call this the stage of *latent failure*. The evolution of the structure is followed by solving the complete equations (9.5). At time t_1, the first bar will break – we call this the time for *initiation of failure*. The creep strain in each bar is noted at this instant and the problem reformulated assuming the first bar to have failed completely. The subsequent evolution of the remaining bars is followed by solving equations (9.6) with $m = 1$ until time t_2 when we must set $m = 2$, and so on. Finally at time t_M the structure fails completely. The period t_1 to t_M is called, for obvious reasons, the *propagation period*; the time t_M is called the time of *final failure* or *fracture*. These phases will be found in any deteriorating structure.

In order to determine the crucial times t_1 and t_M, the equations (9.5) and (9.6) must be solved numerically. To facilitate this, we deform non-dimensional quantities

$$S_i = \frac{\sigma_i}{\sigma_0} \qquad U = \frac{q/L}{\epsilon_0} \qquad \epsilon_0 = \frac{\sigma_0}{E}$$

using the stress $\sigma_o = Q_o/A$ where Q_o is arbitrary, together with the new time scale

$$\tau = E\sigma_0^{n-1} \int B \, \mathrm{d}t$$

The equation for the growth of damage can now be written in the form

$$\frac{\mathrm{d}\omega_i}{\mathrm{d}\tau} = \frac{1}{\tau_0} \frac{1}{1+k} \left(\frac{S_i}{1-\omega_i}\right)^k$$

where we have defined

$$\tau_0 = \frac{EB/D}{(1+k)\,\sigma_0^{1-n+k}}$$

This can be identified as the normalised rupture time of a single bar under the tensile load Q_0. It should be carefully noted that the initial conditions will depend on the load parameter Q/Q_0. Unlike the previous cases that we have considered with a non-deteriorating time hardening power law of creep, the loading cannot be eliminated through redefinition of the time scale. The solution will therefore depend on the material parameters n and k and the loading parameter Q/Q_0; since Q_0 is arbitrary, we choose it such that $\tau_0 = 1$.

As this is a simple problem, we can effectively use any of the time-stepping algorithms discussed in Section 7.4. More importantly, we must pay heed to one particular aspect of the solution procedure. During some time step it is

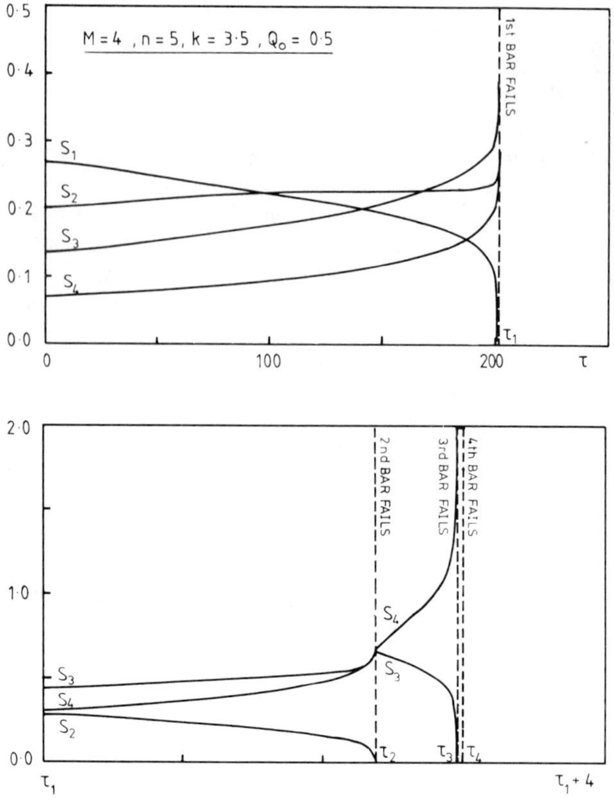

Figure 9.5 Variation of normalised stresses during initiation and propagation of failure

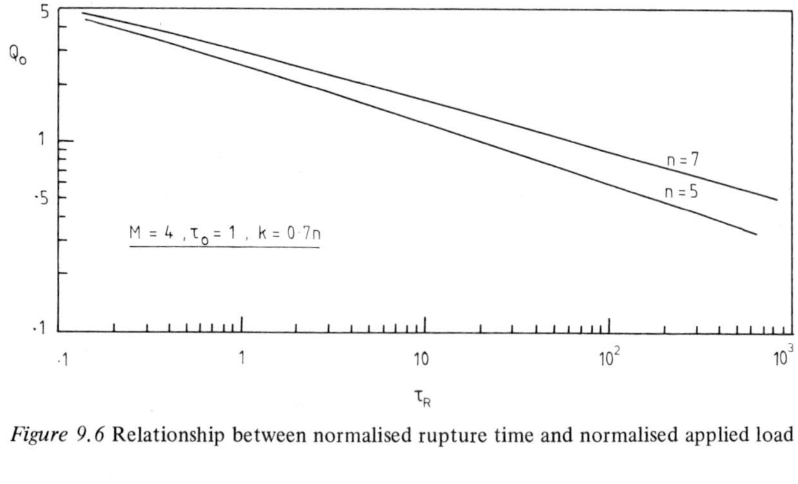

Figure 9.6 Relationship between normalised rupture time and normalised applied load

imperative that a bar does not reach the failure criterion ($\omega_i = 1$). Consequently, if failure of a bar is imminent, the time step must be reduced to avoid this possibility. Complete failure ($\omega_i = 1$) of a bar cannot be achieved in practice; instead failure is assumed to occur once some critical value of damage, say, $\omega_i = 0.9999$, is exceeded.

In *Figure 9.5* we plot the redistribution of stresses in time for a four-bar structure using typical values of the material parameters and load parameter. The stress in the first most highly stressed bar is seen to fall rapidly to zero as damage increases and failure is reached. During this period the stresses in the other bars increase to accommodate the load. Failure of the remaining bars, leading to complete failure of the structure, is seen to occur *very* soon after the initiation time τ_1.

In *Figure 9.6* we plot the final rupture time against the loading parameter Q/Q_0 for typical values of the material parameters. It can be seen that this relation is very nearly linear in a log–log plot.

9.4 Continuum damage mechanics

We need to be quite clear about the progress and effect of material deterioration in a continuum. It is similar to, but subtly different from, that which we observed in the multi-bar structure, which is rather special. As each bar fails, we are faced with a new structural problem, but one which is obvious and readily solved. In a continuous body under load, the constitutive relation tells us how the stress, strain rate and damage are interrelated at a point in the material. Just as various points in the body suffer different stress levels, various points will be at different stages of material deterioration. The constitutive equations remain valid until the specified rupture criterion is reached. At this time, the point under consideration can no longer sustain stress and, at this point, the material has failed. Ultimately a region of failed material will progress through the body until the applied load can no longer be sustained and the body fails as a whole. We can depict this as follows. Consider a body of volume

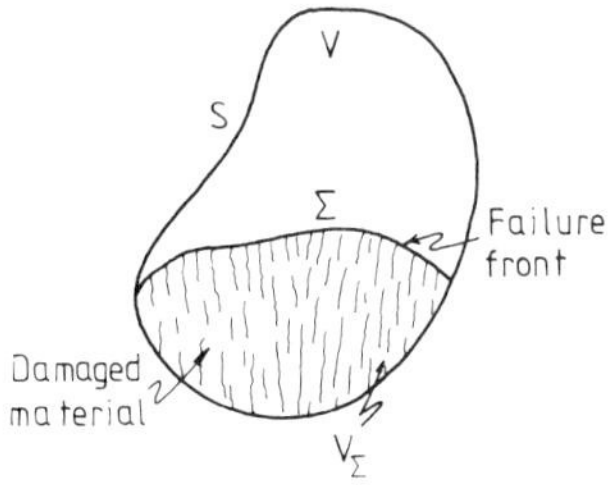

Figure 9.7 Geometry of a deteriorating continuous body

V bounded by a surface S under external loading and supported in some manner (*Figure 9.7*). Up until a time, which we will denote by t_I, no part of the body has failed. As for the multi-bar structure, we call this the stage of *latent failure* (or, sometimes, the *incubation period*). At this time, some point in the body will fail and we call this time t_I the *initiation time*. Subsequently, a surface Σ which separates failed and intact material propagates through the body. This moving surface is called the *failure front*. Finally, the failure front spreads through the body until at a time t_{II} the body fails as a whole. The period t_{I} to t_{II} is called the *propagation period*.

The continuum problem described by the above is different from that which we normally encounter in continuum mechanics. In order to formulate and subsequently solve the problem, we must follow the evolution of the failure front. To differentiate this new class of problem from the old, we call their study *continuum damage mechanics.*

9.4.1 Mathematical formulation of continuum damage mechanics

In order to develop a mathematical formulation of continuum damage mechanics, we return here to the formalism of Section 7.3. The development will be based on the initial value problem for the creep strains, Section 7.3.5, following the discussion for the multi-bar structure of the previous section.

We consider a body of volume V, bounded by a surface S, under the influence of internal body forces. On part of the surface, tractions are applied, whilst on the remainder displacements are prescribed. The material of this body is assumed to deteriorate under these conditions; at some time t_{I} failure will be initiated somewhere in the body and subsequently a failure front Σ will propagate through the body (*Figure 9.7*) until complete failure at a time t_{II}. The region of failed material we denote by V_{Σ}. Both the position of the failure front and the region V_{Σ} will be functions of elapsed time. We consider the two stages of deterioration separately:

9.4.1.1 Stage of latent failure: $0 \leqslant t \leqslant t_{\mathrm{I}}$

In this stage, no failure of the material has occurred, although there is a growth of damage. The evolution of the creep strain tensor $\epsilon_{ij}^{\mathrm{c}}$ can then be described by the initial value problem

$$\frac{\mathrm{d}\epsilon_{ij}^{\mathrm{c}}}{\mathrm{d}t} = f_{ij}(R_{kl}(\epsilon_{mn}^{\mathrm{c}}) + \sigma_{kl}^{\mathrm{o}}, \omega)$$

$$\frac{\mathrm{d}\omega}{\mathrm{d}t} = g(R_{kl}(\epsilon_{mn}^{\mathrm{c}}) + \sigma_{kl}^{\mathrm{o}}, \omega) \tag{9.7}$$

where the functional forms f_{ij} and g can be identified from equations (9.4), together with the initial conditions

$$\epsilon_{ij}^{\mathrm{c}}(0) = 0 \qquad \omega(0) = 0$$

throughout the body.

As previously, σ_{ij}^{o} is the equivalent elastic stress and R_{ij} the tensor-valued residual operator from the residual elastic problem.

These equations are valid up to the initiation time t_{I} when failure first occurs in the body. We denote the creep strain at the initiation of failure by $\epsilon_{ij}^{\mathrm{I}}$ and the distribution of damage by ω^{I}.

9.4.1.2 Stage of propagation of failure: $t_{\mathrm{I}} \leqslant t \leqslant t_{\mathrm{II}}$

Once local failure has been initiated, a failure front, being the boundary of the failed region $V - V_{\Sigma}$, will propagate through the body. Mathematically, we are now faced with a 'moving boundary value problem'. This may be formally described by the initial value problem in $V - V_{\Sigma}$

$$\frac{\mathrm{d}\epsilon_{ij}^{\mathrm{c}}}{\mathrm{d}t} = f_{ij}\,(R_{kl}^{\Sigma}\,(\epsilon_{mn}^{\mathrm{c}}) + \sigma_{kl}^{\Sigma}\,,\,\omega)$$

$$\frac{\mathrm{d}\omega}{\mathrm{d}t} = g\,(R_{kl}^{\Sigma}\,(\epsilon_{mn}^{\mathrm{c}}) + \sigma_{kl}^{\Sigma},\,\omega) \tag{9.8}$$

subject to the conditions

$$\epsilon_{ij}^{\mathrm{c}}\,(t_{\mathrm{I}}) = \epsilon_{ij}^{\mathrm{I}} \qquad\qquad \omega(t_{\mathrm{I}}) = \omega^{\mathrm{I}}$$

Because of the motion of the failure front, the equivalent and residual elastic problems must be reformulated in terms of the instantaneous configuration of unfailed material. Thus σ_{ij}^{Σ} denotes the equivalent elastic stress for the region $V - V_{\Sigma}$ and R_{ij}^{Σ} denotes the residual elastic stress for a prescribed creep strain distribution in $V - V_{\Sigma}$. In both of these boundary value problems, the boundary conditions have to be adjusted to account for the failed region. This will depend on the particular problem in hand.

The motion of the failure front cannot be predicted in advance; it will depend both on the material constitutive model adopted and the level and nature of the loading. For this reason, it is necessary to follow the whole history of deformation of the body. Of course, this must be done numerically.

9.4.2 Numerical solution of problems in continuum damage mechanics

Essentially, the numerical methods which were discussed in Chapter 7 remain applicable to problems in damage mechanics. In practice, the formulation of the problem for numerical solution, and the different stages in this solution, is similar to that described for the multi-bar structure. The most common methods of numerical solution of continuum problems – finite element or finite difference techniques – involve some kind of spatial discretisation. For material creep, this discretisation results in a finite system of ordinary differential equations, as we have seen in various problems. During the stage of latent failure, the complete system of equations is solved as before until failure is initiated. Then the dimension of this system is reduced to remove the (discrete) failed region. A new system of equations must then be formed and the solution procedure is restarted until a further discrete region fails, and so on. In this way a sequence of initial value problems of reducing dimension should be solved. It should be obvious that these initial value problems can be stiff in character so that stiff equation solvers are more appropriate (although, as we discussed in Section 7.4, not necessary). Experience with the multi-bar structure showed us that, as failure is approached, the stress and deformation processes can occur quite rapidly. Because of this, small time steps must be used. As we have mentioned before, the time step should be chosen so that the failure criterion is not satisfied during the increment and, further, the failure criterion itself cannot be satisfied exactly in practice. To overcome these difficulties, it is necessary to reduce the time step as local failure approaches. This can be done by checking if the failure criterion ($\omega = 1$) is exceeded in a time step; if it is, then the numerical time-stepping algorithm should automatically return to the beginning of the step and proceed with a reduced step size (say half its original value). This procedure is followed until a specified upper limit on damage, say $\omega = 0.99$, is exceeded. For complex problems this procedure may require a large number of time steps in the period immediately preceding failure. This may be avoided by reducing the upper limit on damage.

It is fair to say that, at the time of writing, very little investigation of the numerical problems associated with continuum damage mechanics has been made. In particular, the application of the finite element method to damage mechanics, and the unique problems that it would generate, has received very little attention. This will obviously be an area of much future work.

9.5 Example: Creep rupture of a thick cylinder

The concepts described in the preceding section are best understood by way of an example. We consider here the isothermal creep rupture of a thick cylinder of inner radius a, outer radius b, subject to an internal pressure p and in a state of plane strain (*Figure 4.2*). The basic field equations for this problem have been given in Section 4.3, to which the reader is referred for further notation, taking into account only steady creep. In this case, where we assume transient creep caused by material damage according to the constitutive relations (9.4), it can be shown that the initial value problem for creep strains in the stage of latent failure is given by

$$\frac{d\epsilon_r^c}{dt} = B\frac{\bar{\sigma}^{n-1}}{(1-\omega)^n}\left[\sigma_r - \frac{1}{2}(\sigma_\vartheta + \sigma_z)\right]$$
$$\frac{d\epsilon_\vartheta^c}{dt} = B\frac{\bar{\sigma}^{n-1}}{(1-\omega)^n}\left[\sigma_\vartheta - \frac{1}{2}(\sigma_r + \sigma_z)\right] \qquad (9.9)$$

where

$$\bar{\sigma}^2 = \frac{1}{2}\left[(\sigma_r - \sigma_\vartheta)^2 + (\sigma_r - \sigma_z)^2 + (\sigma_\vartheta - \sigma_z)^2\right]$$

The stresses are related to the creep strains according to

$$\begin{aligned}\sigma_r &= R_r(\epsilon_r^c, \epsilon_\vartheta^c) + \sigma_r^o\\ \sigma_\vartheta &= R_\vartheta(\epsilon_r^c, \epsilon_\vartheta^c) + \sigma_\vartheta^o\\ \sigma_z &= R_z(\epsilon_r^c, \epsilon_\vartheta^c) + \sigma_z^o\end{aligned} \tag{9.10}$$

in terms of the equivalent elastic stresses

$$\sigma_r^o = \frac{pa^2}{b^2 - a^2}\left(1 - \frac{b^2}{r^2}\right)$$

$$\sigma_\vartheta^o = \frac{pa^2}{b^2 - d^2}\left(1 + \frac{b^2}{r^2}\right)$$

$$\sigma_z^o = \frac{pa^2}{b^2 - a^2}\, 2\nu$$

and the residual operators

$$R_r = \frac{E}{2(1-\nu^2)}\left(\int_a^r \frac{\epsilon_r^c - \epsilon_\vartheta^c}{\eta}\, d\eta - \frac{r^2 - a^2}{b^2 - a^2}\,\frac{b^2}{r^2}\int_a^b \frac{\epsilon_r^c - \epsilon_\vartheta^c}{\eta}\, d\eta\right)$$

$$+ \frac{E(1-2\nu)}{2(1-\nu^2)}\,\frac{1}{r^2}\left(\int_a^r \eta\,\epsilon_z^c\, d\eta - \frac{r^2 - a^2}{b^2 - a^2}\int_a^b \eta\epsilon_z^c\, d\eta\right)$$

$$R_\vartheta = \frac{E}{2(1-\nu^2)}\left(\int_a^r \frac{\epsilon_r^c - \epsilon_\vartheta^c}{\eta}\, d\eta - \frac{(r^2 + a^2)}{(b^2 - a^2)}\,\frac{b^2}{r^2}\int_a^b \frac{\epsilon_r^c - \epsilon_\vartheta^c}{\eta}\, d\eta\right)$$

$$-\frac{E(1-2\nu)}{2(1-\nu^2)}\,\frac{1}{r^2}\left(\int_a^r \eta\,\epsilon_z^c\, d\eta + \frac{r^2 + a^2}{b^2 - a^2}\int_a^b \eta\,\epsilon_z^c\, d\eta\right)$$

$$-\frac{E}{1-\nu^2}(\epsilon_\vartheta^c + \nu\,\epsilon_z^c)$$

$$R_z = \frac{E\nu}{1-\nu^2}\left(\int_a^r \frac{\epsilon_r^c - \epsilon_\vartheta^c}{\eta}\, d\eta - \frac{b^2}{b^2 - a^2}\int_a^b \frac{\epsilon_r^c - \epsilon_\vartheta^c}{\eta}\, d\eta\right)$$

$$-\frac{E\nu(1-2\nu)}{1-\nu^2}\,\frac{1}{b^2 - a^2}\int_a^b \eta\,\epsilon_z^c\, d\eta + \frac{E}{1-\nu^2}\left[\nu\epsilon_r^c - (1-\nu)\epsilon_z^c\right]$$

noting that, from incompressibility

$$\epsilon_z^c = -(\epsilon_r^c + \epsilon_\vartheta^c)$$

Combining equations (9.9) and (9.10), the required equations can be obtained. In the above form, we can also readily evaluate the stress directly during solution. These equations are completed by the damage relations. In this example, a maximum principal stress criterion is used. For the thick cylinder under constant pressure, the maximum stress is given by σ_ϑ which is always positive. Then we have

$$\frac{d\omega}{dt} = D\left(\frac{\sigma_\vartheta}{1-\omega}\right)^k \tag{9.11}$$

For convenience, the complete set of equations may be non-dimensionalised using an arbitrary stress σ_o on defining

$$S_r = \frac{\sigma_r}{\sigma_0} \qquad E_r^c = \frac{\epsilon_r^c}{\sigma_0/E} \qquad \text{etc.}$$

and using the normalised time scale

$$\tau = E\sigma_0^{n-1}\int B\,dt$$

Then equation (9.11) can be written as

$$\frac{d\omega}{d\tau} = \frac{1}{\tau_0}\left(\frac{1}{1+k}\right)\left(\frac{S_\vartheta}{1-\omega}\right)^k$$

where we have defined

$$\tau_0 = \frac{EB/D}{(1+k)\sigma_0^{1-n+k}}$$

as the normalised rupture time of a uniaxial specimen under the initial stress σ_0. As in the case of the multi-bar structure, we choose $\tau_0 = 1$. Then the solution will depend on the load parameter $p_0 = p/\sigma_0$.

The problem can now be reduced to a finite system of ordinary differential equations suitable for numerical solution by considering only a finite number of points through the thickness of the cylinder in the radial direction, say, r_i, $i = 1, 2, \ldots, M$. The integrals in R_r, R_ϑ and R_z can then be evaluated using the formulae given in Section 7.5.

To begin with, we will re-solve this problem numerically in the stage of latent failure for a particular value of the material parameter n with $k = 0.7n$, and various values of the load parameter p_0. In *Figure 9.8* the redistribution of the stresses S_ϑ and S_r and the damage ω up to the failure initiation time τ_I are shown for a typical case. In that case, it can be seen that failure is initiated at the outer surface and can therefore be expected to propagate inwards. At failure, the stress at the outer surface falls rapidly to zero. In *Figure 9.9* the time to initiation of failure for a range of loading parameters is shown. In this range, failure always initiates at the outer surface. (In general, this will depend on the damage criterion used; failure may propagate from the inner surface or indeed from an internal position.) The problem should now be formulated to allow for the stage of propagation of the failure front.

Let the radius of the failure front be denoted by $c(t)$ (*Figure 9.10*). The material has failed in the region $c < r \leqslant b$. The equations (9.9) and (9.11) remain

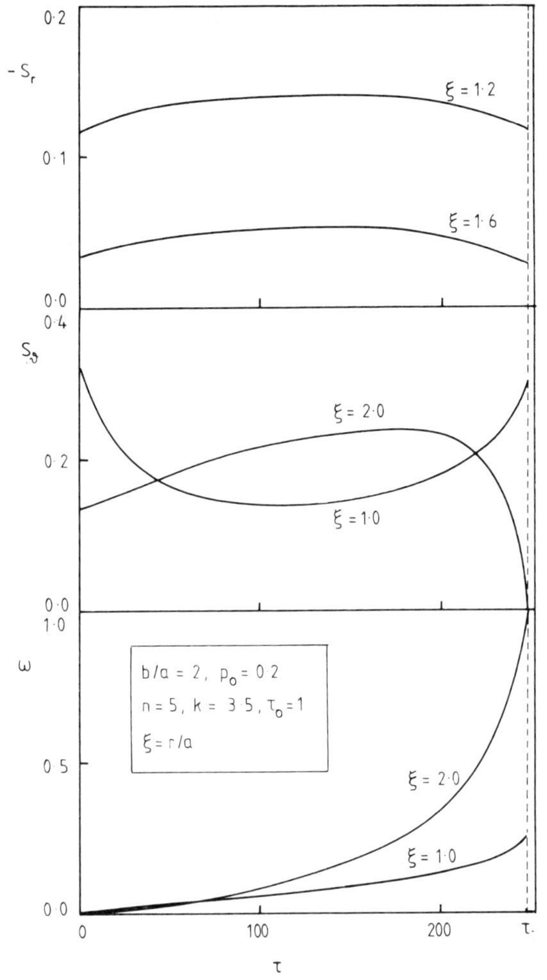

Figure 9.8 Variation of normalised stresses and damage in a thick cylinder during initiation of failure

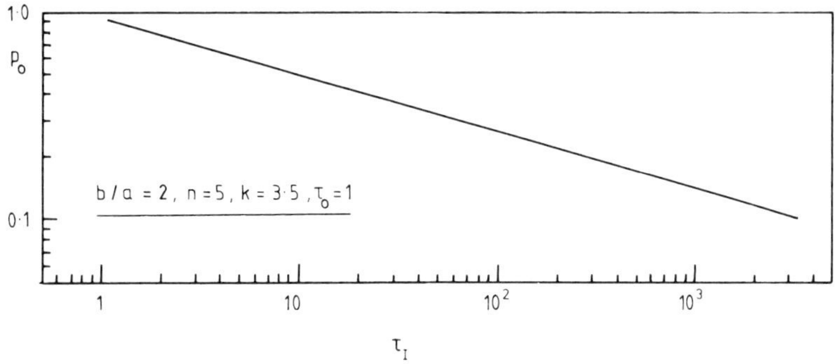

Figure 9.9 Relationship between normalised initiation time and normalised pressure

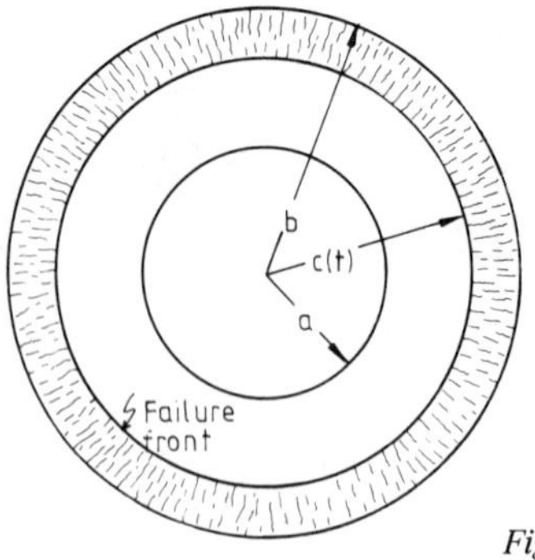

Figure 9.10 Propagation of failure front in thick cylinder

valid in the unfailed region $a \leqslant r \leqslant c$; however, the form of the relations of equation (9.10) must be altered. For a propagating failure front

$$\begin{aligned} \sigma_r &= R_r^{\Sigma}(\epsilon_r^c, \epsilon_\vartheta^c) + \sigma_r^{\Sigma} \\ \sigma_\vartheta &= R_\vartheta^{\Sigma}(\epsilon_r^c, \epsilon_\vartheta^c) + \sigma_\vartheta^{\Sigma} \\ \sigma_z &= R_z^{\Sigma}(\epsilon_r^c, \epsilon_\vartheta^c) + \sigma_z^{\Sigma} \end{aligned} \tag{9.12}$$

The equivalent elastic stresses for the region $a \leqslant r \leqslant c$ are

$$\sigma_r^{\Sigma} = \frac{pa^2}{c^2 - a^2}\left(1 - \frac{c^2}{r^2}\right)$$

$$\sigma_\vartheta^{\Sigma} = \frac{pa^2}{c^2 - a^2}\left(1 + \frac{c^2}{r^2}\right)$$

$$\sigma_z^{\Sigma} = \frac{pa^2}{c^2 - a^2}\, 2\nu$$

The residual elastic operators are

$$R_r^{\Sigma} = \frac{E}{2(1-\nu^2)}\left(\int_a^r \frac{\epsilon_r^c - \epsilon_\vartheta^c}{\eta}\, d\eta - \frac{r^2 - a^2}{c^2 - a^2}\,\frac{c^2}{r^2}\int_a^c \frac{\epsilon_r^c - \epsilon_\vartheta^c}{\eta}\, d\eta\right)$$

$$+ \frac{E(1-2\nu)}{2(1-\nu^2)}\,\frac{1}{r^2}\left(\int_a^r \eta\,\epsilon_z^c\, d\eta - \frac{r^2 - a^2}{c^2 - a^2}\int_a^c \eta\,\epsilon_z^c\, d\eta\right)$$

$$R_\vartheta^{\Sigma} = \frac{E}{2(1-\nu^2)}\left(\int_a^r \frac{\epsilon_r^c - \epsilon_\vartheta^c}{\eta}\, d\eta - \frac{r^2 + a^2}{c^2 - a^2}\,\frac{c^2}{r^2}\int_a^c \frac{\epsilon_r^c - \epsilon_\vartheta^c}{\eta}\, d\eta\right)$$

$$- \frac{E(1-2\nu)}{2(1-\nu^2)}\,\frac{1}{r^2}\left(\int_a^r \eta\,\epsilon_z^c\, d\eta + \frac{r^2 + a^2}{c^2 - a^2}\int_a^c \eta\,\epsilon_z^c\, d\eta\right)$$

$$- \frac{E}{1 - \nu^2} (\epsilon_\vartheta^{\mathrm{c}} + \nu\epsilon_z^{\mathrm{c}})$$

$$R_z^\Sigma = \frac{E\nu}{1 - \nu^2}\left(\int_a^r \frac{\epsilon_r^{\mathrm{c}} - \epsilon_\vartheta^{\mathrm{c}}}{\eta} \, \mathrm{d}\eta - \frac{c^2}{c^2 - a^2}\int_a^c \frac{\epsilon_r^{\mathrm{c}} - \epsilon_\vartheta^{\mathrm{c}}}{\eta} \, \mathrm{d}\eta\right)$$

$$- \frac{E\nu(1 - 2\nu)}{1 - \nu^2} \frac{1}{c^2 - a^2} \int_a^c \eta \, \epsilon_z^{\mathrm{c}} \, \mathrm{d}\eta + \frac{E}{1 - \nu^2} [\nu\epsilon_r^{\mathrm{c}} - (1 - \nu)\epsilon_z^{\mathrm{c}}]$$

which we write in their complete form for clarity

The equations defined by (9.9), (9.11) and (9.12) can be expressed in discretised form. Then the following computational procedure is adopted. At initiation of failure, when the outer surface at radius r_M has just failed, we can eliminate the creep strains at this discrete point from the solution process. The equations are then reformulated for the remaining unfailed points r_i, $i = 1, 2, \ldots,$ $M - 1$. The time-stepping algorithm is then restarted and continues until point r_{M-1} fails. This point is eliminated, the problem reformulated and the solution

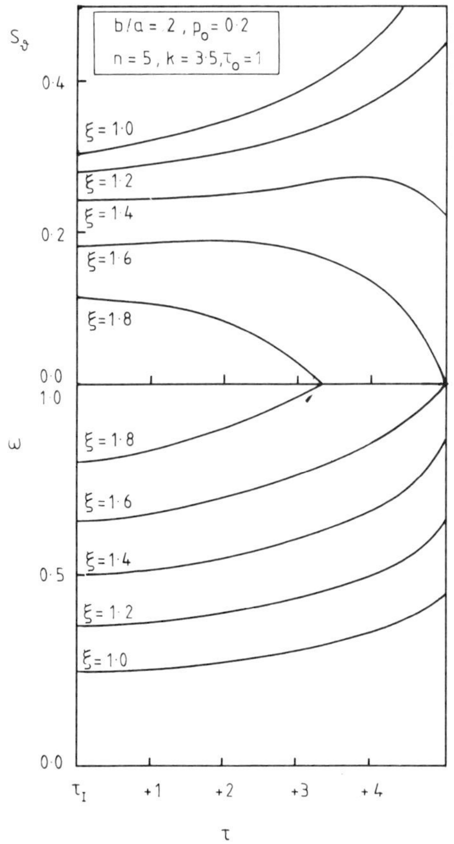

Figure 9.11 Variation of normalised circumferential stress and damage during propagation of failure

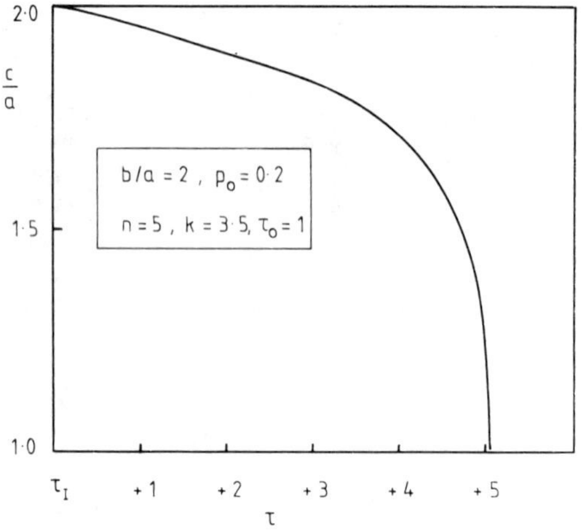

Figure 9.12 Variation of radius of failure front with time

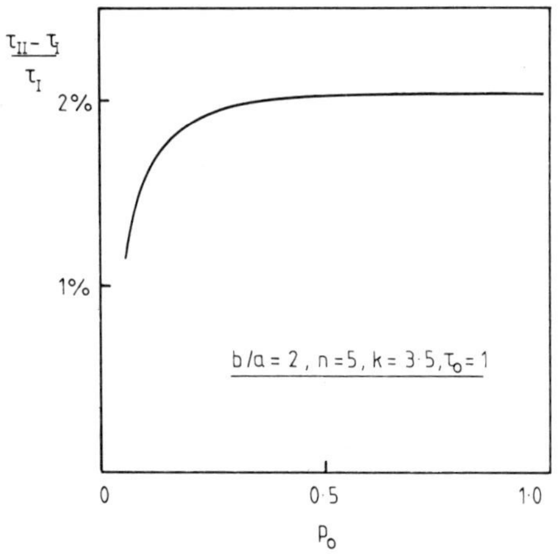

Figure 9.13 Ratio of duration of propagation phase to duration of initiation phase as a function of applied pressure

restarted. This procedure is repeated until complete failure of the cylinder. In this way the propagation of the failure front can be followed; from the nature of our numerical solution, it proceeds in discrete jumps.

In *Figure 9.11* we show the redistribution of the stress S_ϑ and the damage ω during the propagation phase. The motion of the failure front can be clearly seen as the stresses fall to zero. In *Figure 9.12* the radius of the failure front $c(t)/a$ is shown as a function of time in a typical case. The failure front is seen to proceed with increasing velocity as it moves through the cylinder. Finally, in *Figure 9.13* we show the ratio $(\tau_{II}-\tau_I)/\tau_I$, which represents the duration of the stage of propagation, for various values of the load parameter p_0.

9.6 Estimation of failure times in deteriorating structures

We return now to the question posed in the introduction to this chapter – can we estimate the lifetime of a deteriorating structure without resort to a complete damage analysis? We have seen for both the multi-bar structure and the thick cylinder that a complete stress analysis involves considerable detailed computation and further that the numerical realisation of such a computation is fraught with difficulties. If this can be dispensed with, then much expense (and frustration) can be avoided. Ideally, a lower bound on the lifetime of a structure would be wanted by a designer in order to produce a conservative estimate. Not surprisingly for such a complex problem, efforts to obtain such an estimate have been largely unsuccessful. (And it is always worth bearing in mind that any such bound that we may develop relates only to a rather idealised material behaviour, not the actual behaviour!) Nevertheless, it is possible to establish an upper bound on lifetime; this we discuss in the following.

The structural problem which we consider is essentially that of Section 5.4 of a continuous body occupying a volume V acted upon the surface by a certain applied load and supported in the same manner. (Throughout the vector notation with principal stress directions will be used.) In order to establish this bound, a mathematical fabrication is necessary. This is based on the assumption that the functional

$$\Omega_k = D\,\Phi^k(\sigma)$$

appearing in the damage law, equation (9.4), is convex (in fact only the functional Φ need be convex). From *Figure 9.1* this assumption is physically justifiable. We then *define* a hypothetical steadily creeping material with constitutive equation

$$\dot{\epsilon} = \mathrm{d}\Omega_k/\mathrm{d}\sigma \tag{9.13}$$

Then the following result can be established.

Result (*Upper bound on lifetime*) Consider the structural problem as defined composed of a material deteriorating according to equation (9.4). Then an upper bound on the rupture time of a structure is given by

$$t_{\mathrm{II}} \leqslant \frac{V}{(k+1)D\int_V \Phi^k(\sigma_\Phi)\,\mathrm{d}V}$$

where σ_Φ is the steady stress distribution for the same structural problem but with a material described by equation (9.13).

To prove this result, we rewrite the damage relation of equation (9.4), integrating over the volume of the body, as

$$-\frac{1}{k+1}\frac{\mathrm{d}W}{\mathrm{d}t} = \frac{1}{V}\int_V \Omega_k(\sigma)\,\mathrm{d}V$$

on defining the function

$$W(t) = \frac{1}{V}\int_V (1-\omega)^{k+1}\,\mathrm{d}V$$

For the steadily creeping body with constitutive equation (9.13), we can use the theorem of minimum complementary dissipation (Section 5.4.3) in the form

$$\int_V \Omega_k(\sigma_\Phi)\,\mathrm{d}V \leqslant \int_V \Omega_k(\sigma^{\mathrm{s}})\,\mathrm{d}V$$

where σ^{s} is any statically admissible stress. We can of course use the actual stress σ. Then we have the differential inequality

$$-\frac{1}{k+1}\frac{\mathrm{d}W}{\mathrm{d}t} \geqslant \frac{1}{V}\int_V \Omega_k(\sigma_\Phi)\,\mathrm{d}V$$

where the right-hand side is a known, calculable quantity independent of time.

Initially, $\omega = 0$ throughout the body, that is $W(0) = 1$. At final rupture, $t = t_{\mathrm{II}}$, $W = W_{\mathrm{II}} \geqslant 0$. Then integrating the above inequality over the time interval $(0, t_{\mathrm{II}})$ there results

$$t_{\mathrm{II}} \leqslant \frac{V(1 - W_{\mathrm{II}})}{(k+1)D\int_V \Phi^k(\sigma_\Phi)\mathrm{d}V}$$

Since $W_{\mathrm{II}} \geqslant 0$ the result follows.

This upper bound (which, of course, only tells us that the structure will fail *before* this time) has been derived solely on a basis of the damage properties of the material, without any recourse to the strain properties of the material. In effect, it is only a mathematical 'trick'. But it is a very clever one.

If we examine the standard creep curves, Chapter 2, it is seen that the period of tertiary creep, leading to rupture, is preceded by a period of steady-state creep. An obvious estimate of the lifetime of a structure for constant loads would then assume that no stress redistribution takes place and that the stresses before rupture are given by their steady-state values. From the constitutive equation (9.4) the rate of growth of damage at each point in a structure in steady state would be

$$\frac{\mathrm{d}\omega}{\mathrm{d}t} = D\left(\frac{\Phi(\sigma^{\mathrm{ss}})}{1-\omega}\right)^k$$

where $\boldsymbol{\sigma}^{\mathrm{ss}}$ is the vector of principal steady-state stresses. In the above, it is also assumed that there is no coupling between the strain–rate equations and the damage equations. Then the rupture criterion ($\omega = 1$) is first satisfied at a time t_{RS} given by

$$t_{\mathrm{RS}} = \frac{1}{D(1+k)(\Phi_{\mathrm{max}})^k} \tag{9.14}$$

where Φ_{max} denotes the maximum value of $\Phi(\sigma^{\mathrm{ss}})$. We may use the estimate t_{RS} as an estimate of the initiation time t_{I}. Intuitively this estimate is likely to give a lower bound on the initiation time, although this cannot be guaranteed by formal proof.

It is possible also to use this simplification to estimate the duration of the period of propagation. That is, it can be assumed that as the failure front propagates the stresses are approximately the equivalent steady-state stresses for the appropriate moving boundary value problem. However, by the nature of the problem, a propagating failure front would lead to a redistribution of stress into the undamaged material. Thus this simplification is not likely to lead to reliable results, and will not be considered further here.

In the following, we use the above simple estimates to predict the lifetime of the multi-bar structure and thick cylinder and compare the results to the more exact analyses that we have carried out.

9.6.1 Example: Estimates of lifetime for the multi-bar structure

The estimates of lifetime for the multi-bar structure can be evaluated very easily. The steady-state stresses for this problem are obtained as

$$\sigma_i = \left(\frac{M+1-i}{M}\right)^{1/n} \frac{Q/A}{I_n(M)} \qquad i = 1, 2, \ldots, M$$

on defining

$$I_n(M) = \sum_{j=1}^{M} \left(\frac{M+1-j}{M}\right)^{1+1/n}$$

The upper bound is based on this stress field with $n = k - 1$. Then the normalised rupture time is

$$\tau_U = \tau_0 \frac{\Gamma_U(k)}{(Q/Q_0)^k} \tag{9.15}$$

where

$$\Gamma_U(k) = M\left[\sum_{j=1}^{M}\left(\frac{M+1-j}{M}\right)^{k/(k-1)}\right]^{k-1}$$

The estimate based on the steady-state stresses is given in this case simply by

$$t_{RS} = \frac{1}{D(1+k)\sigma_1^k}$$

since the first bar is the most highly stressed and will consequently fail first. The normalised rupture time is then

$$\tau_{RS} = \tau_0 \frac{\Gamma_S(k)}{(Q/Q_0)^k} \tag{9.16}$$

where

$$\Gamma_S(k) = I_n(M)^k$$

The estimates (9.15) and (9.16) are shown in *Figure 9.14* together with the exact solution obtained in Section 9.3. It can be seen that both estimates give acceptable predictions of the lifetime, particularly in view of the difficulty of the problem. The upper bound, however, diverges from the exact solution for lower values of the load parameter.

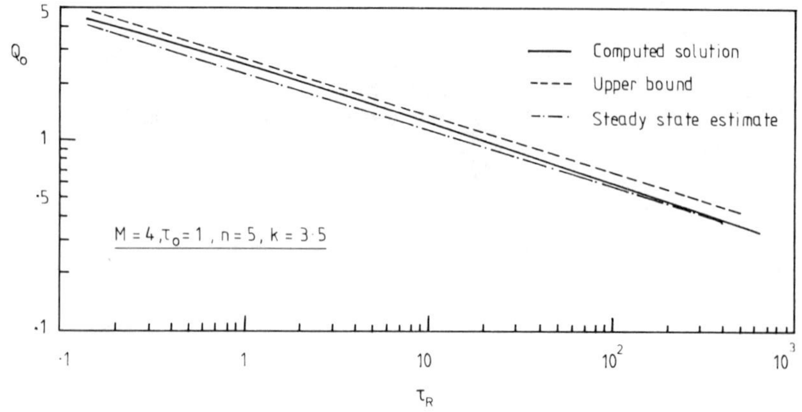

Figure 9.14 Comparison between computed and approximate estimates of rupture time for multi-bar structure

9.6.2 Example: Estimates of lifetime for a thick cylinder

The evaluation of these estimates for the thick cylinder are a little more involved. In Section 4.3 we obtained the steady-state stress distributions for a power law; for $n > 2$ the circumferential stress σ_ϑ is tensile with a maximum at the outside surface $r = b$ given by

$$\sigma_\vartheta \Big|_{\max} = \frac{2/n}{(b/a)^{2/n} - 1}\, p$$

Then the estimate based on the steady stresses, in the normalised time scale, can be evaluated as

$$\tau_{RS} = \tau_0 \frac{\Gamma_S(k)}{p_0^k} \tag{9.17}$$

where

$$\Gamma_S(k) = \left(\frac{(b/a)^{2/n} - 1}{2/n} \right)^k$$

To obtain the upper estimate we must re-solve the steady creep problem with the hypothetical creep law

$$\dot{\epsilon}_r = \frac{\partial \Omega_k}{\partial \sigma_r} \qquad \dot{\epsilon}_\vartheta = \frac{\partial \Omega_k}{\partial \sigma_\vartheta}$$

where the potential Ω_k is given by the rupture surface

$$\Omega_k = D\sigma_\vartheta^k$$

based on the maximum principal stress. That is

$$\dot{\epsilon}_r = 0 \qquad \dot{\epsilon}_\vartheta = Dk\sigma_\vartheta^{k-1}$$

(a very strange material indeed). Following the analysis of Section 4.3 we find for this material

$$\sigma_\vartheta = \frac{k-3}{k-1} \frac{(a/r)^{2/(k-1)}p}{(b/a)^{(k-3)/(k-1)}-1}$$

Then the upper bound is

$$t_U = \frac{V}{(k+1)\int_a^b \Omega_k (2\pi r)\,dr}$$

where $V = \pi(b^2 - a^2)$. In the normalised time scale, this can be rearranged to give

$$\tau_U = \tau_0 \frac{\Gamma_U(k)}{p_0^{\,k}} \tag{9.18}$$

where

$$\Gamma_U(k) = \frac{[(b/a)^{(k-3)/(k-1)}-1]^k}{1-(b/a)^{-1/(k-1)}} \left(\frac{k-1}{k+1}\right)^k \frac{\pi}{2} \frac{(b/a)^2-1}{k-1}$$

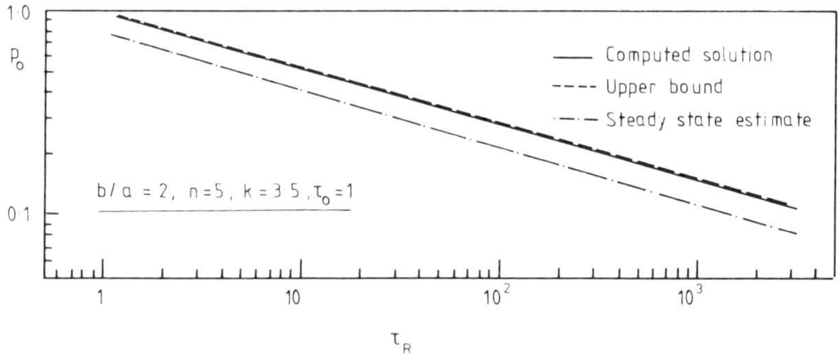

Figure 9.15 Comparison between computed and approximate estimates of rupture time for thick cylinder

The estimates (9.17) and (9.18) are shown in *Figure 9.15* together with the more exact solution obtained in Section 9.5. In this case, the steady-state estimate is much poorer, while the upper bound compares very well with the exact solution.

9.6.3 Discussion

For the two problems that we have considered here, the simple estimates provide a good indication of their likely lifetime. In the load ranges discussed, the steady-state estimate is a lower bound, although as already mentioned this cannot be guaranteed. (It is possible to establish a true lower bound for kinematically determinate structures, but this is not considered here.) There is some divergence between the steady-state estimates and the upper bounds on the rupture times for a given load. However, in the design context where a lifetime is specified, the divergence in the necessary load levels is much less. *For this reason, it is concluded that these simple estimates, which avoid completely the costly damage analyses, are adequate for design purposes*. But it must be emphasised that we have only considered constant-load, isothermal creep damage here. The reliability of these estimates for variable load situations is debatable.

9.7 A reference stress for rupture time

In Section 3.1 and Chapter 6 we discussed the effect of small variations in material parameters and their subsequent effect on steady-state deformation rates in structures. Some elementary calculations show us that predicted creep rupture times can be similarly sensitive to variations in the material parameters. With a little thought, the reference stress method can be extended to deal with rupture times. We give a simple example.

For the multi-bar structure of Section 9.3, the upper bound is given by equation (9.16), which can be expressed in the form

$$\tau_U = \delta \times \tau_{ref} \tag{9.19}$$

on defining a scaling factor δ by

$$\delta = \tau_0 \frac{\Gamma_U(k)}{\alpha_{ref}^k}$$

where $\sigma_{ref} = \alpha_{ref} Q/A$ and

$$\tau_{ref} = \frac{EB/D}{(1+k)\sigma_{ref}^{1-n+k}}$$

is the normalised rupture time from a tensile test at the initial load σ_{ref}.

As in Section 6.6, the reference parameter α_{ref} is chosen in such a way that for the range $k_1 \leqslant k \leqslant k_2$

$$\delta(k_1) = \delta(k_2)$$

that is,

$$\alpha_{ref} = \left(\frac{\Gamma_U(k_2)}{\Gamma_U(k_1)}\right)^{1/(k_2-k_1)} \tag{9.20}$$

We call the stress σ_{ref} the 'reference rupture stress' to distinguish it from the 'reference deformation stress' of Chapter 6.

In *Figure 9.16* we plot the variation in α_{ref} with the damage exponent k taking the limit as $k_1 \rightarrow k_2 \rightarrow k$ in equation (9.20). It can be seen that this reference parameter tends to a limit as k increases. In *Figure 9.17* the variation in the scaling factor with damage exponent k for an estimated mean value $k = 3.5$ is shown. Clearly this variation is small.

It is apparent that all the techniques of the reference stress method can also be applied to the estimation of creep rupture times. Obviously the same benefits are available.

Several attempts have been made to relate the reference stress for rupture to that for deformation: for example, they may be assumed to be approximately

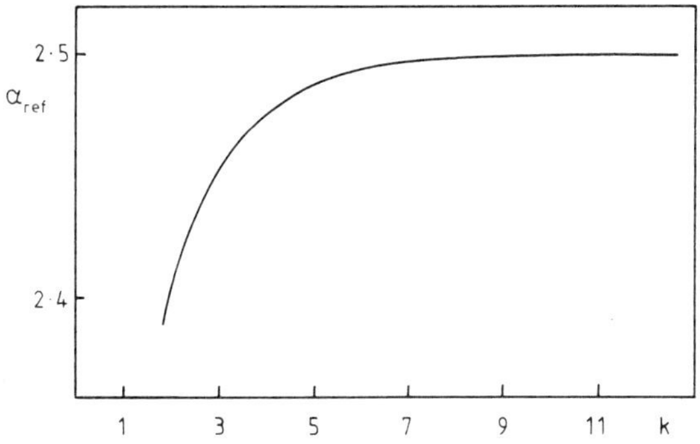

Figure 9.16 Reference parameter for multi-bar structure

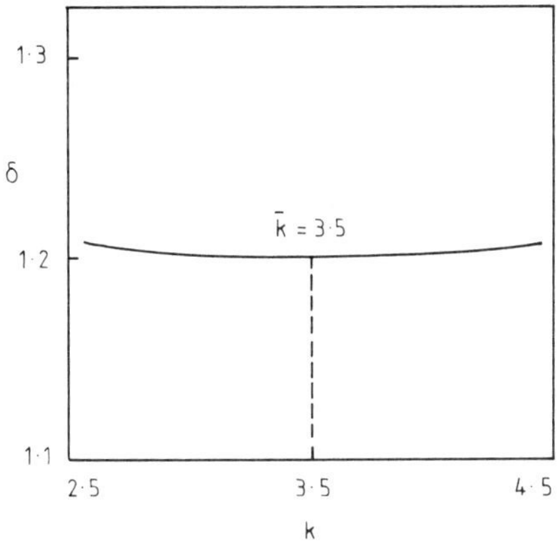

Figure 9.17 Variation of scaling factor with k

the same or that for rupture is scaled to be slightly greater than that for deformation. These are mostly empirical relationships and will not be discussed further here.

Further reading

Kachanov, L.M., *Theory of Creep*, 1960 (transl. National Lending Library for Science and Technology, 1967).
Odqvist, F.K.G., *Mathematical Theory of Creep and Creep Rupture,* Clarendon Press, 2nd Edn, 1974.
Westlund, R., A qualitative evaluation of phenomenological creep rupture theories, *J. Mech. Eng. Sci.,* **18**, 175–8, 1976.
Leckie, F.A., and Hayhurst, D.R., Creep rupture of structures, *Proc. R. Soc. Lond. A*, **340,** 323–47, 1974.
Leckie, F.A., and Hayhurst, D.R., Constitutive equations for creep rupture, *Acta Metallurgica,* **25**, 1059–70, 1977.
Johnson, A.E., Henderson, J., and Khan, B., *Complex Stress Creep, Relaxation and Fracture of Metallic Alloys,* HMSO, 1962.
Garofalo, F., *Fundamentals of Creep and Creep Rupture in Metals,* Macmillan, 1965.
Gittus, J., *Creep, Viscoelasticity and Creep Fracture in Solids*, Applied Science Publ., 1975.
Ashby, M.F., and Raj, M., Creep fracture, *Proc. Conf. on Mechanics and Physics of Fracture, Cambridge,* The Metal Society/Institute of Physics, 1975.

Cocks, A.C.F., and Ashby, M.F., Creep fracture by void growth, *Proc. 3rd IUTAM Symp. on Creep in Structures, Leicester, 1980*, Eds A.R.S. Ponter and D.R. Hayhurst, Springer-Verlag, 1981, p. 368.
Dyson, B.F., and Loveday, M.S., Creep fracture in NIMONIC 80A under triaxial tensile stressing, *Proc. 3rd IUTAM Symp. on Creep in Structures, Leicester, 1980*, Eds A.R.S. Ponter and D.R. Hayhurst, Springer-Verlag, 1981, p. 406.
Hayhurst, D.R., Creep rupture under multi-axial states of stress, *J. Mech. Phys. Solids,* **20**, 381–90, 1972.
Hayhurst, D.R. and Storakers, B., Creep rupture of the Andrade shear disc, *Proc. R. Soc. Lond. A,* **349**, 369–82, 1976.
Trampezynski, W.A., and Hayhurst, D.R., Creep deformation and rupture under non-proportional loading, *Proc. 3rd IUTAM Symp. on Creep in Structures, Leicester, 1980,* Eds A.R.S. Ponter and D.R. Hayhurst, Springer-Verlag, 1981, p. 388.
Leckie, F.A. and Onat, E.T., Tensorial nature of damage measuring internal variables, *Proc. IUTAM Symp. on Physical Non-linearities in Structural Analysis, Senlis, 1980,* Eds. J. Hult and J. Lemaitre, Springer-Verlag, 1981, p. 140.
Murakami, S., and Ohno, N., A continuum theory of creep and creep damage, *Proc. 3rd IUTAM IUTAM Symp. on Creep in Structures, Leicester, 1980,* Eds. A.R.S. Ponter and D.R. Hayhurst, Springer-Verlag, 1981, p. 422.
Chrzanowski, M., Damage parameters in continuum fracture mechanics, *Mech. Teor. i Stos,,* **16**, 151–67, 1978.
Janson, J. and Hult, J., Fracture mechanics and damage mechanics – a combined approach, *J. Mec. Appl.,* **1**, 69–76, 1977.
Chaboche, J.L., Continuous damage mechanics – a tool to describe phenomena before initiation, *Proc. 2nd Int. Seminar on Inelastic Analysis and Life Prediction in a High Temperature Environment, Berlin, 1979; Nucl. Eng. Des.,* **64**, 233–48, 1981.
Kachanov, L.M., Rupture time under creep conditions, in *Problems of Continuum Mechanics*, Ed. J.R.M. Radok, SIAM, 1961.
Hayhurst, D.R. and Krzeczkowski, A.J., Numerical solution of creep problems, *Comp. Meth. Appl. Mech. Eng.,* **20**, 151–71, 1979.
Hayhurst, D.R., Stress redistribution and rupture due to creep in a uniformly stretched thin plate containing a circular hole, *Trans. ASME, J. Appl. Mech.,* **40**, 244–50, 1973.
Hayhurst, D.R., The prediction of creep rupture times of rotating discs using biaxial damage relationships, *Trans. ASME, J. Appl. Mech.,* **40**, 915–20, 1973.
Hayhurst, D.R., Dimmer, P.R. and Chernuka, M.W., Estimates of the creep rupture lifetime of structures using the finite element method, *J. Mech. Phys. Solids,* **23**, 335–55, 1975.
Leckie, F.A. and Wojewodzki, W., Estimates of rupture life – constant load, *Int. J. Solids Struct.,* **11**, 1357–65, 1975.
Wojewodzki, W., Creep rupture of structures at elevated temperature and estimates of rupture life, *Roz. Inzy.,* **26**, 515–40, 1978.
Ponter, A.R.S., Upper bounds on the creep rupture life of structures subjected to variable load and temperature, *Int. J. Mech. Sci.,* **19**, 79–92, 1977.
Goodall, I.W., Cockroft, R.D.H. and Chubb, E.J., An approximate description of the creep rupture of structures, *Int. J. Mech. Sci.,* **17**, 351–60, 1975.
Boyle, J.T., The use of statistical reference stress techniques in nonlinear structural analysis, *Proc. IUTAM Symp. on Physical Non-linearities in Structural Analysis, Senlis, 1980*, Eds. J. Hult and J. Lemaitre, Springer-Verlag, 1981, p. 29.

Chapter 10

Creep buckling

The classical buckling problem in elastomechanics is well known. Some structures are of a shape which cannot always sustain a compressive load: a critical load level is reached where further loading can only be accommodated with a change in shape. Such behaviour is usually associated with 'thin' structures, e.g. slender bars in bending or thin plates and shells in compression. A structure is usually designed for elasticity to a load less than the buckling load. Unfortunately, the presence of creep strains can cause a structure to buckle at a load lower than may be expected from elastic considerations.

Moreover, buckling under creep conditions does not occur at once, but after some period of time. Thus when dealing with creep it is the critical time, *not* the critical load, that must be taken into consideration. The problem is to design such that the expected buckling time is greater than the design life.

Buckling is usually associated with large deformations in a structure. So far, we have only looked at small-deformation theory, although, as we will see, the basic solution techniques remain applicable. In many cases, these large deformations are associated with large strains. Then we should properly include plastic as well as creep strains. It is clear that this problem can be quite involved; indeed, very little information is available on creep in the presence of large strains to assist with the development of a material model for the stress analysis of engineering structures. (For this reason, the possibility of local ductile rupture of thin structures under severe tensile loading, although acknowledged, is not considered here.) We must be content with a rather brief introduction to the basic concepts of creep buckling with small strains. These concepts will be expressed in a form which is largely independent of the assumed material behaviour and, for consistency, includes classical elastic buckling as a special case. Of necessity, we restrict our attention to simple bar frameworks.

10.1 Example: Creep buckling of a shallow Mises truss

Consider the shallow truss depicted in *Figure 10.1* consisting of two rods of length L and area A pinned together as shown under the action of a vertical compressive load Q. The rods are assumed to change length under load, but otherwise they are rigid. The initial height of the point of application of the

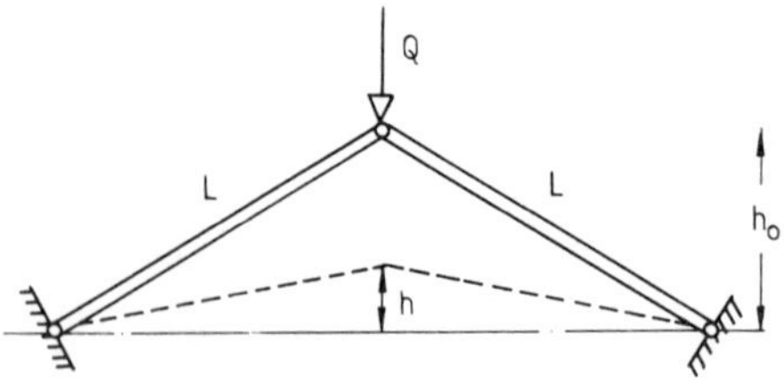

Figure 10.1 Geometry of shallow Mises truss

load is h_0. At any other time the height is given by $h(t)$. If σ is the stress in each bar, then the equation of equilibrium is

$$\sigma h = -QL/2A \tag{10.1}$$

In the case of small deformations, the problem is statically determinate. If ϵ is the strain in each bar, then the second-order strain–displacement relation is

$$\epsilon = \frac{1}{2L^2}(h^2 - h_0^2) \tag{10.2}$$

For the purposes of discussion, the material model is assumed to be given by a constitutive relation with an equation of state in terms of a single state variable, denoted by α

$$\epsilon = f(\sigma, \alpha) \tag{10.3}$$

together with a growth equation for α

$$\mathrm{d}\alpha/\mathrm{d}t = g(\sigma, \alpha)$$

This includes the material models of Chapter 2 where we may identify α with the creep strain and write (10.3) in the form

$$\epsilon = f(\sigma, \epsilon_C) = \sigma/E + \epsilon_C \tag{10.4}$$

as usual.

Buckling of a structure is associated with a loss of 'stability of equilibrium'. In classical mechanics, a position of equilibrium for any system is said to be *stable* when an arbitrary small disturbance does not cause the system to depart far from the position of equilibrium. Otherwise, it is unstable. Obviously this definition needs slight modification for a system in quasistatic equilibrium, for example the slow changes in internal equilibrium in a creeping body. In order to examine the stability criterion, we need to be able to identify equilibrium states of a structure. The current state of the truss is fully described by the couple (σ, h). We must consider a small disturbance to such a state. Such a disturbance is achieved by a perturbation to the couple (Q, α). If the values of these are known, then we can combine equations (10.1)–(10.3) to get the single equation for h

$$h f^{-1}\left(\frac{1}{2L^2}(h^2 - h_0^2), \alpha\right) = -\frac{QL}{2A} \tag{10.5}$$

Given certain trivial conditions on the constitutive function f, this nonlinear equation may be solved for h for a given (Q, α). Then also

$$\sigma = f^{-1}\left(\frac{1}{2L^2}(h^2 - h_0^2), \alpha\right) \tag{10.6}$$

Thus the equilibrium state of the truss is given either by (σ, h) or by (Q, α) since the former is *uniquely* defined by the latter. Since we will consider perturbations to (Q, α), we call this the *control state.* The control state uniquely defines the equilibrium state. (All of this is just jargon so far.)

The stability of the truss is determined from the examination of perturbations to the control state. Suppose that (Q, α) is perturbed by a small amount to $(Q + \delta Q, \alpha + \delta\alpha)$. The state of the truss (σ, h) is changed to $(\sigma + \delta\sigma, h + \delta h)$

From equilibrium equation (10.1), we must have

$$(\sigma + \delta\sigma)(h + \delta h) = -\frac{(Q + \delta Q)L}{2A}$$

i.e.

$$\sigma h + (h\delta\sigma + \sigma\delta h) + (\delta\sigma)(\delta h) = -\frac{QL}{2A} - \frac{\delta QL}{2A}$$

If it is assumed that $\delta\sigma$ and δh are *small*, then we may ignore the second-order term $(\delta\sigma)(\delta h)$; consequently, from equilibrium for the unperturbed state

$$h(\delta\sigma) + \sigma(\sigma h) = -\frac{\delta QL}{2A}$$

Similarly, if the strain is perturbed to a value $\epsilon + \delta\epsilon$, from equation (10.2)

$$\epsilon + \delta\epsilon = \frac{1}{2L^2}\left[(h + \delta h)^2 - h_0^2\right]$$

and, on ignoring the second-order term $(\delta h)^2$, there results

$$\delta\epsilon = \frac{1}{L^2}\, h(\delta h)$$

Finally, from the constitutive relation, equation (10.3)

$$\begin{aligned}\epsilon + \delta\epsilon &= f(\sigma + \delta\sigma, \alpha + \delta\alpha)\\ &= f(\sigma, \alpha) + f_1(\sigma, \alpha)\delta\sigma + f_2(\sigma, \alpha)\delta\alpha + \text{terms of order } (\delta\sigma)^2, (\delta\alpha)^2\end{aligned}$$

using a Taylor series. Hence

$$\delta\epsilon = f_1(\sigma, \alpha)\delta\sigma + f_2(\sigma, \alpha)\delta\alpha$$

where

$$f_1(\sigma, \alpha) = \frac{\partial f}{\partial\sigma} \qquad f_2(\sigma, \alpha) = \frac{\partial f}{\partial\alpha}$$

These equations for the perturbations have the solution

$$\delta h = \left(-\frac{\delta QL}{2A} + \frac{f_2(\sigma, \alpha)}{f_1(\sigma, \alpha)}\,\delta\alpha h\right) \Big/ \left(\sigma + \frac{h^2}{L^2}\,\frac{1}{f_1(\sigma, \alpha)}\right) \tag{10.7}$$

This clearly becomes unbounded, and hence the structure is unstable, if the state (σ, h) satisfies the condition

$$\sigma + \frac{h^2}{L^2} \frac{1}{f_1(\sigma, \alpha)} = 0 \tag{10.8}$$

Since (σ, h) are functions of the control state (Q, α), this condition represents an equation for (Q, α). The control states, and equilibrium states, which satisfy the condition (10.5) we call *critical states.* Since perturbations become unbounded, this case we will call a *limit state.*

It is instructive to examine some particular cases. First, for linear elasticity alone, we have $\alpha \equiv 0$ and

$$f(\sigma) = \sigma/E$$

Thus the control state is given simply by the applied load Q. The equilibrium state (σ, h) is found from the equation (10.5) in the form

$$\frac{1}{2L^2} h(h^2 - h^2{}_0) = -\frac{QL}{2AE} \tag{10.9}$$

then

$$\sigma = \frac{E}{2L^2} (h^2 - h^2{}_0) \tag{10.10}$$

For this simple example, the critical equilibrium state can be evaluated exactly, since on substituting equation (10.10) into (10.8) for the critical states

$$\frac{\sigma}{E} + \frac{h^2}{L^2} = 0 \tag{10.11}$$

there results

$$h_{\text{crit}} = h_0/\sqrt{3} \tag{10.12}$$

and

$$\sigma_{\text{crit}} = -\frac{1}{3E} \frac{h_0^2}{L^2} \tag{10.13}$$

Finally, from equation (10.1) the critical load is

$$Q_{\text{crit}} = \frac{2}{3\sqrt{3}} \frac{h_0^3}{L^3} \frac{A}{E} \tag{10.14}$$

This is the classical picture of *snap buckling* in elastic stability theory. In terms of a load–deflection diagram (*Figure 10.2*), as the load Q increases the deflection $(h_0 - h)$ increases until the critical load Q_{crit} is reached. The truss will then snap through to another geometry.

As a second example, we assume a time hardening creep law of the form

$$\epsilon = \sigma/E + \epsilon_C \qquad \dot{\epsilon}_C = g(t)\,\sigma^n$$

identifying the state variable α with the creep strain ϵ_C. Now the control state is defined by the couple (Q, ϵ_C) and the equilibrium state is given uniquely from a solution of

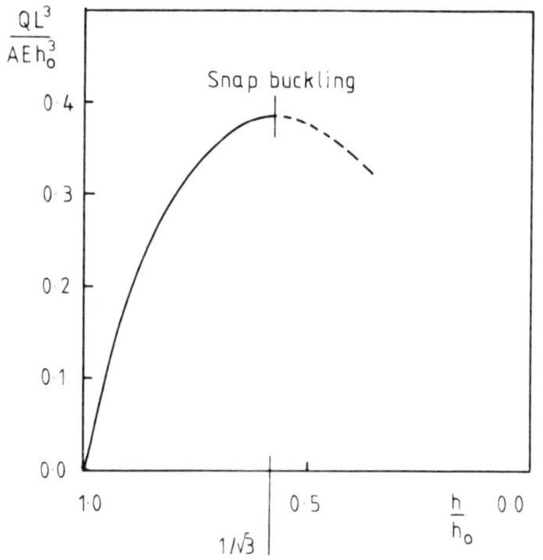

Figure 10.2 Elastic snap buckling behaviour

$$h\left(\frac{1}{2L^2}(h^2 - h_0^2) - \epsilon_C\right) = -\frac{QL}{2A}$$

with

$$\sigma = E\left(\frac{1}{2L^2}(h^2 - h_0^2) - \epsilon_C\right)$$

to give the relation $h = h(Q, \epsilon_C)$, $\sigma = \sigma(Q, \epsilon_C)$.

For this material model, we find that the critical state of equilibrium is again given by equation (10.11) from elasticity, so that the critical control states are given by those couples (Q, ϵ_C) satisfying

$$\frac{\sigma(Q, \epsilon_C)}{E} + \frac{h(Q, \epsilon_C)^2}{L^2} = 0$$

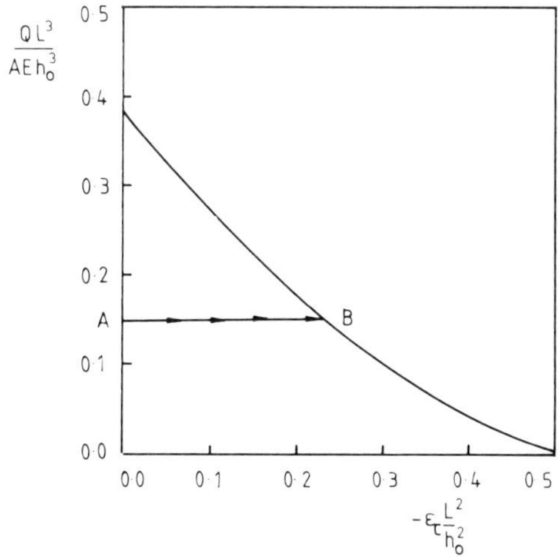

Figure 10.3 Locus of critical states

For this simple example, this can be reduced to

$$\epsilon_C \left(\frac{L}{h_0}\right)^2 = \frac{1}{2}\left[3\left(\frac{1}{2}\ \frac{QL^3}{AEh_0^3}\right)^{2/3} - 1\right] \tag{10.15}$$

This may conveniently be plotted in control state space (Q, ϵ_C) (*Figure 10.3*). For inelastic material behaviour, this diagram is more useful than a load–deflection diagram. If at $t = 0$ the load is below that of the critical elastic load, and is subsequently kept constant, the creep strain rate will increase in magnitude and the truss will follow a path in state space (Q, ϵ_C) (*Figure 10.3*), AB say. When this path meets the locus of critical states, the truss is unstable. The evolution of the truss is described by the equation

$$\frac{dh}{dt} = \frac{g(t)h\sigma^n}{\sigma/E + h^2/L^2} \tag{10.16}$$

where

$$\sigma = -\frac{QL}{2A}\ \frac{1}{h}$$

and we see that at the critical state $dh/dt \to \infty$.

It is of interest to solve this problem numerically. We define normalised variables

$$S = \frac{\sigma}{\sigma_0} \qquad H = \frac{h}{h_0} \qquad \tau = E\sigma_0^{n-1}\int g(t)\, dt$$

where $\sigma_0 = E(h_0/L)^2$. Then equation (10.16) becomes

$$\frac{dH}{d\tau} = \frac{HS^n}{S + H^2} \tag{10.17}$$

$$S = -\frac{1}{2}\ \frac{Q_0}{H}$$

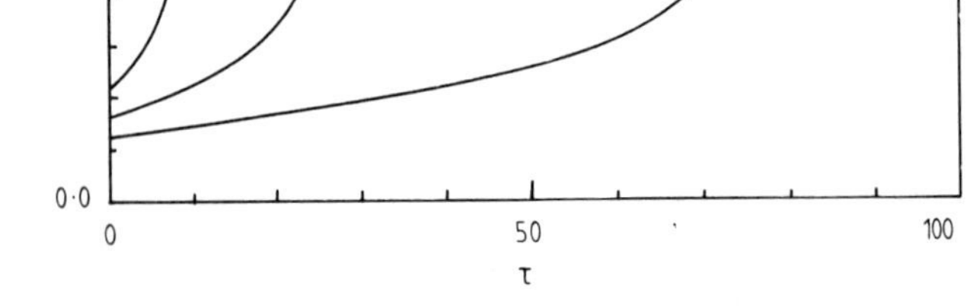

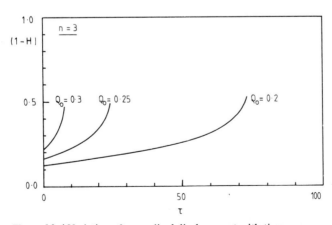

Figure 10.4 Variation of normalised displacement with time

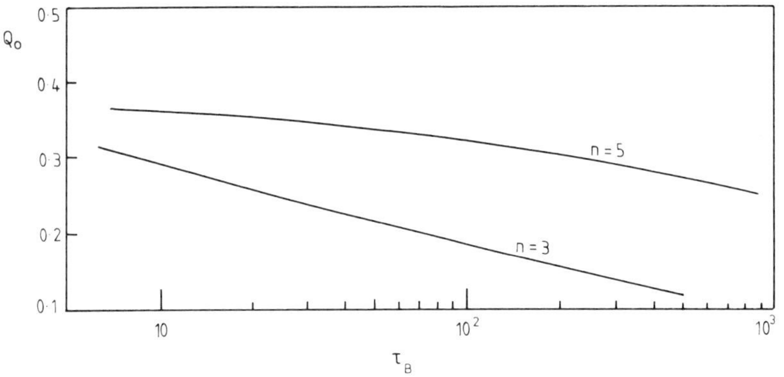

Figure 10.5 Relationship between buckling time and normalised load

The initial conditions are found from a solution of

$$H(0)\,(1 - H(0)^2) = Q_0$$

where $Q_0 = QL^3/AEh_0^3$. The solution will depend on both n and Q_0. Typical plots of the deflection $1 - H(\tau)$ are shown in *Figure 10.4*. We can also plot the load Q_0 against the buckling time τ_B (*Figure 10.5*).

10.2 Example: Creep buckling of an asymmetric arch

The simple shallow truss has only a single degree of freedom and is incapable of demonstrating some more subtle aspects of structural instability. As a second example, we consider the asymmetric arch (*Figure 10.6*) consisting of two rigid rods of length R fixed at points A and D as shown and connected to each other at B and D by a linear spring of length L_0. Rotation of the rods at the hinges A and C is resisted by a spring coil. A vertical load Q is applied at B and D, and is assumed to remain vertical during deformation. The rods are originally inclined

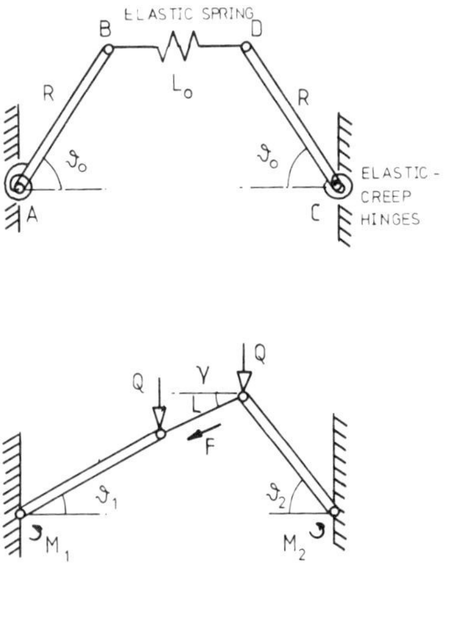

Figure 10.6 Geometry of asymmetric arch

at an angle ϑ_0; during deformation they subtend angles ϑ_1 at A and ϑ_2 at C. The linear spring changes to a length L at an angle γ to the horizontal (*Figure 10.6*).

The change in length of the spring is given by

$$\Delta = L - L_0$$

and the rotation of the rods by

$$\varphi_1 = \vartheta_0 - \vartheta_1 \qquad \varphi_2 = \vartheta_0 - \vartheta_2$$

The linear spring is presumed to deform elastically: if F is the compressive force in the spring, and k the spring constant then

$$\Delta = kF \tag{10.18}$$

The spring coils are presumed to creep (they are called, for obvious reasons, *creep hinges*). We assume an equation of state such that if Q is the rotation of a hinge and M the bending moment causing rotation

$$\varphi = f(M, \alpha) \tag{10.19}$$

where again α is an internal variable dictating the current state of the hinge.

We note the following geometric relations between Δ and φ_1, φ_2:

$$\frac{\Delta}{L_0} = 1 - \frac{\chi}{\cos\gamma}$$

$$\gamma = \tan^{-1}\lambda \qquad \chi = 1 + \frac{R}{L_0}(2\cos\vartheta_0 - \cos\vartheta_1 - \cos\vartheta_2)$$

$$\lambda = \frac{1}{\chi}\frac{R}{L_0}(\sin\vartheta_1 - \sin\vartheta_2)$$

Equilibrium is obtained by taking moments about each hinge, i.e.

$$\begin{aligned} M_1 + FR\sin(\vartheta_1 + \gamma) &= QR\cos\vartheta_1 \\ M_2 + FR\sin(\vartheta_2 - \gamma) &= QR\cos\vartheta_2 \end{aligned} \tag{10.20}$$

The equilibrium state of the arch is given by the variables $(\varphi_1, \varphi_2, M_1, M_2, F)$; the control state is given by (Q, α_1, α_2), which uniquely defines the equilibrium state through a solution of the coupled nonlinear equations

$$\begin{aligned} g(\varphi_1, \alpha_1) + (kL_0R)(1 - \chi/\cos\gamma)\sin(\vartheta_0 - \varphi_1 + \gamma) &= QR\cos(\vartheta_0 - \varphi_1) \\ g(\varphi_2, \alpha_2) + (kL_0R)(1 - \chi/\cos\gamma)\sin(\vartheta_0 - \varphi_2 - \gamma) &= QR\cos(\vartheta_0 - \varphi_2) \end{aligned} \tag{10.21}$$

for a given constitutive function f, with $g = f^{-1}$.

A *symmetrical state* is defined such that

$$\alpha_1 = \alpha_2 = \alpha$$

$$\varphi_1 = \varphi_2 = \varphi$$

in which case equation (10.21) reduces to the single equation

$$\frac{g(\varphi, \alpha)}{kL_0R} - 2\frac{R}{L_0}[\cos\vartheta_0 - \cos(\vartheta_0 - \varphi)]\sin(\vartheta_0 - \varphi) = \frac{Q}{kL_0}\cos(\vartheta_0 - \varphi) \tag{10.22}$$

The critical states of the arch, which affect its stability, are obtained in the following manner. Suppose the control state (Q, α_1, α_2) is slightly perturbed to $(Q+\delta Q, \alpha_1+\delta\alpha_1, \alpha_2+\delta\alpha_2)$. The resulting variation in (φ_1, φ_2) (and hence also in M_1, M_2 and F through the constitutive relations) is given by the equations

$$\begin{bmatrix} A_{11} & A_{12} \\ A_{21} & A_{22} \end{bmatrix} \begin{bmatrix} \delta\varphi_1 \\ \delta\varphi_2 \end{bmatrix} = \begin{bmatrix} b_1 \\ b_2 \end{bmatrix} \tag{10.23}$$

where

$$\begin{aligned} A_{11} &= g_\varphi(\varphi_1, \alpha_1) - QR\sin(\vartheta_0 - \varphi_1) \\ &\quad -kL_0R\Big\{[(\chi_1 + \chi\gamma_1 \tan\gamma)/\cos\gamma]\sin(\vartheta_0 - \varphi_1 + \gamma) \\ &\quad + (1 - \chi/\cos\gamma)(1-\gamma_1)\cos(\vartheta_0 - \varphi_1 + \gamma)\Big\} \\ A_{12} = A_{21} &= (1 - \chi/\cos\gamma)\,\gamma_2\cos(\vartheta_0 - \varphi_1 + \gamma) \\ &\quad - kL_0R\Big\{[(\chi_2 + \chi\gamma_2\tan\gamma)/\cos\gamma]\sin(\vartheta_0 - \varphi_1 + \gamma)\Big\} \\ A_{22} &= g_\varphi(\varphi_2, \alpha_2) - QR\sin(\vartheta_0 - \varphi_2) \\ &\quad - kL_0R\Big\{[(\chi_2 + \chi\gamma_2\tan\gamma)/\cos\gamma]\sin(\vartheta_0 - \varphi_2 - \gamma) \\ &\quad + (1 - \chi/\cos\gamma)(1+\gamma_2)\cos(\vartheta_0 - \varphi_2 - \gamma)\Big\} \\ b_1 &= -\delta\alpha_1 g_\alpha(\varphi_1, \alpha_1) + R\delta Q\cos(\vartheta_0 - \varphi_1) \\ b_2 &= -\delta\alpha_2 g_\alpha(\varphi_2, \alpha_2) + R\delta Q\cos(\vartheta_0 - \varphi_2) \end{aligned}$$

where

$$g_\varphi = \frac{\partial g}{\partial\varphi} \qquad g_\alpha = \frac{\partial g}{\partial\alpha}$$

$$\chi_1 = -\frac{R}{L_0}\sin(\vartheta_0 - \varphi_1) \qquad \chi_2 = -\frac{R}{L_0}\sin(\vartheta_0 - \varphi_2)$$

$$\gamma_1 = \frac{\lambda_1}{1+\lambda^2} \qquad \gamma_2 = \frac{\lambda_2}{1+\lambda^2}$$

$$\lambda_1 = \frac{R}{L_0}\;\frac{\lambda\sin(\vartheta_0 - \varphi_1) - \cos(\vartheta_0 - \varphi_1)}{\chi}$$

$$\lambda_2 = \frac{R}{L_0}\;\frac{\lambda\sin(\vartheta_0 - \varphi_2) + \cos(\vartheta_0 - \varphi_2)}{\chi}$$

If variations from a symmetric state, $\alpha_1 = \alpha_2$, are considered, this reduces to

$$\begin{bmatrix} A_1 & A_2 \\ A_2 & A_1 \end{bmatrix} \begin{bmatrix} \delta\varphi_1 \\ \delta\varphi_2 \end{bmatrix} = \begin{bmatrix} b \\ b \end{bmatrix} \tag{10.24}$$

where

$$A_1 = g_\varphi(\varphi,\alpha) - QR\sin(\vartheta_0 - \varphi) + kL_0R\,[-\chi'\sin(\vartheta_0 - \varphi) - (1-\chi)(1-\gamma')\cos(\vartheta_0 - \varphi)]$$

$$A_2 = kL_0R\,[-(1-\chi)\gamma'\cos(\vartheta_0 - \varphi) - \chi'\sin(\vartheta_0 - \varphi)]$$

$$b = -\delta\alpha g_\alpha(\varphi,\alpha) + \delta QR\sin(\vartheta_0 - \varphi)$$

$$\chi' = -\frac{R}{L_0}\sin(\vartheta_0 - \varphi) \qquad \gamma' = -\frac{R}{L_0}\frac{\cos(\vartheta_0 - \varphi)}{\chi}$$

It is readily seen that equation (10.23) has solution

$$\delta\varphi_1 = \frac{b(A_1 - A_2)}{(A_1 - A_2)(A_1 + A_2)} \qquad \delta\varphi_2 = \frac{b(A_1 - A_2)}{(A_1 - A_2)(A_1 + A_2)}$$

and, provided $A_1 - A_2 \neq 0, A_1 + A_2 \neq 0$, then

$$\delta\varphi_1 = \delta\varphi_2 = \frac{b}{A_1 + A_2}$$

and symmetric states are obtained from perturbations to symmetric states.

However, both $\delta\varphi_1$ and $\delta\varphi_2$ are undefined (become unbounded) if

$$A_1 + A_2 = 0$$

This condition defines a set of *limit states*, for symmetrical states, as in the previous example: these states are defined by the set of control states (Q, α) such that

$$\frac{Q}{kL_0}\sin(\vartheta_0 - \varphi) = \frac{g_\varphi(\varphi,\alpha)}{kL_0R} + 2\frac{R}{L_0}[(\cos\vartheta_0 - \cos(\vartheta_0 - \varphi)\cos(\vartheta_0 - \varphi) + \sin^2(\vartheta_0 - \varphi)] \tag{10.25}$$

where $\varphi = \varphi(Q, \alpha)$ on solving equation (10.22).

But in this example another type of critical state can be achieved: if we have the condition

$$A_1 - A_2 = 0$$

then from equation (10.24) there is no *unique* solution for the perturbed state $(\delta\varphi_1, \delta\varphi_2)$. We call this form of critical state a *bifurcation state* for a symmetric state. This is given by the set of control states (Q, α) such that

$$\frac{Q}{kL_0}\sin(\vartheta_0 - \varphi) = \frac{g_\varphi(\varphi,\alpha)}{kL_0R} + (\chi - 1)\cos(\vartheta_0 - \varphi)\left(1 + 2\frac{R}{L_0}\frac{\cos(\vartheta_0 - \varphi)}{\chi}\right) \tag{10.2}$$

where

$$\chi = 1 + 2\frac{R}{L_0}[\cos\vartheta_0 - \cos(\vartheta_0 - \varphi)]$$

and again $\varphi = \varphi(Q, \alpha)$ from equation (10.22).

In general, a bifurcation state indicates the onset of a non-symmetric equilibrium state with a change to the control state.

As a simple illustration, we consider the familiar equation for a creep hinge in the form

$$\varphi = \frac{1}{\mu} M + \varphi_C$$

where μ is the elastic spring constant and φ_C a 'creep rotation', identified with the variable α and

$$\frac{d\varphi}{dt}C = f(M, t)$$

We will examine the particular numerical case

$$\frac{\mu}{kL_0R} = 0.02 \qquad \vartheta = 15° \qquad \frac{R}{L_0} = 0.5$$

In *Figure 10.7* we plot the locus of the limit and bifurcation states, obtained on solving the coupled equations (10.22) and (10.25) or (10.22) and (10.26)

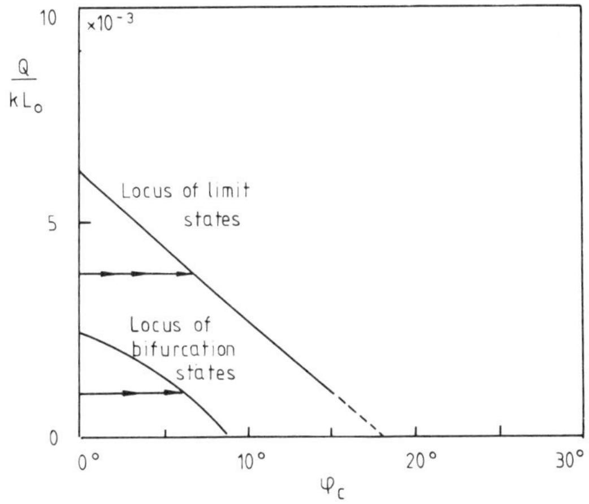

Figure 10.7 Locus of critical states

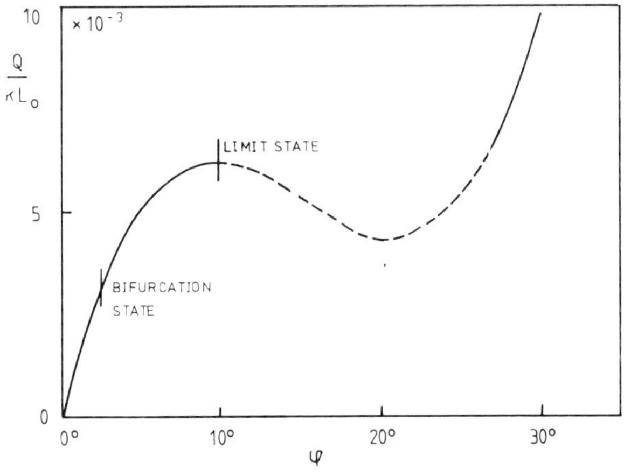

Figure 10.8 Elastic stability

respectively, in control state space (Q, φ_C) for symmetric equilibrium states. The case $\varphi_C = 0$ corresponds to linear elasticity. We see that in general a bifurcation state will be achieved before a limit state, as the load increases, unless the arch is *constrained* to have symmetric states. A typical elastic load–deflection diagram is given in *Figure 10.8*.

If the hinge is allowed to creep so that φ_C increases, we see that from a load less than the elastic bifurcation load the arch will gradually deform until a bifurcation state is reached. Alternatively, if the arch is constrained to have symmetric deformations and allowed to creep, it will eventually reach a limit state.

It can be shown that, for symmetric states, the rate equations for φ_1 and φ_2 are

$$\begin{bmatrix} A_1 & A_2 \\ A_2 & A_1 \end{bmatrix} \begin{bmatrix} \dot{\varphi}_1 \\ \dot{\varphi}_2 \end{bmatrix} = \begin{bmatrix} b \\ b \end{bmatrix} \tag{10.27}$$

as before, equation (10.29), but for a constant load

$$b = \mu \dot{\varphi}_C$$

where

$$\dot{\varphi}_C = f(M, t) \qquad M = QR \cos(\vartheta_0 - \varphi) - kL_0R(1 - \chi) \sin(\vartheta_0 - \varphi)$$

Provided a critical state is not reached, $\dot{\varphi}_1 = \dot{\varphi}_2$ from the above and the arch remains symmetric in creep. However, if a bifurcation state is reached, there is a loss in uniqueness of the rates $\dot{\varphi}_1, \dot{\varphi}_2$. On the other hand, if a limit state is reached, $\dot{\varphi}_1 \to \dot{\varphi}_2 \to \infty$ as before.

This problem is obviously quite complex and we cannot examine it in any detail here. As an example, we assume a power law $\dot{\varphi}_C = g(t)M^n$ and plot the variation in φ in the normalised time scale

$$\tau = \mu(kL_0R \times 10^3)^{n-1} \int g(t)\, dt$$

assuming the arch is constrained to symmetric states, for some typical cases in *Figure 10.9* in terms of the load parameter $Q_0 = Q/kL_0$.

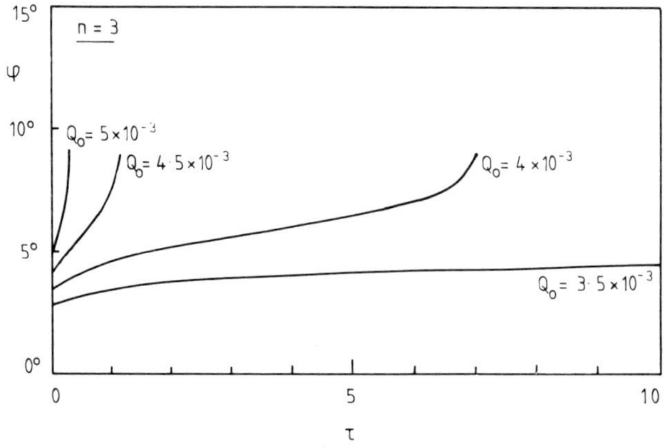

Figure 10.9 Variation of change in angle with time for symmetric deformations

10.3 Creep in the presence of large deformations

The two simple examples that we have considered serve to illustrate the type of behaviour which we can expect if large deformations are included in the formulation of a creep problem leading to compressive instabilities and buckling.

The formulation given above is fairly consistent with modern stability theory since we have defined an equilibrium state and a control state. In elasticity, the control state consists simply of the applied loading; in order to retain this treatment for quasistatic equilibrium due to creep, the inelastic strain has been included as a control parameter. This means that, as the structure creeps, a path is marked out in control space: no external influence is required to alter the control state. A creeping structure may be thought of as self-destabilising (that is, if it is likely to be unstable). Alternatively, natural time may be thought of as a destabilising parameter.

It is important to extend these concepts to a continuum with finite deformations. We will briefly examine this problem here. However, two objections should be noted at the outset. It is common in elasticity to consider only a second-order theory of 'moderately large' displacements while still retaining small strains (as in the shallow Mises truss). This is the common assumption in creep buckling. But, if large strains are present, a generally acceptable continuum theory has only recently appeared. If creep, and worse still plastic, strains are also admitted, the development of a satisfactory continuum theory has hardly begun. This objection notwithstanding, a more serious obstacle is the development of a suitable constitutive relation.

10.3.1 The boundary value problem for creep with large deformations

Consider a body occupying a volume V, bounded by a surface S. As before, we will use a fixed rectangular Cartesian frame to describe the deformation of this body under load. In addition, the undeformed body will be employed as the reference configuration; thus a point $\mathbf{x} \in V$ originally in the undeformed body moves to a point $\mathbf{y} = \mathbf{x} + \mathbf{u}(\mathbf{x})$ where $\mathbf{u}(\mathbf{x})$ is the displacement vector of the point $\mathbf{x}$. Some simplification is possible if we use alternatively a moving reference configuration, which is common in numerical solutions. For clarity, we will adhere to the classical treatment. We can define the *finite strain tensor*

$$\epsilon_{ij} = \frac{1}{2}(u_{i,j} + u_{j,i} + u_{k,i}u_{k,j})$$

which reduces to the usual linear definition if the second-order effects are ignored.

Next we introduce the *first Piola–Kirchoff stress tensor*, denoted by σ^i_j, referred to the undeformed body, such that, if $\bar{b}_i$ are the body forces, referred to unit volume before deformation, the condition of equilibrium is

$$\sigma^i_{j,i} + \bar{b}_i = 0$$

The stress tensor σ^i_j is not symmetric in general, although the finite strain tensor is. For this reason, the constitutive relations are written in terms of the *second Piola–Kirchoff stress tensor*, denoted by σ^{ij}, defined as

$$\sigma^i_j = \sigma^{ik}\frac{\partial y_j}{\partial x_k}$$

which can be demonstrated to be symmetric.

The boundary conditions are

$$u_i = \bar{u}_i \qquad \text{on } S_u$$

$$p_i = \sigma^j_i n_j = \bar{p}_j \text{ on } S_p$$

The boundary value problem is complete if we specify the constitutive equation. We write this as an equation of state relating the finite strain tensor and the second Piola–Kirchoff stress tensor

$$\epsilon_{ij} = f_{ij}\,(\sigma^{kl}, \alpha_{kl})$$

which is assumed invertible

$$\sigma^{ij} = f^{-1}_{ij}\,(\epsilon_{kl}, \alpha_{kl})$$

The second-order tensor a_{kl} represents a number of internal variables governed by growth laws

$$\dot{\alpha}_{ij} = g_{ij}\,(\sigma^{kl}, \alpha_{kl})$$

The equilibrium state of the body is completely described by the couple (σ^{ij}, u_j): as the material creeps, we are interested in the evolution of this state.

The rate form of the above boundary value problem can be summarised as

$$\dot{\epsilon}_{ij} = \frac{1}{2}(\dot{u}_{i,j} + \dot{u}_{j,i} + u_{k,i}\dot{u}_{k,j} + \dot{u}_{k,i}u_{k,j})$$

$$\dot{\sigma}^i_{j,i} + \dot{\bar{b}}_j = 0$$

$$\dot{u}_i = \dot{\bar{u}}_i \qquad \text{on } S_u$$

$$\dot{p}_i = \dot{\bar{p}}_i \qquad \text{on } S_p$$

$$\dot{\epsilon}_{ij} = \frac{\partial f_{ij}}{\partial \sigma^{kl}}\,\dot{\sigma}^{kl} + \frac{\partial f_{ij}}{\partial \alpha_{kl}}\,\dot{\alpha}_{kl}$$

$$\dot{\alpha}_{ij} = g_{ij}\,(\sigma^{kl}, \alpha_{kl})$$

$$\dot{\sigma}^i_j = \dot{\sigma}^{ik}\frac{\partial y_j}{\partial x_k} + \sigma^{ik}\frac{\partial \dot{y}_j}{\partial x_k}$$

which are obtained by partial differentiation with respect to time (an objection is possible here – we will ignore it and assume slow, quasistatic processes).

We will formally write the solution of this boundary value problem (assuming one exists) as

$$\dot{u}_i = R^u_i(u_k, \sigma^{kl}, \alpha_{kl}, \dot{\alpha}_{kl}, \dot{\mathscr{C}})$$

$$\dot{\sigma}^{ij} = R^{ij}_\sigma\,(u_k, \sigma^{kl}, \alpha_{kl}, \dot{\alpha}_{kl}, \dot{\mathscr{C}}) \tag{10.28}$$

$$\dot{\alpha}_{ij} = g_{ij}(\sigma^{kl}, \alpha_{kl})$$

The set $\dot{\mathscr{C}}$ represents the set of rates of change of the boundary data $\mathscr{C} = \{\bar{p}_i, \bar{b}_i, \bar{u}_i\}$.

We will not be more explicit here, but merely note that the equations (10.28) are the finite deformation counterpart of the equations of stress redistribution, Section 7.3.4. They describe a recognisable initial value problem for $(u_i, \sigma^{ij}, \alpha_{ij})$ in the familiar Cauchy form.

10.3.2 Critical states

An examination of the stability of this body involves the proper identification of the critical states. The equilibrium state is described completely by (σ^{ij}, u_i); the control state is given by the set $(\mathscr{C}, \alpha_{ij})$ consisting of the boundary data, internal forces and internal variables, since specification of these uniquely defines the equilibrium state. It should be emphasised here that the sets (σ^{ij}, u_i) and $(\mathscr{C}, \alpha_{ij})$ represent the distributions of these field variables over the whole body.

The stability depends on the effect of small disturbances to the control state. Let the control state $(\mathscr{C}, \alpha_{ij})$ be perturbed by $(\delta\mathscr{C}, \delta\alpha_{ij})$ where $\delta\mathscr{C} = \{\delta\bar{p}_i, \delta\bar{b}_i, \delta\bar{u}_i\}$. The equilibrium suffers a perturbation $(\delta\sigma^{ij}, \delta u_i)$ given by

$$\delta u_i = R_i^u(u_k, \sigma^{kl}, \alpha_k, \delta\alpha_k, \delta\mathscr{C})$$

$$\delta\sigma^{ij} = R_\sigma^{ij}(u_k, \sigma^{kl}, \alpha_k, \delta\alpha_k, \delta\mathscr{C})$$

Thus identification of the critical states depends on the operators R_i^u, R_σ^{ij}. Simply, if these lead to unbounded perturbations δu_i, then we have a limit state; while if R_i^u is not single-valued, we have a bifurcation state. This is an oversimplification, but good enough for our purposes. A more rigorous definition of the critical states can be given in terms of a single variational statement for the rate problem which determines uniqueness and stability (this general framework is due to R.Hill). It also happens that the same general theory will characterise the 'tensile instability' (ductile rupture) of structures under severe loading. The manner in which we have presented the results of this theory here is probably more useful in creep.

10.3.3 Numerical solution of creep problems with finite deformations

A general-purpose solution routine for large-deformation problems in creep, although theoretically plausible, is in practice fraught with difficulties. For simple problems, of course, any of the time incrementation schemes described in Section 7.4 can be used. But, for geometrically complex structures, much attention should be paid to efficiency of solution algorithms. The reason for this is not difficult to see. In Chapter 7 we saw that the solution of a small deformation creep problem essentially reduces to the solution of a sequence of linear elastic problems. If, for example, a finite element discretisation is being used, this involves at the most basic level, computation of some reduced form of the stiffness matrix at the beginning of the calculation which can be used over and over again. For large deformation problems the stiffness matrix *changes* at each step (being dependent on the current deformation) and must be recomputed. This is very costly, even for the simplest time-stepping algorithm. A number of computational strategies have been developed, and are available in the literature, aimed at optimising the cost of each solution step. They are mostly based around applications of an implicit Euler scheme. These are, unfortunately, too specialised to relate here.

10.4 Creep buckling of thin-walled structures

A rigorous treatment of the creep buckling of thin-walled structures, even axisymmetric ones, requires a level of sophistication in mathematics and numerical algorithms which is beyond the scope of this book (and not easily repeated by the reader). For this reason, no further problems will be presented for solution here. Creep buckling is an important phenomenon to be avoided in numerous thin-walled structures – for example, in reactor components and undersea structures. A number of helpful reviews of this topic (by J.C. Gerdeen and V.K. Sazawal, N.J. Hoff and N. Jones) can be found listed at the end of the chapter. In this context, two important points should be made: first, it is well known in elasticity that initial 'imperfections' can entirely change the supposed stability characteristics of a structure. By this we mean shape imperfections induced by manufacture, material anisotropy or inhomogeneity including the presence of welds, or non-idealised loading and support. The neglect of these imperfections can lead the analyst to overlook unexpected stability modes in the real structure. Secondly, stability is concerned with not only the identification of critical states but also the post-buckling characteristics of the structure. These are rarely discussed in creep buckling.

The study of idealised or discretised structures (as we have done with the simple truss and arch) can lead to misleading conclusions, aimed at simplifying the problem. For example, the critical states for the truss, equation (10.15), are independent of the actual creep relation used *if* the total strain can be decomposed into an elastic and creep component. The need for such a simplification can be seen if we examine the general formalism for stability that we have introduced here. The critical states depend not only on the applied loading but also on a distribution of inelastic strain for a real structure. The identification of these states, without recourse to a full analysis, is improbable. Thus the need for a simplified approach is obvious. Unfortunately this has met with little success, except for rather idealised components with idealised material models. It has also led to the proposition of, albeit well intended, physically unrealistic, simplified methods. For example, for certain components it is possible to make the interpretation that creep causes the growth of initial imperfections to a magnitude which would cause elastic buckling. For other components it is possible to make the interpretation that buckling will occur at the same displacement irrespective of the material behaviour (for example, the truss). For still others it may be assumed that buckling will occur at the same (total) strain. Except in those cases where such simplifications have been demonstrated *a priori* to be valid, the reader is advised to show extreme caution. This may sound very sceptical, but the reader should bear in mind that the stability of real structures in elasticity is difficult to determine. It cannot be expected to be any easier in creep.

Further reading

Thompson, J.M.T., and Hunt, G.W., *A General Theory of Elastic Stability*, Wiley, 1973.
Brush, D.O. and Almroth, B.O., *Buckling of Bars, Plates and Shells*, McGraw-Hill, 1975.
Kachanov, L.M., *A Brief Course on the Theory of Buckling*, Solid Mechanics Division, University of Waterloo, 1981.

Brantman, R., On the use of linearised instability analyses to investigate the buckling of nonsymmetrical systems, *Acta Mech.*, **26**, 75–89, 1977.
Rabotnov, Y.N., *Creep Problems in Structural Members*, North-Holland, 1969.
Hoff, N.J., A survey of the theories of creep buckling, *Proc. 3rd US Natl Congr. Applied Mechanics*, **13**, 1958, 29–49.
Hult, J., Creep buckling and instability, *Creep of Engineering Materials and Structures*, Eds A. Bernasconi and G. Piatti, Applied Science Publ., 1980, Chap. 6.
Huang, N.C., Nonlinear creep buckling of some simple structures, *Trans. ASME, J. Appl. Mech.*, **89**, 651–8, 1967.
Hayman, B., Aspects of creep buckling, Parts I and II, *Proc. R. Soc. Lond. A*, **364**, 393–414, 415–33, 1978.
Kulkarni, A.K., Ellin, F., and Neale, K.W., Transient creep analysis with finite deformations, *Trans. ASME, J. Pressure Vessel Techn.*, **97**, 84–9, 1975.
Storakers, B., On uniqueness and stability under configuration dependent loading of solids with or without a natural time, *J. Mech. Phys. Solids*, **25**, 269–87, 1977.
Hoff, N.J., Creep buckling of plates and shells, *Proc. 13th Int. Congr. of Theoretical and Applied Mechanics*, Eds E. Becker and G.K. Mikhailov, Springer, 1973, pp. 129–40.
Gerdeen, J.C. and Sazawal, V.K., *A Review of Creep Instability in High Temperature Piping and Pressure Vessels*, Bulletin No. 195, Welding Research Council, 1974, pp. 33–56.
Jones, N., The creep buckling of shells, *Proc. 3rd IUTAM Symp. on Creep in Structures*, Eds. A.R.S. Ponter and D.R. Hayhurst, Springer, 1981, pp. 308–30.
Bushnell, D., A strategy for the solution of problems involving large deflections, plasticity and creep, *Int. J. Numer. Meth. Eng.*, **11**, 683–708, 1977.
Kranchi, M.B., Zienkiewicz, P.C. and Owen, D.R.J., The visco-plastic approach to problems of plasticity and creep involving geometric nonlinear effects, *Int. J. Numer. Meth. Eng.*, **12**, 169–81, 1978.
Padovan, J. and Tovichakchaikul, S., On the solution of creep induced buckling in general structures, *Computers & Structures*, **15**, 379–92, 1982.
Hoff, N.J. and Levi, I.M., Short cuts in creep buckling analysis, *Int. J. Solids Struct.*, **8**, 1103–14, 1972.
Hoff, N.J., Rules and methods of stress and stability calculations in the presence of creep, *Trans. ASME, J. Appl. Mech.*, **45**, 669–75, 1978.
Chern, J.M., A simplified approach to the prediction of creep buckling times in structures, *ASME Special Publ. on Simplified Methods in Pressure Vessel Analysis*, ASME-PVP-PB-029, ASME, 1978, p. 99.

Chapter 11

Design for creep

In previous chapters we have dealt solely with the techniques of stress analysis. It must be emphasised that this is only a means to an end – that end being design for the safe and reliable operation of a structural component which suffers creep. In this chapter we will describe the main aspects of *design* for creep. We must take cognisance of the fact that this area is enigmatic – of necessity it is continually changing – and consequently that due justice to the topic cannot be adequately given here.

The phenomenon of creep is the source of many problems in design – some of which we are aware of. The principal aim is to ensure that the useful life of a component is not terminated before the operating life required. This can occur due to excessive deformation or failure in some respect. In certain applications in the nuclear industry, even this is not enough for safety reasons: assuming that the likelihood of a failure cannot be completely eliminated, it is often necessary to design for the *consequences* of failure as well.

Design for creep can be explained with reference to the needs, execution and aims of stress analysis, but it must be pointed out that this is not the *raison d'etre* of design. There are three aspects to be considered. First, any stress analysis must be based on an assumed material model, so we should be concerned with *material data requirements*. Secondly, we must have confidence in the stress analysis itself – we must have some idea if it is giving an acceptable answer. This can be partly accomplished by comparison with available *benchmarks.* Thirdly, the results of a stress analysis should be used in such a manner that reliability of a component, within certain limits, is ensured. As we have seen, failure analysis is difficult; the question is, can we ensure safe operation without resort to a failure analysis? In this context the analyst must resort to *design codes* which essentially specify limits on calculated stresses or strains which must be satisfied to ensure a lifetime in excess of the design life. These three aspects will be discussed in more detail in this chapter.

11.1 Material data requirements and constitutive modelling for design

Throughout this book we have assumed that, in order to perform the stress analysis of a structure, an adequate material model, with values for the material parameters for a specific material, will be available. Sometimes, for a particular application, a material model is 'recommended'[1], although this is rare. More often, a suitable material model with appropriate parameter values can be found in the literature. Sadly, the usual situation is that, at best, the raw creep data can be extracted from published information or, at worst, one may have to resort to actual creep testing. In the former situation it is also necessary for the published data to be in a form suitable for defining a material model. Then, given the basic creep data, it is necessary to obtain values for the inherent material parameters in a chosen model. Since the data will be subject to scatter, this must be done in some optimum sense and the fact that these are only best estimates kept in mind *throughout* the analysis. Finally, the available test data will only refer to a limited range of load, time and temperature and some form of extrapolation may be necessary. We will deal with each of these aspects in the following.

11.1.1 Sources and presentation of data

It is obvious that the type of information on basic creep behaviour which will be required largely depends on the use that we are going to make of it. Thus it follows that the information required to characterise fully the simpler creep theories will be less than that required in the characterisation of the more complex unified creep–plasticity theories.

For the simpler theories, we require only the standard creep curves relating to a particular material (Chapter 2, *Figure 2.1*) and the derived cross-plots; that is, we require

(1) standard creep curves (*Figure 2.1*);
(2) isochronous stress–strain rate curves (*Figure 2.3*); and
(3) isostrain stress–time curves (*Figure 2.4*).

in addition to the elastic (and, if necessary, the elastic–plastic) stress–strain curves at temperature. These are sufficient to develop the simplest phenomenological creep theories, i.e. the time hardening, strain hardening or single-parameter damage theories. Unfortunately, raw creep data are not often available in this form. Since creep rupture is usually the governing design consideration for metallic structures and since it is the simplest testing to conduct, a substantial proportion of creep data is available in this latter form. The bulk of this information has been compiled for common pressure-vessel steels and evaluation programs have been undertaken in Britain[2], Germany[3], France[4] Japan[5] and the USA[6] as well as by the International Standards Organisation[7]. With the notable exception of the Japanese NRIM data, strain rates are usually not reported and the information has been presented as creep rupture curves in graphical or tabular form. For the purposes of the development of constitutive equations for stress analysis, this can at best be described as a nuisance. The analyst must resort to an often tiresome search through the published literature and even then the basic data may not be in the required form.

If one of the more complex material models is to be used, then even the standard creep curves are insufficient (perhaps not surprisingly since such models are formulated to include plasticity). In most cases, the additional tests required depend on the theory being used, for example, relaxation and constant-strain-rate elasto–plastic tests[8–10]

11.1.2 The identification problem

Assuming that the raw creep data are given in the required form, the next step is the identification of the numerical values of the material parameters appearing in the assumed constitutive relation. This is called the *identification problem* (or, sometimes in the literature, the *inverse problem*). Throughout, it must be borne in mind that scatter in the data will always be present so that the values that we identify will only be estimates according to some chosen criterion of 'best fit'. In which case, it would be attractive also to have an estimate of the likely variation of the estimates.

Essentially, the parameter identification problem for creep is a special case of the general parameter estimation problem on which a number of standard texts are available[11]. We will be concerned with those techniques which have special application to the creep problem. We can distinguish two different techniques, which we will call here 'direct identification' and 'indirect identification'.

11.1.2.1 Direct identification

Suppose that the material constitutive relation contains only the observables, strain ϵ_C, stress σ and time t, in an equation of state

$$\epsilon_C = f(\sigma, t, p_1, p_2, p_3, \ldots) \tag{11.1}$$

where f is some specified functional form containing certain material parameters $p_1, p_2, \ldots$ which have to be determined. From the standard creep curves, a table of measured values of creep strain can be constructed, which we denote by $\epsilon_C(i, j)$ being the measured creep strain corresponding to a stress σ_i at time t_j where $i = 1, 2, \ldots, S, j = 1, 2, \ldots, T$.

The parameters must now be chosen to provide a 'best fit' to this table of measured values. The general idea is to choose them such that the error term $\epsilon_C(i, j) - f(\sigma_i, t_j)$ is simultaneously rendered as small as possible. This is usually achieved by minimising the sum of squares of the error term (the 'residual sum of squares')

$$R = \sum_{i=1}^{S} \sum_{j=1}^{T} [\epsilon_C(i, j) - f(\sigma_i, t_j)]^2 \tag{11.2}$$

for the available data. In fact, this is a special case of a more general statistical procedure[12] which presumes that the observed values of creep strain are subject to measuring errors, such that $\epsilon_C(i, j)$ is the mean value of a particular measurement. Then it is necessary to minimise the 'weighted' sum of squares

$$R = \sum_{i=1}^{S} \sum_{j=1}^{T} \omega_{ij} [\epsilon_C(i, j) - f(\sigma_i, t_j)]^2 \tag{11.3}$$

such that the weights ω_{ij} are equal to the reciprocal of the expected statistical variance in the measurements, $\omega_{ij} = 1/v_{ij}^2$. This means that if a particular measurement (or set of measurements) is thought to be less accurate than others (and consequently its variance is greater), then less weight is given to it in minimising the sum of squares. In practice, for creep testing the expected variance is presumed to be the *same* in each measurement so that equation (11.3) reduces to equation (11.2).

We may illustrate the above for the particular case of steady creep according to a power law. Then we have a table of measured values $(\sigma_i, \dot{\epsilon}_C(i))$, $i = 1, 2, \ldots, S$, and equation (11.1) takes the rate form

$$\dot{\epsilon}_C = B\sigma^n$$

The unknown material parameters can be found by minimising the sum of squares

$$R = \sum_{i=1}^{S} \omega_i[\dot{\epsilon}_C(i) - B\sigma_i^n]^2$$

This comprises what is called a 'nonlinear least-squares' problem[13]. Suitable algorithms for solving this problem can be found in most computer libraries. However, one advantage of the power law is that it can be *linearised* by taking logarithms of the creep strain rate and stress. If we denote this transformation by

$$S_i = \ln \sigma_i \qquad E_i = \ln \epsilon_C(i)$$

then we must minimise

$$R = \sum_{i=1}^{S} \omega_i^T [E_i - (b + nS_i)]^2 \qquad (11.4)$$

where $b = \ln B$. The minimum can be found from the conditions

$$\frac{\partial R}{\partial b} = 0 \qquad \frac{\partial R}{\partial n} = 0$$

with solution

$$n = \frac{\Sigma_{i=1}^S \omega_i^T E_i S_i - \left(\Sigma_{i=1}^S \omega_i^T S_i\right)\left(\Sigma_{i=1}^S \omega_i^T E_i\right)/S_\omega}{\Sigma_{i=1}^S \omega_i^T S_i^2 - \left(\Sigma_{i=1}^S \omega_i^T S_i\right)^2/S_\omega}$$

$$b = \log B = \frac{1}{S_\omega}\left(\sum_{i=1}^{S} \omega_i^T E_i - n\sum_{i=1}^{S} \omega_i^T S_i\right) \qquad (11.5)$$

$$S_\omega = \sum_{i=1}^{S} \omega_i^T$$

This is the usual procedure adopted for the power law, and it is common to assume that the weights ω_i^T are equal to unity. This is unfortunately incorrect: as mentioned above, ω_i^T should be the expected variance of the measured data. But the original data have been log-transformed: if ν_i^2 is the variance of the untransformed data (assured to be uniform), then it can be verified that the variance of the transformed data is

$$\frac{1}{\omega_i^T} = (\nu_i^T)^2 \simeq \left(\frac{dE}{d\dot{\epsilon}_C}\right)_i^2 \nu_i^2 = \frac{1}{\dot{\epsilon}_C(i)^2}\,\nu_i^2$$

Hence, even if the weights in the original data are taken to be uniform (and can

TABLE 11.1. Steady-state creep data.
For 316SS at 1300° F. See Garofalo, F., *et al.*, *Proc. IMechE Conf. on Thermal Loading and Creep,* London, 1963, Paper 30.

$\sigma(\times 10^3$ psi)	$\dot{\epsilon}_C(\times 10^{-3}$/h)
6	0.0038
6	0.0044
7	0.0079
7	0.0069
9.55	0.027
9.55	0.025
11.23	0.080
11.23	0.055
13.19	0.096
13.19	0.085
15.49	0.17
15.49	0.26
18.63	0.37
18.63	0.49
21.38	1.40
21.38	1.20
23.20	2.30
23.20	3.20
25.20	5.80
25.20	5.10
29.52	14.70
29.52	27.36

TABLE 11.2. Estimation of parameters.

	B	*n*
Unweighted least squares (transformed data)	4.78×10^{-25}	5.06
Weighted least squares (transformed data)	1.61×10^{-38}	8.07
Nonlinear least squares (original data)	1.21×10^{-37}	7.87

thus be taken as unity), the weights in the transformed data should properly be taken as

$$\omega_i^T = [\dot{\epsilon}_C(i)]^2$$

in this case.

As an example, for the data shown in *Table 11.1* the coefficients *B* and *n* have been estimated from a minimisation of equation (11.2) and from equation (11.5) using weights of unity and as above. The results are shown in *Table 11.2*. The most obvious comment is that they give different results, particularly for the coefficient *B*. The best one to take, that is the parameters estimated from the original data, or the transformed data, is not really a valid question. It should be recalled that we have only established the best fit according to *one* criterion (minimum sum of square errors). Others are possible. For example, we could have minimised the 'inverse sum of squares'

$$\bar{R} = \sum_{i=1}^{S} [\sigma_i - f^{-1}(\dot{\epsilon}_C(i))]^2$$

or the 'mean square error'

$$D = \sum_{i=1}^{S} \frac{\delta_i \bar{\delta}_i}{\sqrt{(\delta_i^2 + \bar{\delta}_i^2)}}$$

where

$$\delta_i = \dot{\epsilon}_C(i) - f(\sigma_i) \qquad \bar{\delta}_i = \sigma_i - f^{-1}(\dot{\epsilon}_C(i))$$

The use of these alternatives emphasises different parts of the steady-state creep curve (*Figure 11.1*) and result in different estimates for the various criteria.

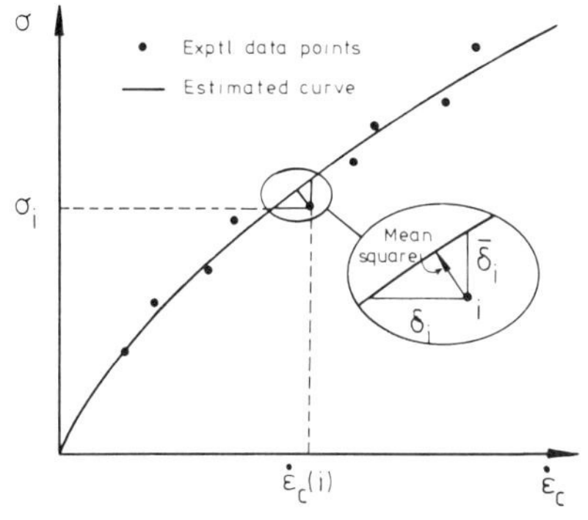

Figure 11.1 Graphical representation of least-squares criteria

Which criterion to use is a matter of personal choice. In fact, if there are a large number of parameters to be estimated and the assumed constitutive equations are highly nonlinear, then the least-squares direct technique can run into difficulties. The analyst must then resort to a second technique, indirect identification.

11.1.2.2 Indirect identification

This technique is the most common in material modelling and identification for creep. It has particular importance for the internal-variable theories where the constitutive equations are not expressed in terms of the observables. The technique requires interactive judgement on the part of the analyst; it can best be described as a method of successive elimination of coefficients and depends on the particular constitutive model being examined. We give two examples.

In the first, the total creep strain is assumed to be decomposed into a primary and a secondary part[14]

$$\epsilon_C = \epsilon_p + \dot{\epsilon}_s t$$

where

$$\epsilon_p = K\sigma^m(1 - a\mathrm{e}^{-bt}) \qquad \dot{\epsilon}_s = B\sigma^n$$

This type of total strain model is often found in finite element applications; it

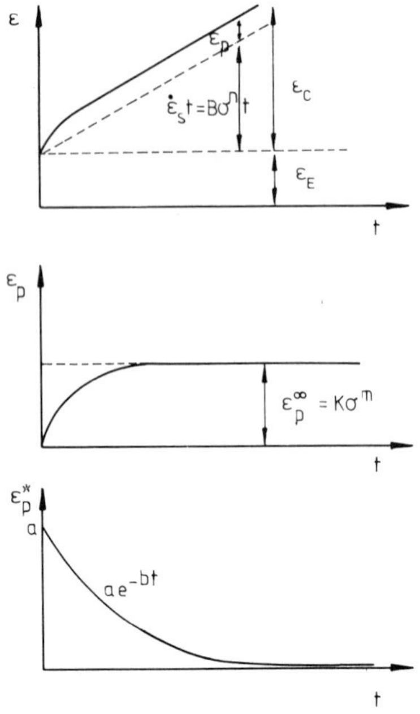

Figure 11.2 Indirect identification for total strain model of creep

requires six material parameters, *k, m, a, b, B* and *n*. They are determined from the standard creep curves as follows (*Figure 11.2*):

(a) Estimate *B* and *n* from the steady-state data.
(b) Form the primary strain component

$$\epsilon_p = \epsilon_C - \dot{\epsilon}_s t$$

It will be found that this tends to a limit as time increases. If the limit, ϵ_p^∞, is plotted against stress in a log–log plot, it is found to be nearly linear. Hence determine *K* and *m* such that

$$\epsilon_p^\infty = K\sigma^m$$

(c) Form the quantity $(1 - \epsilon_p/\epsilon_p^\infty)$. It is found that this decreases exponentially with time and is sensibly independent of the stress. Hence finally determine *a* and *b* such that

$$\epsilon_p^* = 1 - \epsilon_p/\epsilon_p^\infty = ae^{-bt}$$

say, by a least-squares techniques on all the data.

In the second example we examine the Bailey–Orowan model of creep (Section 2.3) which requires four material constants, k_1, k_2, *m* and *n*. These are determined as follows[15]:

(a) Estimate $B = k_1$, k_2 and *n* from the steady-state data.
(b) Perform a sequence of creep recovery tests as follows. Let the stress in the steady state be reduced by an amount $\Delta\sigma$. With reference to *Figure 11.3*, if this occurs at point A then there will be a period of duration Δt where the creep strain rate is zero followed by an increasing strain from point

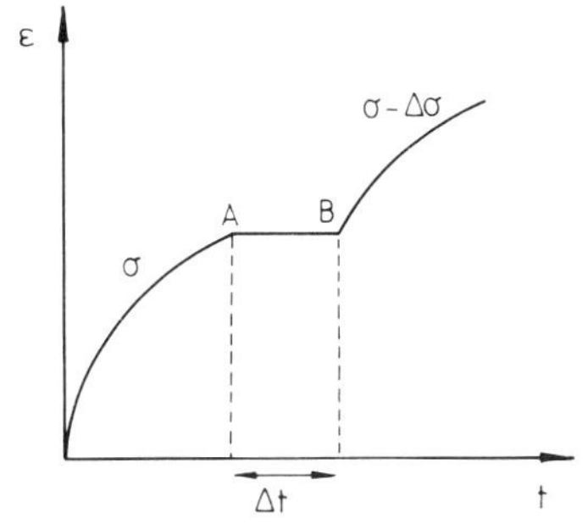

Figure 11.3 Indirect identification for Bailey–Orowan model of creep

B to a reduced steady static rate. During this time, the flow stress R reduces from σ to $\sigma-\Delta\sigma$ and

$$\mathrm{d}R/\mathrm{d}t = -r(R) = k_1|R|^{n-m}$$

The function r can then be determined by taking the limit as $\Delta\sigma \rightarrow 0$ and will be a function of the initial stress level σ with $|\sigma| = R$. If this is plotted against stress in a log–log plot, it is found to be nearly linear; then the parameters k_1 and $(n-m)$ may be determined using least squares.

It should be clear to the reader that direct identification involves number crunching, pure and simple. No judgement on the part of the analyst is required and, as already mentioned, it is inapplicable to the internal-variable theories. The indirect method takes account of the processes involved and is to be preferred. However, it suffers from the disadvantage that estimates of the likely variation of the parameters cannot be made; this is possible through standard formulae using the direct method[12].

11.1.3 Correlation and extrapolation of data

There is one obvious drawback in our approach so far – the available creep data may be insufficient. For example, in the case of creep strain data, they might not have been reported or, in the case of creep rupture curves, they may only have been produced for a limited range of stress and temperature. And, as always, there is the business of scatter. In order to be able to proceed with a stress analysis of some structural component, we have devised a best-fit, but approximate, mathematical model of the material behaviour: we must be content with this approximation. Because of this, many design methodologies involve a final reference back to the original creep data. For example, design may simply be based on limits on calculated elastic stress (thus avoiding any need for this text) which, if satisfied, should ensure structural integrity throughout the specified lifetime. Alternatively, we may use a reference stress approach (Chapter 5, Section 9.7) which automatically relates component behaviour to the basic creep data. Either of these approaches could involve a stress level outside the available range (which is not unexpected since we would essentially need a creep test of duration comparable to the design life). To overcome these difficulties, we need to be able to extrapolate the basic data confidently to lower stress levels and temperatures. But even this may not be sufficient if there are not enough data on which to base the extrapolation, as is often the case for creep strains. Then we would like to be able to correlate strain data with the more readily available rupture data. The techniques for correlation and extrapolation are similar; we will examine correlation first.

11.1.3.1 Correlation of strain and rupture data

The stages of secondary and tertiary creep are not independent; rather, they are each one component part of the same overall creep process. It follows that there should exist some discernible relationship between the point at which rupture occurs and the preceding portions of the standard creep curve (although there is no reason to assume that this will be a simple algrebraic expression).

If we refer to the standard creep curve, *Figure 11.4*, then we see that the minimum creep rate can be expressed by

$$\epsilon_{min} = (\epsilon_4 - \epsilon_1)/t_R$$

Consequently, provided the quantity $\epsilon_4 - \epsilon_1$ remains sensibly constant with stress, there exists an inverse proportionality of minimum creep rate with rupture

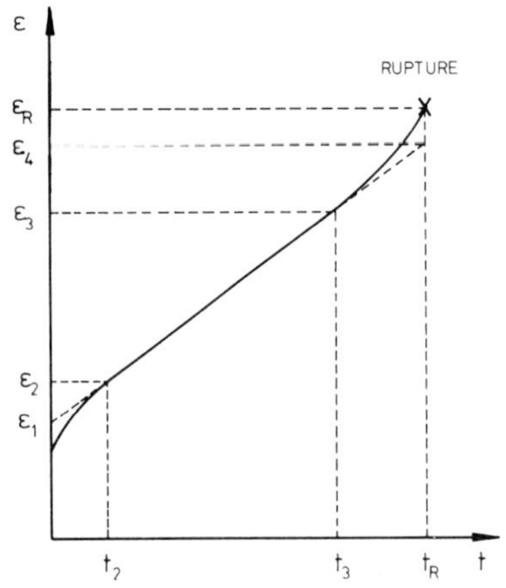

Figure 11.4 Standard creep curve

time. In fact, such an approximate relationship can be seen in a variety of metals[16]; further, a better correlation can be had if a second constant is introduced, so that the above becomes[17] in logarithmic form

$$\log t_R + m \log \dot{\epsilon}_{min} = C \qquad (11.6)$$

where m and C are constants. In practice, the constant m has a value very close to unity. A relationship like equation (11.6) is satisfied for some metal alloys quite well, but in others the parameters m and C are both stress- and temperature-dependent. As an improvement upon equation (11.6), which reduces this dependence, a further empirical relation of the form[18]

$$\log (t_R/\epsilon_R) + m^* \log \dot{\epsilon}_{min} = C^* \qquad (11.7)$$

has been proposed in terms of the creep strain at rupture ϵ_R and new constants m^* and C^*. As an example, we reproduce some results of Lonsdale and Flewitt[19] for 2.25% Cr–1% Mo steels (*Figure 11.5*), where the variation of rupture time

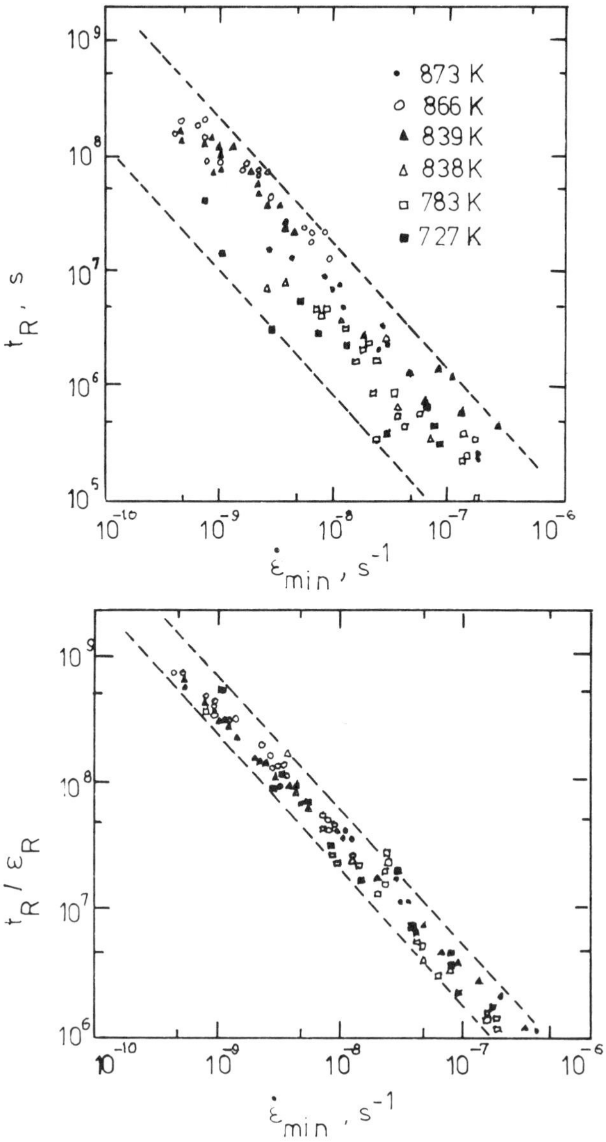

Figure 11.5 Variation of time to faiiure and ratio of time to failure to strain at failure with minimum creep rate (see ref. 19, Figure 1(a) and (b))

with minimum creep rate and the variation of the ratio t_R / ϵ_R with minimum creep rate are both shown. Clearly the correlation is better using equation (11.7).

The use of a relationship like (11.7) would allow, for example, the estimation of steady-state creep material parameters from the rupture data. Some attempts at this have met with fair success[20]. Nevertheless, such a simple empirical relationship as equation (11.7) cannot be expected to be generally applicable. Although the secondary and tertiary stages of creep belong to the same process, the physical mechanisms controlling these stages are different. These mechanisms are usually determined by duration and temperature and it is therefore possible that a correlation which does not explicitly include time would be more successful. One possibility which has arisen is based on the observation[21,22] that there exists a correlation between rupture strength, the stress to produce rupture in a specified time, and creep strain, in the same time and at the same temperature. As an example, we reproduce some comparisons made by Goodman[23] (*Figure 11.6*) for 9% Cr–1% Mo steel in the temperature range 475–625°C. This shows that for this material a nonlinear correlation is necessary.

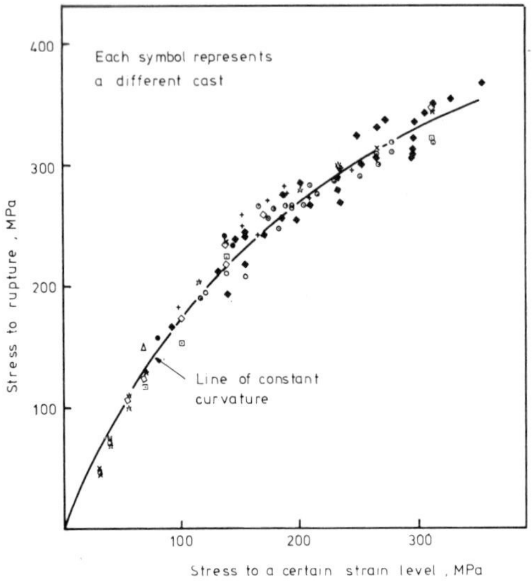

Figure 11.6 Correlation between stress to rupture and stress to 0.5% strain for 9% Cr–1% Mo steel (see ref. 23, Figure 7)

11.1.3.2 Extrapolation of creep rupture data

The purpose of extrapolation is to allow the designer to use short-term (high-stress, high-temperature) creep tests in the assessment of the long-term life of a component. Of course, it follows as a rather fundamental assumption that such a short-cut is physically possible (we will return to this question briefly at the end).

Most extrapolation techniques are based on an attempt to reduce the family of stress against rupture time curves for different temperatures to a single line. Then high-temperature tests can be used to predict the results of lower-temperature tests. How can this be done? As an elementary example, we can presume the following relationships from equations (2.1) and (11.6).

$$t_R = A/\dot{\epsilon}_{min} \qquad \dot{\epsilon}_{min} = Cf(\sigma)\exp(-\Delta H/RT)$$

We may combine these and rearrange to give

$$\log t_R - \frac{\Delta H}{2.3RT} = \log\left(\frac{A}{Cf(\sigma)}\right) \tag{11.8}$$

with the assumption that A and C are independent of stress and temperature. This means that if we plot stress against the parameter

$$P_1 = t_R \exp(-\Delta H/RT)$$

then a *single* line, independent of temperature, should result. If we make the further assumption that $f(\sigma) = \sigma^n$, then a plot of $\log P_1$ against $\log \sigma$ should give a straight line, which could be used for extrapolation (*Figure 11.7*).

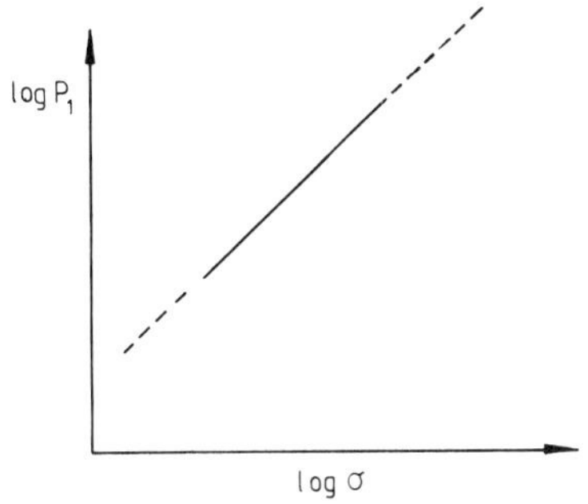

Figure 11.7 Correlation between time–temperature parameter P_1 and stress

The applicability of this technique is based on the assumption that in a plot of $\log t_R$ against the reciprocal of absolute temperature, $1/T$, for a given stress a family of straight, parallel lines results (referring to equation (11.8)). In practice, the real data may not conform exactly to this requirement, which is shown in *Figure 11.8(a)*, and may exhibit a tendency to converge to a focal point as in *Figure 11.8(b)*.

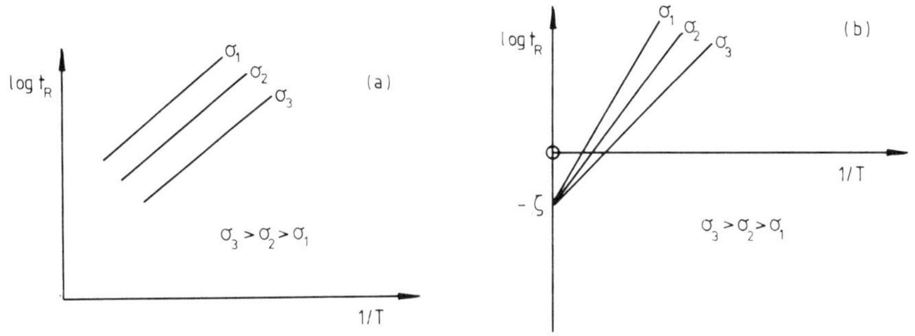

Figure 11.8 Variation of time to failure with reciprocal of temperature at different stress levels

In the latter case, we must infer from equation (11.8) that the grouping log $(A/Cf(\sigma))$ is independent of stress (with a value $-\zeta$, say) while ΔH is stress-dependent. Then a new parameter

$$P_2 = T(\log t_R + \zeta) = \Delta H/2.3R$$

should be used to correlate with stress.

In fact the parameters P_1 and P_2 above are only two examples of a class of so-called 'time–temperature parameters', which aim to collapse a family of rupture curves into a single curve for extrapolation purposes. A full review can be found in the text by Conway[24] or the more recent historical perspectives[25,26].

The two that we have developed here are due to Orr-Sherby and Dorn (P_1)

and Larson and Miller (P_2). The majority of these approaches depend on a unique functional relationship between rupture time, stress and temperature which can be written in the form

$$P(T,\ t_R) = f(\sigma)$$

In fact, many of these assume the special form

$$P(T,\ t_R) = G(\log\sigma)$$

where the time–temperature parameter is

$$P(T,\ t_R) = \log t_R + Ap(T)\log t_R + p(T) \qquad (11.9)$$

with A a constant and $p(T)$ a prescribed function of temperature. For example, in the case of the Orr-Sherby–Dorn parameter (P_1)

$$A = 0 \qquad p(T) = -\ \Delta H/2.3RT$$

In fact, the form given by equation (11.9) can be related to most of the well known time–temperature parameters[27]. The question remains of which parameter to use; in practice, it is found that no single parameter is suitable for all materials. Some idea of the difficulties involved can be seen in *Figure 11.9(a)* reproduced

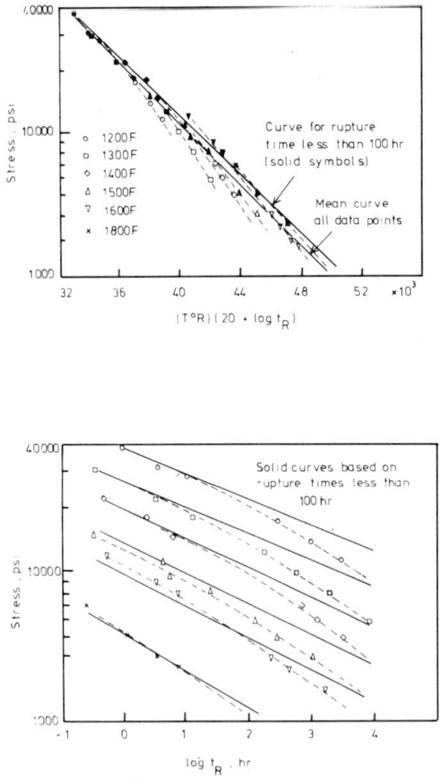

Figure 11.9 Master rupture plot for 18% Cr–8% stainless steel based on Larson–Miller parameter and derived stress–rupture plot (see ref. 28)

from Manson and Haferd[28]. In this, a master rupture plot for a stainless steel (18% Cr–8%) based on the widely used Larson–Miller parameter is shown, together with two mean curves – one derived from short-term (less than 100 h) and one derived from long-term data. Clearly the actual data from long-term tests are not well described by the mean curve based on the short-term tests. This is also illustrated on a conventional plot of stress against rupture time, in *Figure 11.9(b)*. It is also worth noting from *Figure 11.9(a)* that the deviation from a mean curve is systematic, suggesting that, for this material, the linearity of the log t_R versus $1/T$ plot is in question.

In order to overcome these difficulties, the following approach has been recommended as a standard[29]: a procedure should be adopted which automatically chooses the best time–temperature parameter for a particular material. If the Orr-Sherby–Dorn or Larson–Miller parameters were best for that material, then the final result derived from the procedure would appear in the appropriate form: if another parameter proved best, then the procedure would identify it. Such a systematic procedure has been called a 'minimum-commitment method', implying that the analyst need not initially be committed to any single parameter, but could adapt to whichever form emerged.

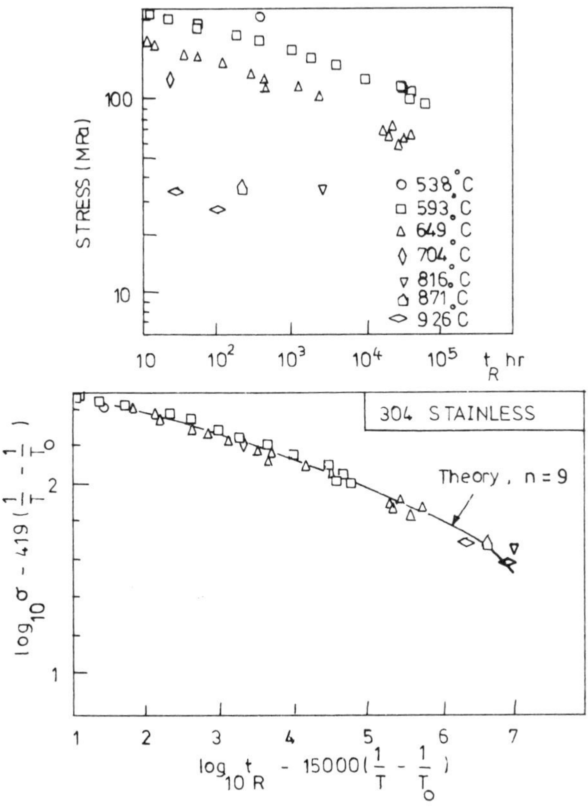

Figure 11.10 Creep rupture data for 304 stainless steel as raw data and in normalised form (see ref. 32, Figure 6)

Of course, such an approach to extrapolation is based on purely phenomenological considerations. Ideally, some cognisance of the physical fracture mechanisms should be taken into account. Early attempts at this[30,31] required subjective estimates from the analyst, or high-quality data, and did not suggest any systematic method of extrapolation. However one possibility which has recently emerged[32] is based on a study of the microstructural changes accompanying the growth of damage.

Essentially this study predicted that a single master curve should result when basic rupture data at different temperature levels are replotted as

$$P_1 \simeq \log \sigma - H\left(\frac{1}{T} - \frac{1}{T_0}\right)$$

against

$$P_2 \simeq \log t_R - J\left(\frac{1}{T} - \frac{1}{T_0}\right)$$

where T_0 is a reference temperature (that of the centre of the range of data, for instance) and the qualities H and J are chosen to minimise the scatter. As an example of this, we reproduce a result from Cocks and Ashby[32] for 304 stainless steel in *Figure 11.10*. Also shown is a smoothed line obtained from theorctical values of H and J predicted from a power law of creep and from consideration of the mechanisms of void growth. This smoothed line can be used as a guide for extrapolation.

11.2 Verification and qualification of stress analysis

So far we have examined techniques of stress analysis, creep data requirements and material modelling. We can no longer avoid the crucial question: how well does our stress analysis of some component compare with an actual creep test of that component?

We have seen in previous chapters that a detailed stress analysis of the time history response of a creeping structure under load can involve a fairly demanding numerical analysis. Whether this numerical analysis is specialised for a particular component or one aspect of a general-purpose finite element computer program, there is the question of validation and final acceptance of the analysis results. For design we must have confidence in the stress analysis. There are two factors to be taken into consideration in validation: verification and qualification. Verification is simply the demonstration that a stress analysis actually does correctly what it is supposed to do. That is, there have been no fundamental errors in its formulation. Given that an analysis has been verified (and this may not be so easy in a general-purpose computer program), there arises the more important question of qualification; that is, does the analysis coupled with the material model given an acceptable solution to the real problem? In relation to this one, very important point should be made: the result of a stress analysis can be no better than the quality of the basic creep data on which it is based. Of necessity, the material model used in a stress analysis is an approximation to the real behaviour. Therefore, it is to be expected that there will always be differences between predicted and measured component behaviour. Indeed, any analysis which does give good quantitative agreement is suspect. Thus, rather than exact predictions which can be expected in the elastic case, a creep stress analysis can do no better than qualitatively indicate the basic characteristics and features of inelastic response. Qualification then involves a judgement as to the acceptability of an analysis.

Verification and qualification of a creep analysis can perhaps best be carried out by benchmark calculations; that is, by solving a limited number of carefully chosen problems, ideally representing good experimental results. The selected benchmark comparisons should demonstrate the validity of an analysis and give an increased level of confidence in its further use. However, it is not as simple as this, and the reader should be aware of a number of factors which complicate the validation.

(a) A good comparison with a benchmark is only a necessary but not a sufficient condition for the validity of an analysis. This means that all we can conclude is that, if the comparison is bad, then the analysis is somehow deficient (i.e. there is a mistake in the computer program, or there is a poor material model, or indeed that the mechanical model poorly reflects the real situation). What we cannot say is that, if the comparison is good, the analysis is validated.

(b) Usually in creep stress analysis an accurate mechanical model is too expensive to realise in practice, and we must resort to either a simpler model or an approximate analysis (cf. Chapter 8). When we compare this simplification to a benchmark, it will be impossible to isolate those errors due to the simplification from those due to the material model. This may cause, for example, a simplified model to be rejected when it is the material model which is deficient.

(c) It must be recognised that creep stress analysis of a real structure is very much an art and can be done successfully only by suitably experienced analysts. This can be a drawback since, given *a priori* the result of a benchmark, the analyst will strive for the best comparison. A true validation of both an analysis *and* the analyst can only be done with 'blind' benchmarks.

Recognition of the need for benchmarking a stress analysis has led to an international effort aimed at identifying suitable benchmark problems[33–37]. Many of these problems are typical of those which can arise in nuclear power plant so that transient thermal effects, plasticity and large deformations are also present. The illustrative examples which we give below are not all taken from this international effort, but rather they have been selected from the past literature on experimental creep testing. In the simplest of these, the reader will be able to verify the stress analysis for himself; hence sufficient information will be provided to allow this. In the more complex examples, we will limit ourselves to demonstrating the comparison between experiment and theory; in each case the reader is strongly advised to refer to the original publications.

11.2.1 Creep of a three-bar structure[38]

A schematic representation of this test, with appropriate dimensions, is given in *Figure 11.11*. The structure consists of three parallel bars, rigidly fixed at each end and subject to a tensile load. The centre bar has twice the length and twice

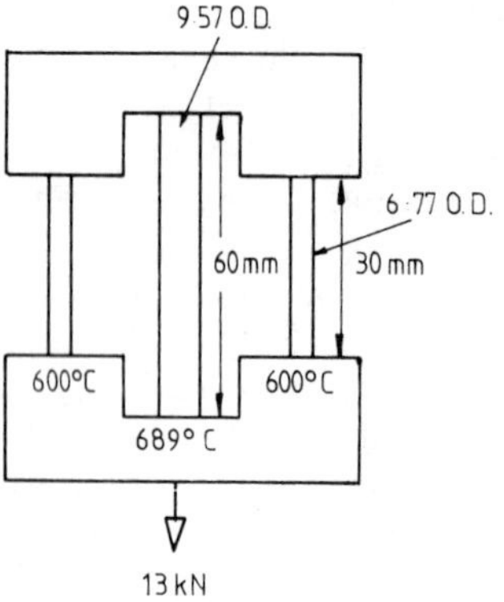

Figure 11.11 Geometry and loading on a three-bar structure (see ref. 38)

the area of the outer bars and is at a higher temperature. The bars are manufactured from the same material, 316 stainless steel. A secondary creep power law for this material has the form

$$\dot{\epsilon}_{min} = \sigma^n \exp\left[\frac{\Delta H}{R}\left(\frac{1}{T} - \frac{1}{T_o}\right)\right]$$

where $\Delta H = 38.6RT_o$, $T_o = 600°$ C, $n = 6.4$ and rate is in h^{-1}; the coefficient of thermal expansion is given as $\alpha_T = 18.8 \times 10^6 /°C^{-1}$ while Young's modulus at 600° C is $E = 1.53 \times 10^5$ MN m^{-2}. The steady-state displacement rate measured during the test was 7.64×10^{-5} mm h^{-1}. The theoretically calculated value is found to be 6.29×10^{-5} mm h^{-1}. The comparison is clearly acceptable.

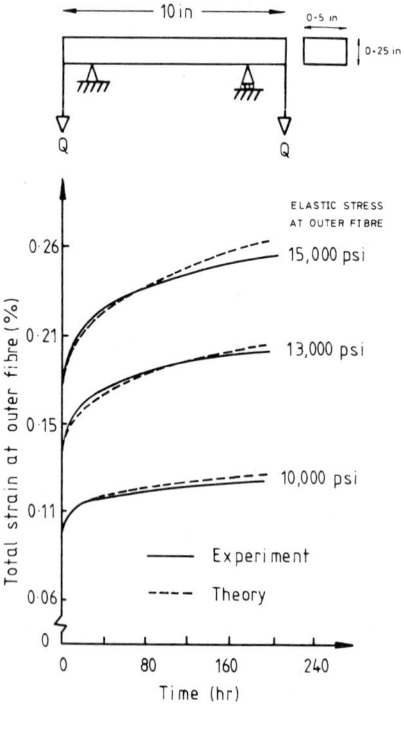

Figure 11.12 Creep of a beam under bending: comparison of experimental and theoretical results (see ref. 39, Figure 9)

11.2.2 Creep of a beam in bending[39]

In this test, the conditions of pure bending for a rectangular beam are obtained using a standard four-point loading system: this is shown schematically, with dimensions, in *Figure 11.12*. The beam is made from a commercially pure aluminium which exhibits creep at room temperature. A suitable creep law for this material is given by the time hardening form

$$\dot{\epsilon}_C = Bt^{-m}\sigma^n$$

where $B = 6.65 \times 10^{-16}$, $n = 3.2$, $m = 0.7$ and stress is in psi, time in hours; Young's modulus is given by $E = 10.5 \times 10^6$ psi. In *Figure 11.12* we reproduce experimental results for the measured strain in the outer fibre as it varies in time for three different load levels corresponding to elastic stresses at the outer fibre of 10000, 13000 and 15000 psi. Also shown are analytic solutions obtained with the above creep law using the methods of Chapter 7. Comparison is quite good, although it is less acceptable for higher loads.

11.2.3 Creep of a thick, pressurised cylinder[40]

A schematic diagram of this test is given in *Figure 11.13*, representing an internally pressurised thick tube of 0.19% carbon steel at 450°C. The tube is

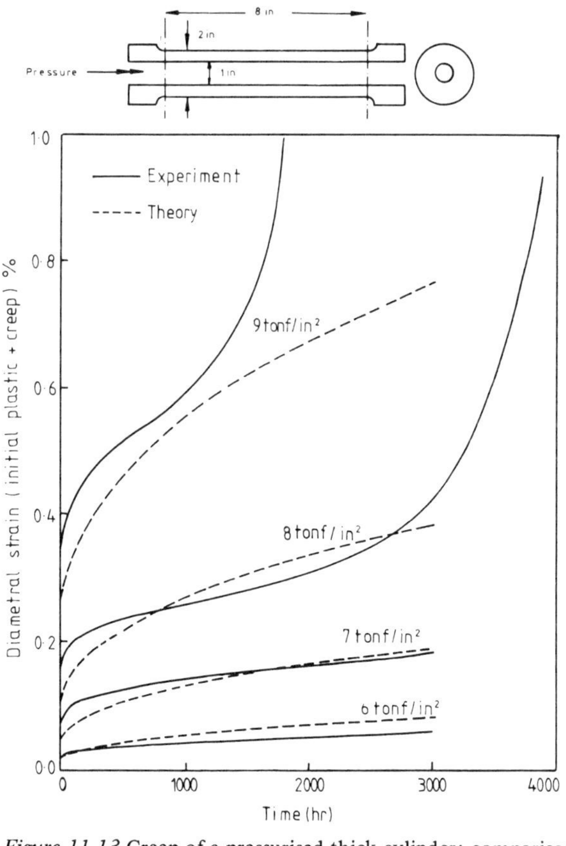

Figure 11.13 Creep of a pressurised thick cylinder: comparison of experimental and theoretical results (see ref. 40, Figure 7.2)

constrained not to move in the axial direction so that conditions of plane strain can be assumed. In *Figure 11.13* we reproduce experimental results for the variation of the diametral creep strain (i.e. the strain based on changes in the measured tube diameter) with time for three different levels of pressure. Also shown is a comparison with an analytical solution, found using the methods of Chapter 7, for the creep law

$$\dot{\epsilon}_C = Bt^{-m}\sigma^n$$

where $B = 1.225 \times 10^{-8}$, $n = 2.99$, $m = 0.55$ with stress in psi, time in hours. An averaged value for Young's modulus is $E = 24.53 \times 10^6$ psi. The comparison is acceptable, although again it gets worse as the load increases. It should be noted that the stated creep law was obtained from short-term (less than 200 h) tensile data; if longer term data were used, the comparison should improve.

11.2.4 Creep of a rotating disc[41]

This is an old, but well established, test on a rotating disc with a central hole; it is shown in *Figure 11.14* together with necessary dimensions. The disc rotates at 15 000 rpm (thus, in the notation of Section 5.3, the load $Q = 1810$ lb in^{-4}), and is composed of a 12% chrome steel (called Allegheny 418); the test was conducted at 1000° F. In *Figure 11.14* we reproduce the strain distributions

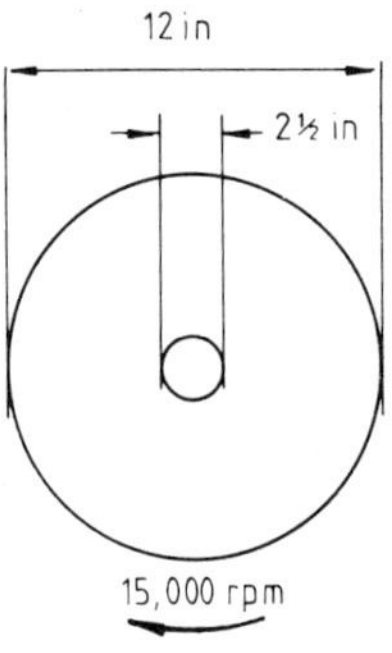

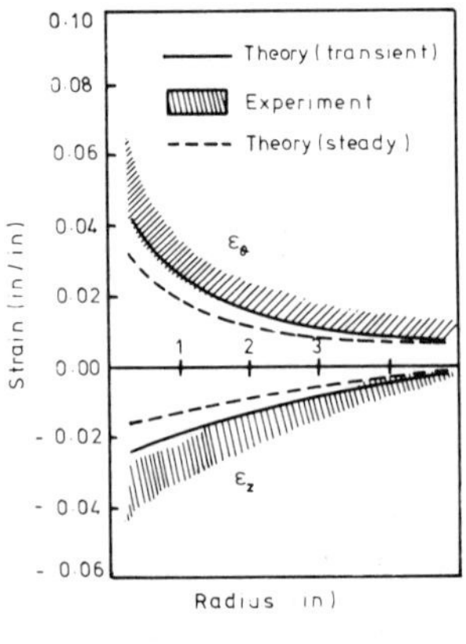

Figure 11.14 Creep of a rotating disc: comparison of experimental and theoretical results (see ref. 42)

which were obtained at 180 h. Also shown are calculated values[42] obtained from a transient analysis using the creep law

$$\dot{\epsilon}_C = Bt^{-m}\sigma^n$$

where $B = 10^{-32}$, $n = 6$, $m = 0.33$ and Young's modulus $E = 18 \times 10^6$ psi. It can be seen that the computed results lie at the limit of the scatter band obtained from the test. In fact, at 180 h the disc thickness had started to reduce, eventually leading to ductile rupture.

11.2.5 Creep of a nozzle–sphere intersection[43]

The problem of localised creep straining in pressure-vessel intersections and discontinuities has been recognised for a long time. In order to assess the applicability of various techniques of stress analysis, Penny and Marriott[43] conducted a series of creep tests on some small model nozzle–sphere intersections, one of which is shown schematically in *Figure 11.15.* The model vessels were manufactured from the aluminium alloy HE15WP and the tests were conducted at 180° C with a steady internal pressure of 220 psi. The measured maximum circumferential strains at two gauge locations on the inside surface are shown in *Figure 11.15*. An attempt was also made to predict the behaviour of the model shells using a finite difference analysis based on rotationally symmetric, small deflection, thin-shell theory. In the transient creep analysis, an approximate

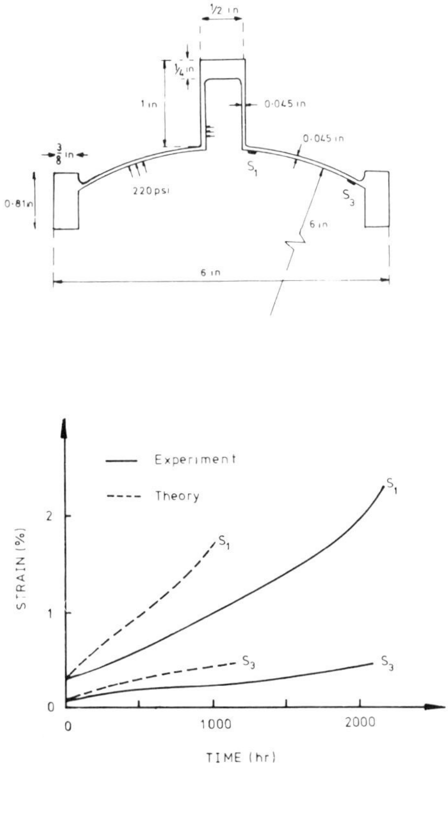

Figure 11.15 Creep of a nozzle–sphere intersection: comparison of experimental and theoretical results (see ref. 43, Figure 204–9)

time hardening power law was assumed. The results of this analysis are also shown in *Figure 11.15* for comparison. Clearly the thin-shell analysis seriousl over-estimates the measured strain at the intersection. The reason for this discrepancy lies principally with the fact that the theoretical junction assumed in the rotationally symmetric thin-shell analysis does not model the real intersection well. A more detailed analysis in the region of the junction, for example using finite elements, should improve the comparison.

11.2.6 Creep and relaxation of a circular plate under prescribed transverse deflections[44,45]

A simply supported plate, loaded at its centre, is one of the simplest possible types of structural test and is particularly valuable as a benchmark. The test[44] presented here relates to a simply supported, type 304 stainless steel, circular plate to which a program of central deflection was applied at 1100° F. The geometry of the plate tested, together with the deflection history, is shown in

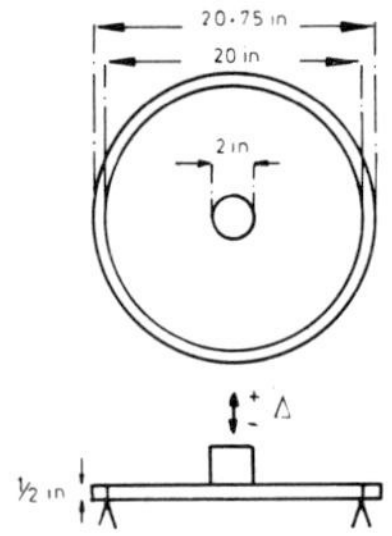

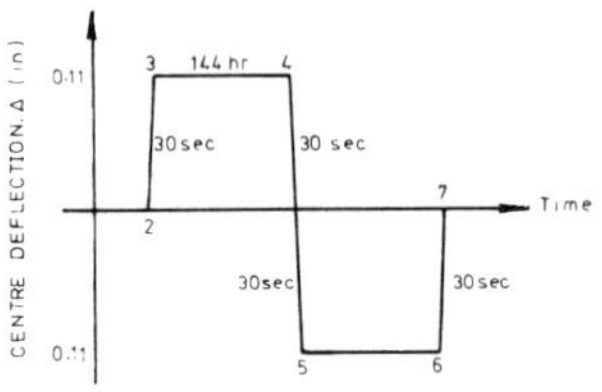

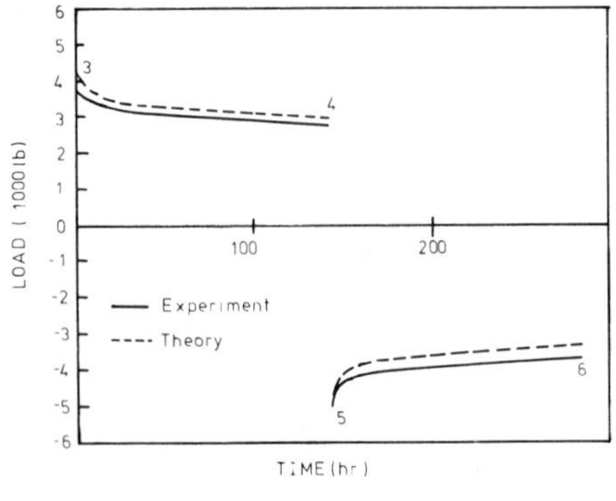

Figure 11.16 Relaxation of a circular plate under prescribed transverse deflections: comparison of experimental and theoretical results (see ref. 45, Figure 10)

Figure 11.16. Similarly, we reproduce the measured load response during the two hold periods also in *Figure 11.16.* This plate was also analysed[45] using the finite element program PLACRE; the model consisted of 784 three-node axisymmetric, triangular ring elements, giving in total 464 nodes. The material constitutive relation assumed consisted of a bilinear kinematic hardening plasticity model, coupled with a strain hardening creep model. The results of this analysis are also given in *Figure 11.16*. The comparison is clearly very good.

11.2.7 Creep rupture of a tension plate containing a central circular hole[46]

The problems that we have considered so far relate to creep in its primary and secondary phases; as our next example, we describe a series of creep rupture tests on long, narrow plate specimens containing small central holes under uniaxial tension (*Figure 11.17*). Two specimens were tested, one of copper with an applied tensile stress of 31.03 MN m^{-2} and one of aluminium with an applied tensile stress of 70.67 MN m^{-2}. The tests were conducted at temperatures of 250° C and 210° C respectively. The measured extension of the tension plates (shown as a ratio

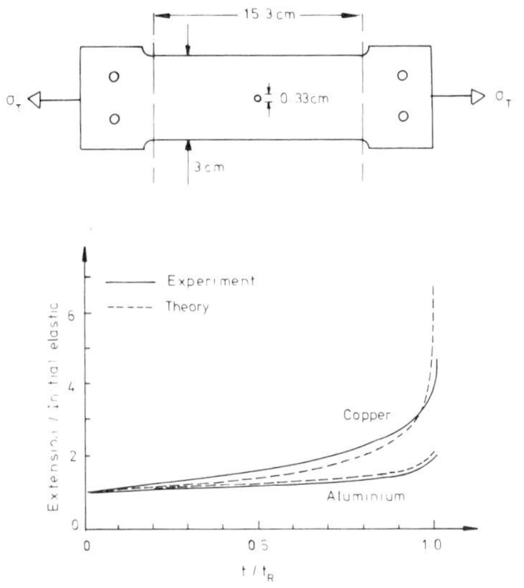

Figure 11.17 Creep rupture of a tension plate containing a central circular hole: comparison of experimental and theoretical results (see ref. 46, Figure 9)

with the initial elastic extensions) are given as smoothed curves in *Figure 11.17* for both tests in terms of a normalised time scale of real time divided by measured rupture time. These experimental results were compared to those of a special-purpose finite element solution for a quadrant of the plate split up into triangular elements of different sizes. A simple damage law of the type described in Chapter 9 was assumed as the material model. Results of these computations are also given in *Figure 11.17.* Comparison in this case is also very good.

11.2.8. International benchmark project: the Oak Ridge pipe ratchetting experiment[47,48]

Finally, we briefly describe one of the most widely studied benchmark problems, which is aimed at realistically simulating the effect of transient thermal loading in

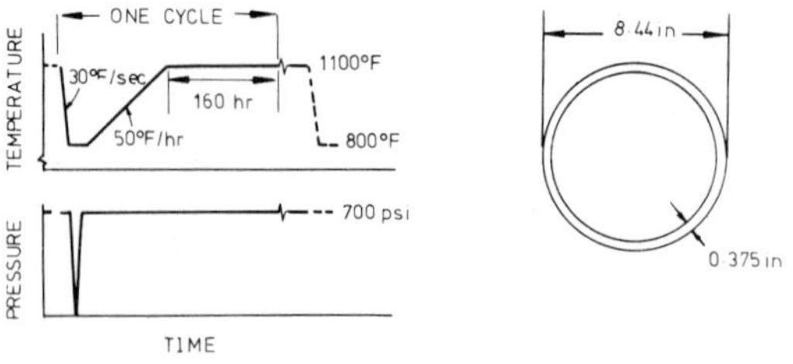

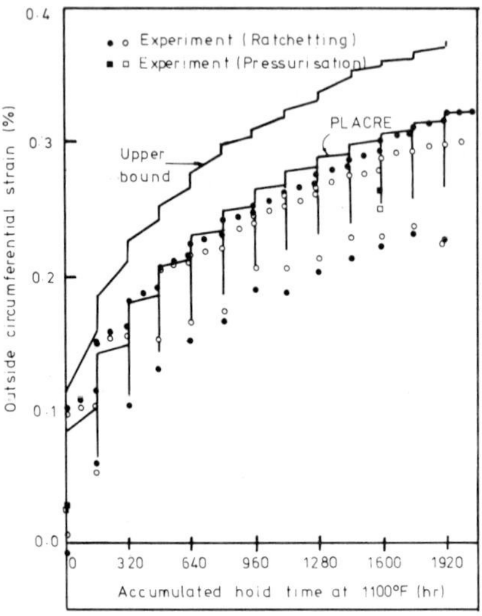

Figure 11.18 The Oak Ridge pipe ratchetting experiment (see refs 47 and 48)

a pipe carrying liquid sodium[47]. The pipe, which was machined from a well documented heat of type 304 stainless steel, was subjected to a series of thermal downshocks (sudden loss of sodium temperature and pressure) followed by sustained periods under internal pressure loading at a nominal temperature of 1100°F. These intermittent thermal downshocks can be expected to produce progressive inelastic straining (ratchetting). The dimensions of the test pipe, together with an idealised temperature–pressure histogram for a typical cycle, are shown in *Figure 11.18*. The measured accumulated circumferential strain at two positions 90° apart is reproduced in *Figure 11.18;* initial pressurisation data and data corresponding to the minimum strain during each transient are included. Thermal ratchetting is clearly shown. Also shown are the results of two analytical solutions. The first[45] is a finite element solution obtained using the program PLACRE with a special pipe-wall bending element which performs both a heat-transfer analysis and an elastic–plastic–creep structural analysis. The material model is the same as that used in Section 11.2.6. The agreement is excellent, particularly in the later stages of the test. The second is a simplified analysis based on the bounding methods described in Chapter 8. The comparison is acceptable and provides an upper bound, as expected. This problem has also been the subject of an enthusiastic international collaboration, which was reported in depth by the program coordinator, Dr H. Kraus in ref. 48.

11.2.9 Discussion

Component testing at elevated temperatures is difficult, time-consuming and, to the rigorous standards set by conventional elastic testing, not completely satisfactory. In general, this has resulted in a lack of usable high-temperature experimental test data on real structures. What is then perhaps surprising about these illustrative examples is how good the comparison between analysis and experiment can be. This is particularly surprising considering the inadequacy of basic creep data. Indeed, in those cases wherein the material creep behaviour is well documented (e.g. Sections 11.2.6–11.2.8), the comparison is especially good. The conclusion must be drawn that conventional techniques of stress analysis are adequate.

11.3 Design methodology

The aim of this book has been to introduce the reader to the basic techniques of stress analysis for creep. Thus, given an appropriate model of material behaviour, he should be able to predict the deformation and stress histories in a component for a prescribed history of loading. So far we have not discussed how a designer would use this information to produce a 'safe' design. That is, the designer must also be concerned with such questions as: how much creep deformation is too much, or what factor of safety on a calculated lifetime should he assume for assured structural integrity during service? How these questions are answered will depend on the application. For example, in aero gas-turbine design, it is imperative to maintain critical dimensions such as minimum clearance between rotating parts and static parts and to avoid excessive creep damage which would lead to rupture. But the service life of such a component is sufficiently short for real-time testing to be conceivable, as are techniques for in-service monitoring of turbine blade creep damage. Petrochemical plants have longer lifetimes but critical components can usually be identified and repaired or replaced if necessary. It is in the area of power generation plant that service lives are longest and for nuclear plant this is coupled with strict safety requirements and the difficulty of access for inspection and repair. It is in the nuclear field that the need for reliable techniques of stress analysis is most compelling; thus we will consider solely the problems of design in this field. Thus we will be concerned with the behaviour of austenitic heat-resistant steels at high temperatures in the range 425–675° C subject to steady or cyclic loading and required for a long operational lifetime.

It must be realised that any design methodology for high-temperature plant will not deal exclusively with problems deriving from material creep; it will also include the avoidance of failure modes associated with plasticity and fatigue. In the following, we will attempt to highlight the requirements of current design codes with respect to creep. Of course, the most obvious question is: when does a component operate in the creep range? In many instances the anticipated service temperature and chosen material will clearly indicate that its behaviour is likely to be governed by creep. But there are occasions when it may be difficult to decide when the design should be based on creep (time-dependent) or plastic (time-independent) considerations. For example, the plant may normally operate at a temperature where creep is negligible, although there may be short-duration excursions into the creep range. One possible method of identifying these

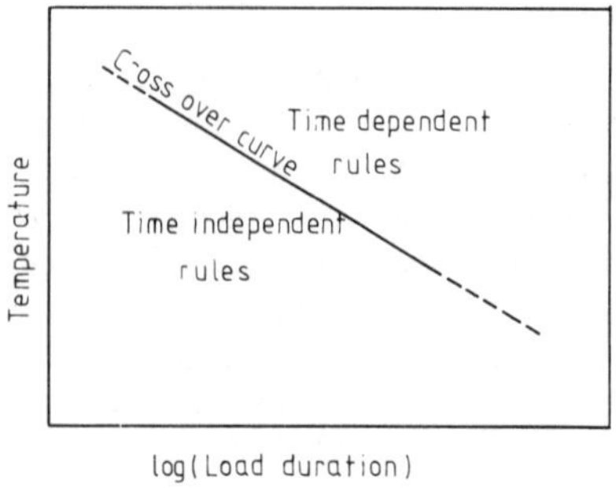

Figure 11.19 Creep range cross-over curve for design

excursions where a time-dependent analysis would be necessary is to adopt the concept of a 'creep range cross-over curve' (*Figure 11.19*), which has been recommended for the construction of pool-type French LMFBR reactors[49].

In the following we will develop three aspects of design methodology: the criteria by which a design must be assessed, the nature of some existing design codes of practice and finally the problems which arise from a creep–fatigue interaction.

11.3.1 Design criteria

The design process is concerned with the identification and prevention of several modes of failure. We have examined two of these in some detail – creep rupture (Chapter 9) and creep buckling (Chapter 10) – with respect to simple structures under steady loading. If a designer only needed to account for the behaviour of such structures in order to avoid these two failure modes, then design calculations would be pleasantly straightforward. Unfortunately, he also needs to exercise his judgement on real structures of complex shape subject to a variety of stress concentrations and gradients. In real life, it would not be immediately clear where or when failure is likely to occur or, indeed, if the emphasis had been placed on the correct modes of failure.

In the first instance, it will be useful to list, and briefly explain, the structural failure modes which can occur. For steady loading, these are:

(a) Extensive short-term yielding, which could lead to a tensile instability (ductile rupture).
(b) Creep rupture from long-term loading. In many cases, if it is assumed that short-term yielding is preventable, then much of the design effort is focused on this mode.
(c) Loss of function due to excessive deformation. This can manifest itself either as a creep distortion which can obstruct moving parts, for example in pumps or valves, or as creep or plastic buckling. It is also possible for thin pipes or vessels to experience gross distortion for quite moderate strains.

For cyclic loading, the principal failure modes are:

(d) Gross distortion due to incremental collapse or ratchetting. The latter can occur when plastic behaviour is such that during cyclic loading the deformation progressively increases. This may be enhanced by creep.
(e) Combined creep and low-cycle fatigue damage. This area is not yet fully understood.

Finally, two other modes of failure may occur for which adequate design rules are not available; these are:

(f) High cycle fatigue. This should be taken into consideration if it cannot be realistically assumed that a component is free from defects, so that small cracks may rapidly initiate and propagate to eventual failure (for example, in a vibrating structure). The absence of defects cannot be ensured, but these can be kept under control by quality assurance during manufacture.

(g) Progressive deterioration of a material from environmental effects caused by the presence of a combination of deleterious fluids and elevated temperature.

These will not be considered further here.

11.3.2 The nature of design codes

Most existing design codes, aimed at preventing the occurrence of the above failure modes, have general application. Thus they can all be assumed to be deficient in some respect, particularly in relation to optimum design. In addition, care is needed to understand why they have become accepted through experience and how they could prove unsatisfactory in some future situation. Most countries have at least one national design code for boilers, pressure vessels and associated piping for general application. Only one, the American Society of Mechanical Engineers[50] (ASME) Code Cases N-47, represents any formal attempt to regulate design practice for both steady and cyclic loading at elevated temperatures. In the following, we will describe the main aspects of one national code, the British Standards Institute[51] Specification BS 5500, which is similar in principle to many others, as well as the ASME Code Case.

11.3.2.1 BS 5500: 1976. Specification for unfired fusion-welded pressure vessels

This British Standard specifies requirements for the design, construction, inspection, testing and certification of unfired fusion-welded pressure vessels. It was developed primarily for components which are expected to operate below some creep threshold. But, in the absence of any specific criteria for time-dependent effects, it does permit the application of its rules to certain components in creep under steady loading. The main feature of this code is that an attempt has been made to eliminate the need for stress analysis of any sort. Thus design is carried out by the use of simple formulae and by reference to charts; indeed the same charts and formulae are to be used both above and below the creep range, although they are partly derived from experiment and theoretical analysis outside the creep range.

The main body of the code consists of five sections covering scope, materials, design, manufacture and workmanship, and inspection and testing. The two sections of interest here are Section 2, which deals with the selection of materials and provides basic material data on a wide range of ferritic and austenitic steels, and Section 3, which covers design and details formulae and charts for the common pressure vessels and various stress concentration features. In addition, there are eleven appendices which are non-mandatory and of an explanatory nature; of most interest is Appendix K on the derivation of material nominal design strengths. The code provides rules for the determination of the minimum thickness or dimensions of a vessel to ensure protection against gross plastic deformation, incremental collapse or buckling. Thus formulae or charts are provided to relate minimum thickness to a permitted design stress for the chosen material and operating temperature. For example, for a cylindrical vessel under pressure loading

$$\text{minimum thickness, } e = \frac{pD_1}{2f - p}$$

where p is the design pressure and D_1 the inside diameter: the nominal design stress is denoted by f. It is chosen as the lesser of the design stress f_E, for short-term tensile strength and f_F, corresponding to *creep* characteristics. For austenitic steels above 100° C

$$f_E = \text{lesser of } \frac{R_{p(t)}}{1.35}, \frac{R_m}{2.5}$$

where $R_{p(t)}$ is the proof stress at temperature t and R_m is the minimum tensile strength at room temperature. The denominators are some factors. The design stress f_F in the creep regime is defined as

$$f_F = \frac{S_{RT}}{1.3}$$

where S_{RT} is the mean value of stress required to produce creep rupture in a time T and a temperature t. This stress generally corresponds to those obtained from the ISO rupture curves[7].

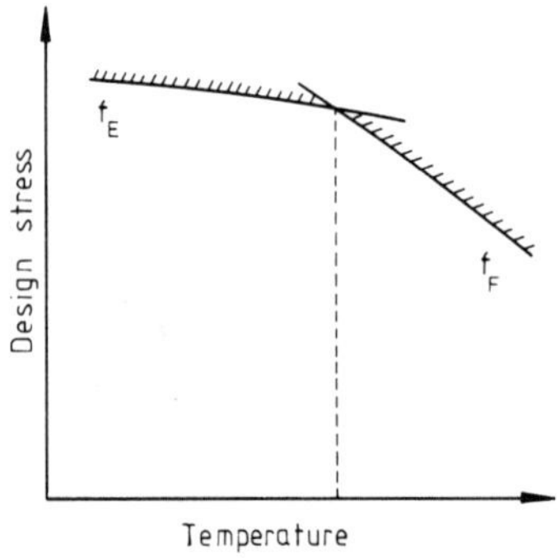

Figure 11.20 Dependence of design stress on temperature according to BS 5500

The choice of design stress for a given material and lifetime (say, 100000 h) can be depicted schematically (*Figure 11.20*); f_F will come below f_E above a certain 'cross-over temperature' which has no particular physical significance.

11.3.2.2 ASME Boiler and Pressure Vessel Code, Case N-47-14: Class 1 components in elevated temperature service

The design of elevated temperature components has a long history in the ASME Boiler and Pressure Vessel Code. Although the design procedure typified by BS 5500, commonly referred to as 'design by rule', can still be found in some parts of the ASME Code, it recognises that improvements are possible through 'design by analysis' which allows a reduction in factors of safety if the results of stress analysis are used. The Code Case N-47-14 is an outgrowth of Section

III of the ASME Code which applies to water-cooled nuclear power plant. It is intended to provide rules for materials, design and certification of components whose temperatures during service exceed 800° F for austenitic and high alloy steels and 700° F for ferritics. The Code document comprises three sections dealing with scope, materials and design. Paragraphs 2000 describe elevated temperature requirements for materials and those products which are permissible. Paragraphs 3000 discuss the general requirements for design by analysis followed by the special requirements for vessels, pumps, valves and piping.

The Code Case is complex and its contents are extensive. A complete description is outside the scope of this book; we can only outline the general features. Before design can proceed it is necessary for the plant owner to provide a 'design specification' which essentially categorises the expected loading conditions into several 'levels' of service loading. Different rules apply for the various levels of loading. Design then proceeds by the comparison of calculated stresses with appropriate design limits which depend on the 'stress category', whether 'primary' membrane (P_m), bending (P_b) or local peak (P_c) stress or 'secondary' thermal stress (Q). The limit on each stress category must be satisfied by the maximum 'stress intensity' which is calculated from the principal shear stresses, and is necessarily rather complex. Essentially, for normal service conditions, in order to avoid certain creep failure modes the stress intensities for primary membrane and primary membrane plus bending should be compared to a time-dependent material stress limit S_t, defined as the lower of

(a) the minimum stress to rupture in time t divided by 1.5,
(b) the minimum stress to 1% creep strain in time t,
(c) 80% of the minimum stress to reach tertiary creep in time t.

The Code Case also considers that excessive deformation may prove a prior limitation; these Rules are detailed in Appendix T of the Code Case. The strain and deformation limits in Appendix T are directed at preventing structural failure due to cyclic loading and buckling. These limits require that in regions expecting elevated temperatures maximum accumulated inelastic strain should not exceed

(a) strains averaged through the thickness, 1%,
(b) linear strain distributions, 2% at the surface,
(c) localised strain concentrations, 5%.

For steady loading, improved design can be achieved through inelastic analysis and the calculation of the steady state; although it is possible to use calculated elastic stresses, this would not include the beneficial effects of stress redistribution. For cyclic loading it has been recognised that the cost of inelastic analysis is high so that Appendix T also outlines simplified methods, based on elastic analysis, which are used in conjunction with material data curves given in the Code Case to estimate the total inelastic strain in the stationary condition at the end of service life. Appendix T provides four screening tests, any of which may be used to show compliance with the requirements of ratchet. These tests, coupled with the strain limits, represent one of the more controversial aspects of the Code Case, and will not be developed further here.

We have contrasted the British and American pressure-vessel codes merely as illustrations of two different design methodologies. It should be pointed out that

BS 5500 does not represent any serious attempt to provide comprehensive design rules for high-temperature service. Other methodologies are possible – the most detailed of these are honestly aimed at eliminating some of the deficiencies in the ASME Codes, for example, the notorious elastic route. The French RAMSES Committee[49] has been set up to examine the pessimistic design procedures implied by the elastic route as well as to extend the rules to a wider range of materials and to include the effects of environmental influences on material strength. An altogether novel approach has been suggested by the English CEGB based on reference stress techniques[52,53]. For example, in the design of steadily loaded components a limit on excessive deformation can be satisfied using the global reference stress described in Chapter 6 together with an approximation for the scaling factor. Further, an approximate 'representative rupture stress' given by the empirical formula

$$\sigma_{R0} = \sigma_R\left[1 + 0.13\left(\frac{Q_L}{Q_y} - 1\right)\right]$$

where σ_0 is the global reference stress under a single load (equation (6.6))

$$\sigma_R = Q\sigma_y/Q_L$$

and Q_y is the load to first yield, can be used directly with the basic creep curves, for a component limited by ductile creep rupture. The CEGB recommendations also provide a revised ASME N-47 elastic route for cyclic loading, with amended stress limits based on reference stress arguments. Again these modifications are the subject of some discussion.

Each of the above approaches to design can be seen as simply an *initial* attempt to resolve a highly complex problem and should not be treated in any way as being final.

11.3.3 Creep–fatigue interaction

The ASME Code Case N-47 pays considerable attention to combined creep and material fatigue under cyclic load. Thus, in order to close our discussion of design for creep, it is worth while briefly outlining the approach which is used. We cannot provide an adequate background to fatigue here, so the reader is referred to standard works[54]. Fatigue data are commonly presented in the form of plots of

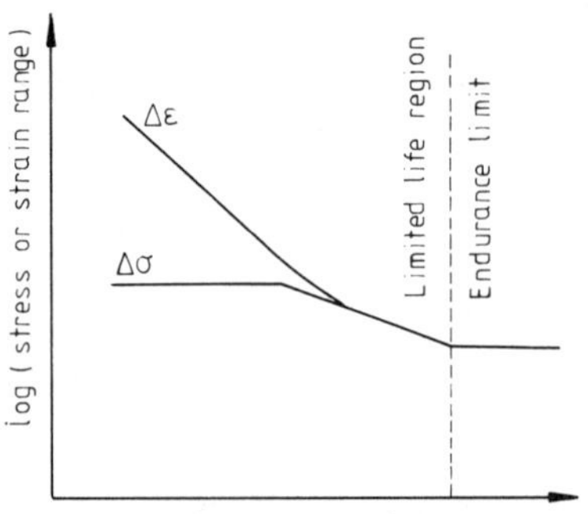

Figure 11.21 Endurance curves for material fatigue

stress range $\Delta\sigma$ or strain range $\Delta\epsilon$ in a cycle against the number of cycles to failure N_f (*Figure 11.21*). These are known as 'endurance curves', usually denoted by (S–N) or (ϵ–N) curves respectively. In real structures, a constant stress or strain range is not imposed; rather, a number of different cycles can be identified. If creep effects are ignored, the common approach in design is to perform a 'damage summation' according to the Palmgren–Miner rule. According to this rule, the fatigue damage for a number of cycles is defined as the ratio of number of cycles of amplitude $\Delta\epsilon_i$, N_i to the endurance N_{fi} for this range, assumed constant throughout the life of the component. If there are p different strain ranges, then the total damage is given by the linear sum

$$S = \sum_{i=1}^{p} (N_i/N_{fi})$$

Failure would occur when $S = 1$. A similar approach can be used to assess creep damage, in the absence of material fatigue.

For steady loading, the Kachanov damage parameter ω can be identified with the ratio of current time to rupture time. Thus for a programme of different load levels, the following approach is suggested: the creep damage incurred during a load level σ_i of duration t_i can be defined as $\omega_i = t_i/t_{ri}$ where t_{ri} is the rupture time under a constant load σ_i throughout the life of the component. If there are q load levels, then the total damage is approximated by the linear sum

$$R = \sum_{i=1}^{q} (t_i/t_{Ri})$$

Failure due to creep rupture would occur when $R = 1$; this is known as Robinson's cumulative damage rule.

The approach used in N-47 for combined creep–fatigue sums the damage for these two processes, which are assumed to be independent. The design criterion is given in the form of the damage summation

$$S + R \leqslant D$$

where D is a specified damage ratio.

This approach assumes that the stress and strain history of a component can be calculated; if not, simplified rules based on the use of elastic stresses are prescribed in Appendix T.

The damage summation approach has had some success, although again its usefulness is limited by the quality of the available creep–fatigue data.

References

1. Corum, J.M., *et al., Interim Guidelines for Detailed Inelastic Analyses of High Temperature Reactor System Components*, Report ORNL-5014, Oak Ridge Natl Lab., December 1974.
2. British Steelmakers Creep Committee *BSCC High Temperature Data*, Iron and Steel Institute, 1973.
3. *Ergebrisse dutscher Zeitstandversuche Langer Dauer*, Verein Deutscher Eisenhutten-leute, Verlag Stahleisen MBH, 1969.
4. *Results of High Temperature Creep Tests in French Steels* IRSID, 1972.

5. National Research Institute for Metals *NRIM Creep Data Sheets: Elevated-Temperature Materials Manufactured in Japan*, NRIM, 1973.
6. ASTM–ASME, *Compilation of Available High Temperature Creep Data,* ASTM–ASME Special Technical Publications Series, 1938 *et seq.*
7. International Standards Organisation *ISO-Data Number 1,* 1975.
8. Hart, E.W., Constitutive equations for the non-elastic deformation of metals, *Trans. ASME, J. Eng. Mat. Techn.,* **98**, 193–202, 1976.
9. Miller, A., An inelastic constitutive model for monotonic, cyclic and creep deformation, *Trans. ASME, J. Eng. Mat. Techn.,* **98**, 97–113, 1976.
10. Larsson, B. and Storakers, B., A state variable interpretation of some rate-dependent inelastic properties of steel, *Trans. ASME, J. Eng. Mat. Techn.,* **100**, 395–401, 1978.
11. Eykhoff, P., *System Parameters and State Estimation,* Wiley, 1971.
12. Deming, W.E., *Statistical Adjustment of Data,* Chapman & Hall, 1946.
13. Fletcher, R. *Practical Methods of Optimisation*, Vol I, Wiley, 1980.
14. Conway, J.B., *Numerical Methods for Creep and Rupture Analyses,* Gordon & Breach, 1968.
15. Mitra, S.K. and McLean, D., Work hardening and recovery in creep, *Proc. R. Soc. A,* **295**, 288–99, 1961.
16. Garofalo, F., *Fundamentals of Creep and Creep-Rupture in Metals,* Macmillan, 1965.
17. Monkman, F.C. and Grant, N.J., An empirical relationship between rupture life and minimum creep rate, in *Deformation and Fracture at Elevated Temperature,* Eds N.J. Grant and A.W. Mullendore, MIT Press, 1965.
18. Dobes, F and Milicka, K., The relation between minimum creep rate and time to fracture, *Metal Sci.,* **10**, 382–4, 1976.
19. Lonsdale, D. and Flewitt, P.E.J. Relationship between minimum creep rate and time to fracture for 2¼ Cr–1 % Mo steel, *Metal Sci.,* **12**, 264–5, 1978.
20. Brathe, L. and Joscfsun, L., Estimation of Norton–Bailey parameters from creep rupture data, *Metal Sci.,* **13**, 660–4, 1979.
21. Goldhoff, R.M. and Gill, R.F., A method for predicting creep data for commercial alloys on a correlation between creep strength and rupture strength, *Trans. ASME, J. Basic Eng.,* **94**, 1–6, 1972.
22. Murphy, M.C., Rating the creep behaviour of heat resistant steels for power plant, *Metals Eng. Q.,* **13**, 41–50, 1973.
23. Goodman, A.M., Materials data for high temperature design, in *Creep of Engineering Materials and Structures,* Eds G. Bernasconi and G. Piatti, Applied Science Publ., 1979, Chap. 11.
24. Conway, J.B., *Stress–Rupture Parameters: Origin, Calculation and Use,* Gordon & Breach, 1969.
25. Goldhoff, R.M., The evaluation of elevated temperature creep and rupture stress data: an historical perspective, in *Characterisation of Materials for Service at Elevated Temperature*, ASME Special Publ, 1978.
26. Manson, S.S. and Ensign, C.R., A quarter century of progress in the development of correlation and extrapolation methods for creep rupture data, *Trans. ASME, J. Eng. Mat. Techn.,* **101**, 317–25, 1979.
27. White, W.E. and Le May, I., On the minimum-commitment method for correlation of creep rupture data, *Trans. ASME, J. Eng. Mat. Techn.,* **100**, 333–5, 1978.
28. Manson, S.S. and Haferd, A.M. A linear time–temperature relation for extrapolation of creep and stress–rupture data, NASA TN No. 2890, NASA, 1953.
29. Goldhoff, R.M., Towards the standardisation of time–temperature parameter usage in elevated temperature data analysis, *J. Testing Evaluation,* **12**, 387–424, 1974.
30. Glen, J., A new approach to the problem of creep, *J. Iron Steel Inst.,* **189**, 333–43, 1954.
31. Grant, N.J. and Bucklin, A.G., On the extrapolation of short time stress rupture data, in *Deformation and Fracture at Elevated Temperatures*, Eds N.J. Grant and A.W. Mullendore, MIT Press, 1965.
32. Cocks, A.C.F. and Ashby, M.F., Creep fracture by void growth, *Proc. 3rd IUTAM Symp. on Creep in Structures*, Eds A.R.S. Ponter and P.R. Hayhurst, Springer, 1981, pp. 368–87.
33. Cocks, A.C.F. and Ashby, M.F., *Pressure Vessels and Piping: 1972 Computer Programs Verification – An aid to developers and users,* ASME Special Publ., 1972.
34. Cocks, A.C.F. and Ashby, M.F., *Pressure Vessels and Piping: Verification and qualification of Inelastic Analysis Computer Programs*, ASME Special Publ., 1975.

35. Yamada, Y., Research and development of inelastic analysis procedures for reactor component design in Japan, *Nucl. Eng. Des.*, **51**, 85–96, 1978.
36. Yamada, Y., Subsequent activities in Japan of inelastic structural analysis for elevated temperature design, *Proc. 2nd Int. Seminar on Inelastic Analysis and Life Prediction in High Temperature Environment, Proc. 5th Int. Conf. on Structural Mechanics in Reactor Technology, Berlin,* 1979.
37. Yamada, Y., *Compilation of Piping Benchmark Problems: Cooperative International Effort, Ed. W.J. McAfee,* Rep. No. IWGFR-27, IAEA Intl Working Group on Fast Reactors, 1979.
38. Ainsworth, R.A., An experimental study of a three-bar structure subjected to a variable temperature, *Int. J. Mech. Sci.*, **19**, 247–56, 1977.
39. Sim, R.G. and Penny, R.K., Some results of testing simple structures under constant and variable loading during creep, *Exptl Mech.*, **10**, 152–9, 1970.
40. Skelton, W.J. and Crossland, B., Correlation of tension and thick-walled cylinder creep based on experimental data, *Proc. I. Mech. E.*, **182**, Pt 3 C, 159–65, 1967.
41. Wahl. A.M. *et al.*, Creep tests of rotating discs at elevated temperature and comparison with theory, *Trans. ASME, J. Appl. Mech.*, **76**, 225–35, 1954.
42. Mendelson, A. *el al.*, A general approach to the practical solution of creep problems, *Trans. ASME,* **81D**, 585–589, 1959.
43. Penny, R.K. and Marriott, D.L., Creep of pressure vessels, *Proc. Int. Conf. on Creep and Fatigue in Elevated Temperature Applications,* I. Mech. E. Paper C204/73.
44. Corum, J.M. and Richardson, M., Elevated temperature tests of simply supported beams and circular plates subjected to time-varying loads, in ref. 34, pp. 13–26.
45. Clinard, J.A., Corum, J.M. and Sartory, W.K., Comparison of typical inelastic analysis predictions with benchmark problems, in ref. 34, pp. 79–98.
46. Hayhurst, D.R. *et al.*, Estimates of the creep rupture lifetime of structures using the finite element method, *J. Mech. Phys. Solids,* **23**, 335–56, 1975.
47. Corum, J.M. *et al.*, Thermal ratchetting in pipes subjected to intermittent thermal downshocks at elevated temperatures, in ref. 34, pp. 47–58.
48. Kraus, H., *International Benchmark Project on Simplified Methods for Elevated Temperature Design and Analysis: Problem 1 – The Oak Ridge Pipe Ratchetting Experiment,* Bulletin No. 258, Welding Research Council, 1980.
49. Jakubowicz H. *et al.*, Design methods and criteria recommended by the RAMSES Committee, *Proc. Int. Conf. on Engineering Aspects of Creep.* I. Mech. E., Paper C297/80. 1980.
50. ASME Code Case N47 – 15 (1592–15) *Class 1 Components in Elevated Temperature Service,* ASME, 1979.
51. BS 5500, *Unfired Fusion Welded Pressure Vessels,* British Standards Institute, 1976.
52. Ainsworth, R.A. and Goodall, I.W., *Proposals for Primary Design Above the Creep Threshold Temperature – Homogeneous and Defect-free Structures,* CEGB Rep. RD/B/N4394.
53. Goodman, A.M., Design related aspects of creep and fatigue, *Creep and Fatigue in High Temperature Alloys,* Ed. J. Bressers, Applied Science Publ., 1981, Chap. 6.
54. Osgood, C.C., *Fatigue Design,* Wiley, 2nd Edition, 1981.

Index